Time

THE COMPARATIVE STUDIES IN SOCIETY AND HISTORY BOOK SERIES

Raymond Grew, Series Editor

Comparing Muslim Societies:
Knowledge and the State in a World Civilization
   Juan R. I. Cole, editor

Colonialism and Culture
   Nicholas B. Dirks, editor

Constructing Culture and Power in Latin America
   Daniel H. Levine, editor

Time: Histories and Ethnologies
   Diane Owen Hughes and Thomas R. Trautmann, editors

Time

Histories and Ethnologies

edited by
Diane Owen Hughes
and
Thomas R. Trautmann

Ann Arbor
THE UNIVERSITY OF MICHIGAN PRESS

Copyright © by the University of Michigan 1995
All rights reserved
Published in the United States of America by
The University of Michigan Press
Manufactured in the United States of America
⊚ Printed on acid-free paper

1998   1997   1996   1995   4 3 2 1

A CIP catalogue record for this book is available from the British Library.

Library of Congress Cataloging-in-Publication Data

Time : histories and ethnologies / edited by Diane Owen Hughes and
Thomas R. Trautmann.
    p.   cm.—(Comparative studies in society and history book series)
Includes bibliographical references and index.
   ISBN 0-472-09579-X (alk. paper).—ISBN 0-472-06579-3 (pbk. :
alk. paper)
   1. Time—Cross-cultural studies. 2. Time perception—Cross-
cultural studies. I. Hughes, Diane Owen. II. Trautmann, Thomas R.
III. Series.
GN469.5.T56   1995
304.2′3—dc20   95-15095
                                                                            CIP

Contents

Foreword   vii
  Raymond Grew

Acknowledgments   xiii

Introduction   1
  Diane Owen Hughes

Part 1. Local Time

The Pasts of an Indian Village   21
  Bernard S. Cohn

The White Man's Book: The Sense of Time, the Social Construction of Reality, and the Foundations of Nationhood in Dominica and the Faroe Islands   31
  Jonathan Wylie

Time and Memory: Two Villages in Calabria   71
  Maria Minicuci

Part 2. Big Time

Remembering the Future, Anticipating the Past: History, Time, and Cosmology among the Maya of Yucatan   107
  Nancy M. Farriss

Chronology and Its Discontents in Renaissance Europe: The Vicissitudes of a Tradition   139
  Anthony T. Grafton

Indian Time, European Time   167
  Thomas R. Trautmann

Part 3. Time and the Story of the Past

Time and Historical Consciousness:
The Case of Ilparakuyo Maasai   201
  Peter Rigby

Time and the Sense of History in an Indonesian Community:
Oral Tradition in a Recently Literate Culture   243
  R. H. Barnes

Ruins of Time: Estranging History and Ethnology
in the Enlightenment and After   269
  Peter Hughes

Contributors   291

Index   293

Foreword

For more than thirty years the quarterly issues of *Comparative Studies in Society and History* have published articles about human society in any time or place written by scholars in any discipline and from any country. Those articles, inevitably reflecting the changing methods and interests within the specialized fields of research from which they grew, have presented new evidence and new techniques, challenged established assumptions, and raised fresh questions. Now this series of books extends and refocuses the comparisons begun in some of the most stimulating of those essays.

The editors of each volume identify a field of comparative study and then consider the kinds of essays that will exemplify its range and excitement, beginning with articles that have appeared in *CSSH* from October 1958 to the present, plus the scores of new manuscripts currently under consideration. The conception of the book thus builds on a group of articles that are part of a continuing dialogue among scholars formed in different disciplines, traditions, and generations. In addition, each volume in this series includes essays never before published and specially commissioned to suggest more fully the potential range of the larger topic and new directions within it. The authors of the previously published articles are also given the opportunity to revise their essays in the light of this project. Each volume is therefore a new work in the specific sense that its chapters are abreast of current scholarship but also in its broader purpose, a cooperative enterprise reconsidering (and thereby reconstructing) a common topic.

Having established the theme to be addressed and identified the scholars to do it, the editors then invite these colleagues to join in exploring the ramifications of their common interest. In most instances this includes a conference in Ann Arbor, attended by contributors and by many other scholars, where issues of conceptualization, interpretation, and method can be debated. Sometimes the volume's topic is made the basis of a graduate course, with contributors giving a series of lectures in a seminar lasting a term or more and attended by a variety of interested specialists. The book, which starts from an indirect dialogue in the pages of *CSSH,* thus takes form through direct exchanges. In open-ended and lively discussion, individual manuscripts are criticized and new suggestions tried out, common concerns identified and then matched against the criteria of different disciplines and the experience of different societies. Reshaped by the community it creates, each volume becomes a statement of where scholarship currently stands and of questions that need to be pursued. Through the process in which individual chapters are reconsidered and revised, general problems are reformulated.

In this way this series extends the tradition that *CSSH* represents. A scholarly quarterly is a peculiar kind of institution, its core permanently fixed in print, its rhythmic appearance in familiar covers an assurance of some central continuity, its contents influenced by its past yet pointing in new directions. *CSSH* seeks to create a community without formal boundaries, a community whose membership is only loosely determined by subject, space, or time. Just as footnotes and references embed each article in particular intellectual traditions while stretching beyond them, so the journal itself reaches beyond editors, contributors, and subscribers, speaking in whatever voice unknown readers respond to whenever and wherever they turn its pages. The resulting dialogues are not limited to any single forum, and the journal itself changes from within while balancing between venturesomeness and rigor as old debates are refined and new problems posed.

The books in this series further in another form aspirations acknowledged in the opening editorial to the first issue of *CSSH* in which Sylvia Thrupp declared her belief that "there is a definite set of problems common to the humanities, to history, and to the various social sciences." Changes in the way these problems are conceived and in the vocabulary for expressing them have not lessened the determination to reject "the false dilemma" between "error through insularity and probable superficiality." Insistence upon thorough, original research has been the principal defense against superficiality, emphasis upon comparison the means for overcoming insularity. Many of the articles published in *CSSH* are systematically comparative, across time and between societies, and that is always welcome; but many are not. Each published article is independently chosen for its qualities of scholarship and imagination as well as for its broader implications. For the contributors to and readers of that journal, comparison has come to mean more a way of thinking than the merely mechanical listing of parallels among separate cases. Articles designed to speak to scholars in many disciplines and to students of different societies are recognized as intrinsically comparative by the nature of the problems they pose, the structure of their argument, and the effect of their conclusions.

Every piece of research deserves to be seen in many contexts: the problems and concerns of a particular society, the immediately relevant scholarly literature with its own vocabulary and evidence, the methods and goals of a given discipline, a body of theory and hypotheses, and sets of questions (established, currently in vogue, or new). Nor can any prescription delimit in advance how far subsequent comparisons of similar problems in different contexts may reach. For the past twenty years, *CSSH* has placed articles within rubrics that call attention to a central comparative theme among adjacent studies. In addition an editorial foreword in each issue notes other sets of connections between current articles and earlier ones, inviting additional comparisons of broad themes, specific topics, and particular problems or methods. A variety of potential discourses is thus identified, and that open-ended process has culminated in this series of books. Some of the volumes in the series are built around established themes of comparative study, subjects known to require comparison; some address topics not always recognized as a field of study, creating a new perspective through fresh questions. Each volume is thus an autonomous undertaking, a discussion with its own purposes and focus, the work of many authors and its editors' vision of the topic, establishing a field of knowledge, assessing its present state, and suggesting some future directions.

The goal, in the quarterly issues of *CSSH* and in these books, is to break out of received categories and to cross barriers of convention that, like the residual silt from streams that once flowed faster, have channeled inquiry into patterns convenient for familiar ideas, academic disciplines, and established specialties. Contemporary intellectual trends encourage, indeed demand, this rethinking and provide some powerful tools for accomplishing it. In fact, such ambitious goals have become unnervingly fashionable, for it no longer requires original daring nor theoretical independence to attack the hegemony of paradigms—positivism, scientism, Orientalism, modernization, Marxism, behavioralism, etc.—that once shaped the discourse of social science. Scholars, however, must hope that the effort to think anew can also allow some cumulative element in our understanding of how human societies work; and so these books begin their projects by recognizing and building upon the lasting qualities of solid scholarship.

Questions about time are, as Diane Owen Hughes demonstrates in the introduction to this volume, central to the modern study of culture and society, crucial to any anthropology of history, at the core of any history of anthropology, the very essence of poststructuralism. Discussions about time quickly become debates about the meaning of modernity, but they can also be about so much else—because time presents itself in so many guises: astronomical, geological, biological, historical, philosophical, theological, mythical, and ritual—that time seems a topic without any natural organizing principle, one that is ubiquitous, like ether or human nature, and that, like them, baffles comparison. Perhaps there are facts too fundamental and concepts too grand to give mere scholars much guidance. The flow of time does transgress both these extremes, and some of the excitement of this volume arises from the challenge of how to structure discussions about how human beings experience time. Diane Hughes makes it clear from the beginning, with her call for rethinking myth and history, that in this book coherence will not be purchased at the price of some contrived constriction of the topic. Rather, she and Thomas Trautmann have designed an organization that guides the reader through the ways of comparing times.

The book begins with three essays that are explicitly comparative, each a local study in which a small world's explicit context facilitates comprehension and exposes some of the multiple meanings of time. Bernard Cohn compares the different pasts nurtured within a single Indian village, each of them offering its own avenues of adaptability. Traditional and modern are not dichotomous. Jonathan Wylie compares two fishing communities that experience time, remember history, and employ the past in almost opposite ways. Time, like memory, is irregular, its reliability independent of accuracy, its thinness or depth an expression of social relations. Maria Minicuci compares two adjacent and similar towns. In both, time is tied to language and place, land and lineage, but it serves as a steady referent in one and is only feebly recalled in the other. The past builds boundaries but also domesticates the intrusions of change.

The next three essays explore the ways in which cosmologies work and the intense concern they invoke. Nancy Farriss traces in Mayan thought the ligaments that tie the everyday to the universal, crops and politics to the gods, and historical to cosmic time. Anthony Grafton connects clocks and iconography to early modern Europe's reconstructions of the ancient world and its conquests on other continents. Thomas Trautmann follows the faith-filled logic of Christian scholars of India who sought God's plan in language and whose contributions to a new science include the ironies of the erroneous. In each of these essays we see time become the object of systematic and learned thought, define the Other, and facilitate comprehension through misunderstanding. What Europeans and Mayans thought they knew, like what Caribbean fisherman and Calabrian peasants forgot, helped keep them apart from strangers.

The final three chapters are, like this book itself, about ways of using time past (refracted through cultural time) to shape present change (understood as a process in historical time). Peter Rigby, writing on the Ilparakuyo section of the Massai; R. H. Barnes, writing on Lamalera in Lembata, Indonesia; and Peter Hughes, writing on some of modern Europe's most famous authors, all have much in common. These essays contemplate the cultural meanings of time and in each case find internal contradictions fruitful and complexity an aid to creativity. Each study rejects an easy distinction between myth and history, theory and practical experience. Each is led to reassess broad issues of scholarship. Rigby criticizes much of modern anthropology, Barnes reconsiders the role of religion in keeping tradition flexible, and Peter Hughes shows that history and literature have shaped each other from Giambattista Vico to Marc Bloch. These studies of how human beings on four continents place themselves in time has led scholars from different disciplines to fresh understanding of specific societies, to broader categories of cultural analysis, and to stimulating reflection on our own ways of knowing.

Raymond Grew

Acknowledgments

This volume was first formed for students of a course of the same name given in the fall of 1992. The editors benefitted from the critical intelligence of the participants of the course: Marc Baer, Javier Barrios, Katherine Brophy, Laurent Dubois, Jason Fink, John Hamer, Matthew Herbst, Elizabeth Horodowich, Alexander James, Vivian Katz, Chris Ogilvie, Manish Patel, Juan Javier Pescador, and Anupama Rao.

The essays in this volume were presented at a conference held on 20–22 November 1992 at the University of Michigan under the sponsorship of *Comparative Studies in Society and History*. We are grateful to Robert McKinley, Walter D. Mignolo, Timothy H. Bahti, and Rudolf Mrazek for commenting on the papers; to John Hamer for program graphics; to Carin MacCormack for administrative assistance; to John C. Dann and Arlene Shy for the hospitality of the William L. Clements Library; and, above all, to Raymond Grew and *CSSH*.

Comparative Studies in Society and History has graciously given permission to reproduce the following articles in this volume.

Bernard S. Cohn, "The Pasts of an Indian Village," *CSSH* 3:3 (1961): 241–49.

Jonathan Wylie, "The White Man's Book: The Sense of Time, the Social Construction of Reality, and the Foundations of Nationhood in Dominica and the Faroe Islands," *CSSH* 24:3 (1982): 438–66.

Nancy M. Farriss, "Remembering the Future, Anticipating the Past: History, Time, and Cosmology among the Maya of Yucatan," *CSSH* 29:3 (1987): 566–93.

Peter Rigby, "Time and Historical Consciousness: The Case of Ilparakuyo Maasai," *CSSH* 25:3 (1983): 428–56.

Introduction

Diane Owen Hughes

What, then, is time? If no one asks me, I know. If someone asks me to explain, I cannot tell him.
—St. Augustine

The problem of time, in both history and anthropology, continues to haunt the disciplines that grew out of attempts to solve or resolve it. As they have grappled with it, and as the studies gathered here continue to suggest, historians and anthropologists have begun to converge after a long period of divergence provoked by differing ways of narrating and analyzing the otherness of the past and the pastness of the other. This divergence, and the birth of modern anthropology, grew out of nineteenth-century history's expulsion of the "strange time" of myth, of superstition, of numinous narratives that resisted dating according to standard chronology as stubbornly as they resisted attempts to interpret them according to modern ideology. The "expulsés" were not only the tribal tellers of such narratives but also the ethnographers impelled into exile in order to hear and retell their stories.

The claim to truth made by the historian was explanatory, as in Ranke's attempt to narrate the past "wie es eigentlich gewesen," and its explanatory power was both increased and narrowed by appeals to general laws and to archival evidence. Both appeals were embodied in a narrative the temporality and coherence of which depended upon (and was limited to) datable facts and events. The anthropologist's claim to truth, by contrast, was from the start based on a narrative and narrative analysis that sought to represent the consciousness that underlay the events narrated. Such narratives might displace the problem of time by invoking mythic tenses of primacy or oneness, but the problem remained in the sequential ordering of events within the narrative itself. Sequence distinguishes temporal events from structural elements, and the taxonomic enterprise of structural anthropology might be described as the boldest of several attempts to settle the problem of time once and for all by "displacing" it, by translating time into space.

Such translations, in which what is exotic can be likened to what is past, the spatially to the temporally remote, mark the common origin or shared prehistory of historical and anthropological narrative, and several of the following studies show how recent work in both fields has returned, with some radical differences, to this discourse and dialogue. The *histoire des mentalités* set as a goal by the Annales school is clearly very close in its scope, if not in its methods, to the narrative of underlying consciousness offered by anthropology. But the analogy between time and space should not be confused with symmetry or identity.

Recent and debated studies of temporal awareness in Indonesian—especially Balinese—communities show how uncertain are attempts to uncover stable structures of ritual and workaday time. Clifford Geertz opened the debate by positing a sharp, even exclusionary, distinction in Bali between ritual time represented as spatial pattern and workaday time represented as recurrent flow—a distinction that excludes historical events.[1] In the course of a larger argument, Maurice Bloch called this study or position into question, proposing instead that two distinct kinds of temporality operate in Bali.[2] One is linear and durational, applied to the world of praxis; the other, cyclic and static, is applied to the world of ritual. The central question posed by Geertz and Bloch is the question posed by thinkers as diverse as Augustine, Kant, and Marx. How real is time? How far does our experience undermine it? Geertz concentrates on the patterned behavior that seems to subvert the reality of temporal change, while Bloch insists on the coeval and practical awareness that change is real, that we cannot step twice into the river's stream at the same point or reverse the arrow of time. Bloch's dualist distinction is attacked by Leopold Howe who argues that neither pure linearity nor cyclicity is possible: both are combined in a unified Balinese temporality.[3] What underlies Howe's argument is a moderate recognition that while all cultures experience duration each culture records or expresses its experience in different ways.

It is at this point that R. H. Barnes's contribution takes up and shifts the grounds of dispute. His earlier book influenced Howe's argument by showing —as Anthony Grafton does here in a Renaissance historical context—that our western sense of time depends upon our instruments of measurement, and that what those instruments measure results from our need to know different, usually more precisely delimited, things than those Indonesian societies need to know.[4] The world of Lamalera, Lembata, presented by Barnes is a village community whose "daily life" is at once influenced by "mythical or legendary events" yet critically and literately aware that legend is a pragmatic "index of the insecurity of inherited arrangements." He insists that, however nuanced a society's figuring of time may be, "the underlying experience of duration is irreversible." That implies, in his village and everywhere, that the spatial analogy is limited and finally inadequate. We can't go home again, not only because that home may be changed or even no longer exist, but because we cannot reverse the path that leads from there and then to here and now. Barnes also concludes, as a further limitation to spatial analogies, that his findings do not allow any neat association of the linear, historical, and "modern" as opposed to the cyclic, legendary, and indigenous. But, once we move beyond questions of duration and try to deal with responses to those questions cast in the form of anthropological or historical discourse, we realize how much remains uncertain. Innovations and crossovers in narrative that show what historians and anthropologists can learn from one another (and from their colleagues in sociology) leave unresolved most of the issues and problems posed by our attempts to reason and deal with temporal reality or reality as temporal.

We must acknowledge from the start that some kinds of certainty are beyond us. There is a philosophical sense in which we can *demonstrate* the reality of present experience and objects of sense but only *argue* about historical events or narratives of the other. It is of course true that many more matters are open to argument than to demonstration, but certainties about historical proof or ethnological analysis are both shaped and warped by our consciousness of time. It is this consciousness, and not the chronometric order imparted by clocks and calendars, that makes most of our discussions, including those that follow, at once valid and unverifiable. Anthony Grafton closes his richly detailed inquiry into earlier chronometric schemes by warning us that "a real history of early modern chronology would evidently not make a pretty sight" and would hardly prove to be "a triumph of historicist virtue." The source of muddle in schemes to regulate time is also the source of our historical sense, and that is our need to give meaning to a process that is in itself meaningless, even vertiginous. As Peter Munz has observed,

> The process of time by itself does not allow us to identify anything as an event, either small or large. The observation of the passage of time can only create an impression of an undifferentiated continuum bound to create giddiness and a blurring of vision in the observer. If the continuum is to be differentiated we first have to imagine that we remove such time structure it

has—as if we were filleting a fish. Thus we get something like a mollusc—a wobbly, still undifferentiated mass, now even deprived of its time skeleton. To reassemble this mass we have to select and to construct. We can do this only by introducing nonempirical factors, where "nonempirical" does not mean metaphysical but something *not* derived from the observation of the passage of time.[5]

The nonempirical factors selected by the anthropologist to pattern time have long differed from those chosen by the historian, but until recently both choices have been masked by rarely questioned assumptions, which have been further exempted from question by terms such as *method, structure,* or *description.* It could even be said that the historian has tended to construct linear and sequential patterns of time, which the anthropologist deconstructs through patterns of depth and simultaneity. Hence the double role of taxonomic structure, which at once confers the stability of strata to anthropology's temporal discourse and destabilizes the patterns of change and event constructed by historians.

To destabilize change and mutability is a paradoxical goal that has animated anthropology from its origins. As Johannes Fabian has polemically expressed it, "Winning the taxonomic game consists of demonstrating synchronic relations of order beneath the flux and confusion of historical events and the expression of personal experience."[6] The growing and critical awareness that such discourse is paradoxical or even duplicitous, exemplified by Fabian's book, animates much that is fresh in recent ethnographic discourse, exemplified by a number of the following studies that show how historical sequence and anthropological depth may change places and sides. Jonathan Wylie, for example, shows through his research into two very different island societies that insularity gives no guarantees of a separation from history or ability to float above its currents; that the temporal sense itself may be shallow or deep; and that its depth depends upon whether or not a community distinguishes (or can distinguish) between history and society. Literacy is certainly a factor in this deepening and shaping of memory, but, as other contributors have suggested, clear distinctions between the oral and the literate in patterning time do not seem borne out by the evidence. The interaction of orality and literacy may be in this respect sinuous or inconclusive.

At the core of this consciousness, whether historical or ethnological, we find narratives through which we express the temporal nature of our experience. For the historian, temporality is characterized by change, sequence, and coherence, and these are the qualities usually foregrounded in historical narratives. The first of these, more exactly described as our first awareness of time through our recognition of change, is crucial to historical narratives. Without change, there is no temporal reality; and it is no accident that the most drastic recognitions of change—the fall of Rome, the French Revolution—seem to evoke the most powerful historical narratives. The narrative may even be described as an attempt both to represent change and to present a way of overcoming it, if not retroactively, then at least retrospectively. R. W. Southern is not alone in thinking that this attempt is most strongly evoked by change that seems strong enough to disrupt our connections with the past, as this possibility or threat, posed by the French Revolution, underlies in his opinion the growth of modern historical narrative.[7] Change cannot be total, however, if it is to be the spur to narrate: we must be simultaneously aware of what remains and of what has been lost. Hence the importance to the historical sense of time, as Peter Hughes suggests in his essay, of ruins, which in the world of experience of the Enlightenment and Revolution present both what lasts and what perishes in one structural whole—or fragmented structure—through which we perceive the double power of memory and time to preserve and lay waste. Hence, too, he suggests, the importance to the historian or thinker of experienced ruin or catastrophe: the fall of France in 1940 was not only the end of Marc Bloch's idea of France's modernity but also the impulse in ethnologists and thinkers such as Claude Lévi-Strauss and Simone Weil to create an alternative to historical disaster in a new structural and timeless order.

Sequence, the second property of temporality, looks at first glance simpler than change or coherence. One thing follows another, one event comes after another. But even in the world of objects everything depends on where something—a statue, a hieroglyph—comes in a series. In material culture there is no such thing as pure and simple repetition. The same hat can be the last word in 1937 and a piece of retro nostalgia in 1994. Sequential repetition underlies boredom and cultural fatigue as well as expectation and surprise. And in the world of ethnographical narrative, as more than one of our contributors has observed in recording a return some years later to the scene of an earlier study, even a sequence of two involves a reciprocal impact. The earlier narrative clearly elicits the later, and the later now interprets the earlier—accounts for it in the ethnographer's terms—whereas before the earlier narrative explained its subject—sought, that is, to account for a situation or group in its own terms.

A further feature of both historical and ethnological time serves to complicate sequence in all the narratives with which we deal. Our culture so values origins that earlier terms we select in creating historical sequences tend to be described as causes of later terms, just as the primal in an ethnological sequence tends to be valued as the explanation or mythic account of what follows. It could even be argued that ethnographic narrative has for this Rousseauistic reason long tended to be set in a strange time that is *before,* and it is impressive to see how many of our contributors have involved themselves in narratives that deal so explicitly with times and senses that come *after.* Since historical sequence, in its turn, is generally aligned with the chronometric time of dates and hours, we may not notice the implications of where the historian chooses to begin and end. The fact that Gibbon goes on from the fall of Rome to the fall of Constantinople alters our reading of his narrative as certainly (if silently) as any sequence of events within it. As a historian of art and archaeology, George Kubler recognizes that made things, starting with stone tools, exist in a sequential continuity that resists division more stubbornly than historical narrative.

> Everything made now is either a replica or a variant of something made a little time ago and so on back without break to the first morning of human time. This continuous connection in time must contain lesser divisions. The narrative historian always has the privilege of deciding that continuity cuts better into certain lengths than into others. He never is required to defend his cut, because history cuts anywhere with equal ease, and a good story can begin anywhere the teller chooses.[8]

Many historians would recoil from this view of themselves as storytellers, but insofar as they are it is sequence and coherence that make them so—and draw their narratives close to those of anthropologists.

The narrator's dream is to give presence and truth to what is absent and imagined, past and forgotten. Both historian and anthropologist narrate in a waking dream clarified and validated by documents and observations. The analogy with dreams is also an analogy with fictions and plots. Just as there are categories of dreams, there are—as Hayden White has shown[9]—categories of historical narratives, and these categories replicate the tropes of rhetoric, from metaphor (*The Waning of the Middle Ages*) to irony (*Time on the Cross*). In trying to narrate time past or the behavior of the other, both historian and ethnographer start from a claim or hope that narrative will, in R. G. Collingood's phrase, "re-enact" its subject, but in the course of narrating both are pulled away from this repetition by the power of pastness and strangeness toward something more like a recreation or reimagining of the subject. The narrative pull is from one kind of truth to another, from truth by correspondence to truth by coherence. Since neither account can correspond to a lost or forgotten original, it must cohere (or at least not conflict) with what others have said or written. And it is this coherence within the narrative and outside it to other narratives that constitutes the truth of history and anthropology. What is offered as explanation and correspondence turns inexorably into interpretation and consistency. Historical "causality" and anthropological "consciousness," both of which are alleged to bridge the distance into the past, are in fact coherent narratives backed by the recorded memories and narratives of others.[10] In both cases imaginative power is essential. It is not for nothing that Aristotle described memory as the imagination plus time.

Even this reduced claim of coherence—rather than correspondence—for our narrations of time is bold and far-reaching. Arching over the differing approaches of historians and anthropologists, and including both, is the grand theory proposed by Paul Ricoeur in *Time and Narrative*. In what has been justly described as "the most important synthesis of literary and historical theory produced in our century,"[11] Ricoeur argues that temporality marks us and our condition as human, and that only through narrative can this temporality and our humanity be expressed and realized.[12] And yet, any attempt to narrate, whether by an ethnographer or a historian, by a native informant or a contemporary witness, must overcome obstacles to satisfy the desire to *relate*, in the double sense of telling and making coherent, of acting out and "configuring," our otherwise baffled and anguished understanding. Ricoeur, appealing to tragic and metaphysical expressions of this need to relate, finds in birth and death the spur to redeem through narration what would otherwise be lost in obscurity and oblivion. Myths of origin or genealogical accounts, like histories of decline and fall, underlie the temporal consciousness studied here by Nancy Farriss in Mexico, for example, Thomas Trautmann in India, and Maria Minicuci in Calabria. But those accounts of time seem to be provoked by a dread even more pervasive than death: a fear of disorder, of loss, of strangeness and the stranger. Minicuci's research revealed startling contrasts in temporal awareness between two neighboring villages but records their further and perhaps equally surprising unanimity in staving off real or imagined threats to their coherence and order. In both Fitili and Zaccanopoli the sense of community is expulsive and claustrophobic.

> [It is not] enough to have been born and raised in the village to be considered part of it. One's forebears, as far back as genealogical memory goes, have to have been born there, too. People whose parents or grandparents came from even a neighboring community, say Alafito, which is little more

than a mile from both Fitili and Zaccanopoli, are still called strangers. People who have moved away continue to be considered locals, but descendants born elsewhere are not. Conversely, people born in Argentina or Milan, say, are considered Argentinian or Milanese, even if their forebears are local.

What this extended comparison makes plain is the interplay between space and time that distinguishes anthropological from historical temporality. History dates and times, while myth places and spaces our shared sense of past, present, and future—and our less fully shared sense of sequence and memory.

Myth and memory are often still thought of as dealing with natural objects, but these studies cast doubts on that assumption. Too many of their discoveries and conclusions escape common sense or naturalist explanations. Since Minicuci's two villages exchange spouses and live in contiguous space, for example, why should they assess the past so differently? If nature is common, why should memory differ? The answer may lie in a rethinking of the question, turning it against the natural. We might say instead that there is no such thing as natural behavior and that memory and myth, time and history, fall under this general rule. Farriss, Grafton, and Trautmann show, in addition, how complex are the temporal systems elaborated within the societies they have studied. Why should we assume simple causes for such complex results? We might be wiser to extend to our other terms what Jonathan Z. Smith has concluded about myth.

> As we have come to learn, myths are not about nature. They are not best understood as "primitive" attempts to explain natural phenomena. Myths often think *with* natural objects; they are almost never about them. Their focus is not on the genealogy of things but on the topography of relationships.

Smith expands that succinct distinction by citing the view of Lévi-Strauss that it is a mistake "to think that natural phenomena are *what* myths seek to explain, when they are rather the *medium through which* myths try to explain facts which are themselves not of a natural but a logical order."[13]

From this perspective we may be able to position our rethinking of myth and history, of time and the temporal, at the center of a larger rethinking in culture and philosophy. This rethinking might best be identified by echoing Richard Rorty's call for "a repudiation of the very idea of anything—mind or matter, self or world—having an intrinsic nature to be expressed or represented."[14] Our understanding of myth and time, and of our concepts of myth and time, should start to think of them as *sources* of knowledge rather than as representations of nature. This pragmatic regard is clearly one we would share with those who create temporal discourse, whether it be Calabrian villagers recounting genealogical memory or Mayan astronomers creating cosmologies. We would, like them, hope and try to *learn* something by reflecting on duration and change. The discourses of myth and history thus become tools, like Rorty's philosophical language, intended to help us cope with shared situations and shared events. Our temporal discourse would and should come to possess the value that James Fentress and Chris Wickham attribute to social memory, which they describe as

> a source of knowledge. This means that it does more than provide a set of categories through which, in an unselfconscious way, a group experiences its surroundings; it also provides the group with material for conscious reflection. This means that we must situate groups in relation to their own traditions, asking how they interpret their own "ghosts", and how they use them as a source of knowledge.[15]

Taken together, our various contributors show how forcefully we and they *construct* time and how expansive that spatial term has been in the making of human temporality. Myth and history are both directed toward explaining how things got the way they are, or are said to be, and our cultural shaping of time leans heavily toward creating significance and heavily against deconstructing meaning. We have few narratives, and even fewer rituals, that celebrate the power of time to erode or trash our lives, that give focus to the ways in which duration and repetition drain out value and let boredom in. Themes and events signifying breakdown and ruin become, on the contrary, as Hughes suggests, provocations to create a new sense of time: the fall of France in 1940 provoked a shift from historical *longue durée* to structural ethnography. In exploring the ways in which the Maya, through their calendars and chronicles, participated in the making of their history, Farriss casts doubt on structuralist suppositions about the ahistorical or even atemporal features of cyclic myth and consciousness.

> The Maya did not therefore assume the attitude of helpless spectators. They had a crucial role to play in the cyclical drama, through their calendrical computations, their related rituals, and even their wars and revolts. Theirs was the task of helping the gods to carry the burden of the

days, the years and the *katuns* and thereby to keep time and the cosmos in orderly motion.

Edmund Leach has gone so far as to argue that our sense of time as repetitive and irreversible, a sense as strong in Marxism as it is in the Judeo-Christian tradition, is ideological or religious, "one of those categories which we find necessary because we are social animals rather than because of anything empirical in our objective experience of the world."[16] Peter Rigby's insistence that the Maasai and his account of them exist within a unified historical order is faithful to the Marxist pledge to deliver humankind from false consciousness, to bring about a fully historical existence—even though Marx himself considered British colonialism a brutal necessity to abolish feudal and tribal institutions in India and Africa on the path toward modernization. This pledge, which emerges from Fabian's critique of the alienating impulses of anthropology, seeks to change the atemporal and undated into the present and located, to stop using our privileged temporal judgments as barriers to fence off the other. Our present judgments about human time themselves result from a salvage operation staged in the mid-nineteenth century after discoveries of human bones alongside those of long-extinct animals exploded Mosaic chronology, which limited human history to a few thousand years squeezed inside a biblical frame. This "revolution in ethnological time," as Trautmann has called it, opened a deep backward abyss of time in western thought. Most of human time now lay somewhere outside providential chronology, in a much longer "archaeological time span" closer in scale to the *yugas* of Hindu narrative than to the chronicles of Moses. And all this happened at the very moment of Darwin's ascent, during which social evolutionism became intertwined with a chronological trajectory stretching back to a vanishing point in the remote past. Just as European voyages of five hundred years before tried first to confirm a squeezed cartographic vision of the globe, then to salvage its remains, and finally to exploit a new and dizzying spatial order that arose out of its ruins, so did social and physical anthropology develop outward and "backward" into a new time frame that was at first as expansive and dangerous to faith as Copernican space or Columbus's Ocean Sea. Throughout these revolutions in time and space the energy that was increasingly and finally directed toward drastic and irreversible change was originally directed toward the restoration of prophetic and providential time. The structure of time has repeatedly been consumed by the zeal of a western quest to confirm it: empirical discoveries about time begin in attempts to renew worn faiths and mythologies.

Read in sequence, the contributions by Trautmann and Bernard S. Cohn, show this process developing in nineteenth- and twentieth-century confrontations with Indian narrative and society. After partial and vexed attempts to set Hindu cosmogonies beside Judeo-Christian and Enlightenment accounts, western scholars came to acknowledge, if not to accept, the otherness of Indian time. Always, however, with an

> intensity of feeling . . . that there is more at issue than time alone. It is a question, we sense, of time being an integument that joins the individual to the object of a quest, a source of knowledge and meaning, that lies beyond time: the pilgrim's progress toward salvation, or the nation's progress from rudeness to civilization. It is this sense that is baffled and affronted before the immensity of Indian time.[17]

We are only slowly coming to see, as Trautmann goes on to point out, that this "immensity of Indian time" may account not only for some gaps and lacks in that culture—biography and geography, for example—but also for some of its great achievements in algebra, structural linguistics, and now in computer science.

If we turn back from this large-scale, "big time" sense of chronology to the "local time" of Cohn's Indian village, we should be struck by analogies as much as by contrasts. The villagers of Senapur, as Cohn makes clear in a series of elliptical accounts, have not one but many pasts, according to their grouping in one of the village's twenty-three castes, which in turn often differ there from their significance in other parts of India, as well as their membership in a variety of sects and political movements. All this is complex enough, but it becomes even more so when viewed in the light of Cohn's thesis that Senapur has not one but two narrative pasts. One is the "traditional past" of the dominant Thakurs, derived from the *Ramayana* and *Mahabharata,* which are at once their epic, their genealogy, and the history of their dominance over the other villagers who derive their past in turn from this traditional account. The other is a "historic past"—though one much less conclusive than the term might suggest—that recounts a sort of clan chronicle of ancestors from the sixteenth to the middle of the nineteenth century. Although only hinted at in his article's allusions to Gandhi and nationalism, there would also seem to be traces of what we might call a "modern past" of new crops, technical change, and political division. Taken together, these pasts argue for a temporal sense every bit as intricate as the "big time" of high Indian culture and a social order more Byzantine than bucolic. The excess of structure in this village society seems to have been there still when Cohn returned in 1970 and again in 1990—manifest through new paths, canals, and metric systems. Senapur is not "modern," Cohn concluded a generation ago, because it participates in many traditions and pasts rather than in one; and that multiplicity is as complex as the works of the great sixteenth-century clock of Strasbourg, offered by Anthony Grafton as an emblem of Renaissance chronology, which itself seems to respond to energies and needs analogous to those of Senapur.

Although worlds apart in many respects, Renaissance time and Indian village pasts share an appetite for proliferating systems that marks so many expressions of temporality. Human attempts to measure or remember time show many qualities from one century to another and from China to Peru. Simplicity has never been one of them. The multiple pasts of Indian villages, like the many faces of Renaissance escapement clocks, bear witness to impulses and patterns closer to chaos and complexity theory than to principles of parsimony or conservation of energy. "Gorgeous and complex as a Mannerist grotto," in Grafton's description, the Strasbourg clock was an icon and index of time and eternity, of astronomical revolutions and of temporal allusions: "everything from history and poetry, sacred texts and profane ones," claimed its creator Dasypodius, "in which there is or can be a description of time." Its system is a chronometric analogy to metric structures and histories like Dante's *Divine Comedy* or Milton's *Paradise Lost*, which also aim at an inclusive and encyclopedic history of human time, one in which the metronomic movement of verse parallels the ticking of the clock, at once dividing time and driving it into the future. Such architectonic structures, literally towering in Strasbourg, remind us of a point made by Farriss in analyzing the complex relations among the Maya between human history and cosmic time: "people do not record their past so much as construct it, with an eye to the present, and at the same time use that past in molding the present."

A striking example of this structuring and creative sense of time is what Jonathan Wylie named and sketched in our closing discussion as "radial time." Although not the explicit subject of any one paper, a recurrent topic in our discussions and in the studies that follow has turned out to be the composite and mixed character of patterns Wylie identified by this name. In a widespread process led by works of art, figures and events from different moments of sacred and profane time can be mingled or composed in paintings and narratives like segments or panels separated by radial divisions that nonetheless run out to a circumference or frame that unites the whole. Saints and Madonnas appearing together with contemporary patrons in medieval triptychs; calypso songs recounting the deeds of Columbus along with those of movie stars and astronauts; war memorials showing dead airmen borne aloft by Winged Victory; all are examples of radial time. We might add to these examples most records of religious experience that combine ordinary time with numinous states. And, as a member of our audience pointed out, by contrasting our academic constructs of historical and ethnographic time with shamanic time, we limited ourselves to conceptualizing rather than venturing or entering into other ways of *experiencing* time.

Time is always pressing, on the world of ethnology no less than on the world of history. The pressure on the present, whether to recount Mayan *katun* rounds of 260 years, to calculate in India that eighteen twinklings of an eye make a second, or for a Portuguese navigator to date a cross raised on the African coast "in the year 6685 of the creation of the earth and 1485 after the birth of Christ," was a pressure to make redemptive order out of the wastage of hours and waste of ages. Taken separately, unmeasured time and measureless time may rob human history of value and significance. But taken together, multiplied and calculated into chronological systems, they confer meaning on the void, rather as multiplying negative numbers yields positive results. In the present of Grafton's early modern Europe, in which scientific curiosity blends with reforming zeal, the structuring and measuring of time keeps it from getting lost, from slipping away wasted and unused,

> for the guilty, secret obsession of early modern society was neither sex nor money but the desperate desire to use time well and the pervasive fear that wasted time would waste those who abused it.

The ethical import of ethnological time is expressed even more forcefully by Farriss.

> The key, I think, to the Mesoamericans' conception of time and to their entire cosmology is their preoccupation with order and above all with cosmic order. Paul Ricoeur links this type of preoccupation to a perception of evil as a force external to humans. Evil is disorder, represented by the chaos that preceded creation and that constantly threatens to reassert itself. The drama of creation is therefore an ongoing one, for the cosmic order must continually be reaffirmed in the face of this ever-looming chaos. Time is part of the cosmic order. Indeed, for the Maya and the rest of Mesoamerica, time is cosmic order, its cyclical patterning the counterforce to the randomness of evil.

Just as the coherence of historical narrative comes to be valued as a triumph over meaninglessness—a victory often gained at the expense of truth—so does the intricate precision of chronologies come to be valued as a good in itself, as a triumph over that evil the Maya "characterized as deviation from expected patterns."

Such precision about time was in several senses an externalizing or "outing" of disorder as evil, so that it could be perceived as external and projected onto the other. This externalizing or expulsive process took a long time in both human history and ethnological time. It could also be described with some bitter irony, as a process of modernization. Lévi-Strauss once distinguished two basic attitudes toward the other by contrasting societies that drove out their enemies from those that ate them up.[18] Certain modern societies might in this spirit be characterized as those that tend to disapprove of incest or cannibalism rather more forcefully than they do of ethnic cleansing or mass murder. And it is in growing skepticism about intricate external chronologies, in the search for something more persuasive, if less precise, that another kind of modernity, one we identify with the Enlightenment, begins to develop. Both Grafton and Hughes point to Vico's *Scienza Nuova* as a critique of the chronic inability of the chronologers to get their dates right and, beyond that, as a work that advanced the psychic time of the imagination, tellable through the "archive of the Past" of law and philology, of legal codes, folk customs, and poetic texts. This active temporal imagination, replacing the theater of memory and cosmic mensuration of the Renaissance, did something decisive to bring what was to become anthropological time within the historical awareness of western culture. It did this by changing the Mayan and Alsatian paradigm of cosmic time and human history, macrocosm and microcosm, big time and local time, to Vico's paradigm of the "return" of *ricorso*—in itself a temporal rather than a spatial concept—through which the individual psyche and speaker reenact the experience of time of the childlike and the primal, among them Homer's warriors, preserved in the language of the tribe and the texts of the past. The Enlightenment should not be mocked for its figure of the noble savage because that figure brings "otherness" into the time of our own lives, projects our experience of time into that of others apparently very different from ourselves. In so doing it expands the field of temporal and historical inward consciousness as decisively as Darwin, Boucher de Perthes, and the excavations in Brixham Cave were to transform external time frames a century later.

If one impulse toward modernity was to create and then isolate the "other" in a strange time of the primitive, a second, more recent impulse has been to follow Vico and Fabian in seeing the other as coeval with ourselves. That perception, as Rigby shows in the postscript to his article, involves a spatial and material cognition and cognation that he and we, "Pita" as observer and Maasai as co-observer and observed, share a historical order of time and space as well as the sublimated distance of ethnographic production. Although Rigby would hardly accept Geertz's larger thesis that distinguishes so sharply those who share a social space and consciousness from those who share a temporal and historical awareness, he could rightly be described as a "consociate" as well as a contemporary of the nomads he has observed and been observed by. The cast of the play of anthropological time is larger than the number of actors and native informants. And Rigby is surely right to extend Fabian's charge against the "tendency to place the referent(s) of anthropology in a Time other than the present of the producer of anthropological discourse" to criticize further the tendency to place those referents in "another epistemological 'Space.'" Now that the "strange time" of imperial anthropology has faded away, along with virtually all of the "primitive" societies that provided that time's separate space and ghetto, we must strive to restore the Enlightenment's awareness that humankind lives within a turbulent stream of time. Within it reflection and self-analysis have the power to make our own, or at least akin to us, what reading and observation set at a spatial and temporal distance. We need this power above all in turning, as Rigby does, to African societies increasingly bracketed and made remote by the camera lens, by the elongated perspectives of famine and AIDS, by the infrared gunsight.

What Marx read in Vico was not only that the past lies on us all like the Alps but also that the conflicted textual struggle of the past should enter into and animate our present studies in anthropology and history. Ending the temporal exile of the ethnological world puts within reach, but also within harm's way, groups and societies that have long been fenced off behind walls and palisades. Many of these were clearly fragile or nonexistent, like Geertz's ceremonial Balinese, who could not keep out the slaughter in 1965 of thousands by anti-Communist violence that shared the same temporal and spatial world as its victims. Others were much higher and thicker than even Cold War rhetoric could admit, like the Wall separating East and West Germany. Its collapse revealed not only differing time frames and conflicting interpretations of events, which historians and ideologues had prepared us for, but also a yawning gap in language and consciousness that needs bridging by ethnographers and sociologists.

The world we live in has extended outside the world of art and into our daily lives the coexistence of tenses, of radial time. Here historians could learn from anthropologists whose recent work—here represented by Barnes and Wylie—juxtaposes and presents this composite sense of time in actual communities. Indonesians and Faroese live this composite as concurrent, like the festivals and rituals of societies that are telecast from within their societies by their own broadcasters onto the journalistic scene and screen of CNN. Historians, for all they may pride themselves on a more supple sense of time, have shown much less willingness to admit the often conflicting testimony of present witness into the archival record of the past, to admit the populist or fanatical significance of yellow ribbons and *fatwas* into the historical record. There is a past sense in which Grafton is utterly right to say of the Renaissance, "The sense that the past is another country had to arise before intellectuals could develop techniques for mapping that country's borders and features in accurate detail." The anachronism of traditional thought had to give way to the historical. But there is another present sense in which we have to recognize that the past is not only another country but also *this* country; that we are all, temporally speaking simultaneously at home and strangers here. The phrase and notion that "the past is another country" has passed so quickly into the language that we are likely to forget where it comes from and what it implies. It echoes, with a telling change of one word, the first sentence of a nostalgic English novel, L. P. Hartley's *The Go-Between:* "The past is a foreign country: they do things differently there." The past it refers to is not the historian's past but the time remembered of childhood recalled in old age—another kind of radial time.[19]

As we reflect on the studies gathered here, we might close by sharing some thoughts of the anthropologist Janet Hoskins who concludes a recent ethnographic study of the Kodi of Indonesia, read through their perception and representation of time, by reminding us that

> the particular consciousness of time realized in Western historical writings is, of course, culturally constructed; it is especially evident in the forms of history that assume a collective unity encompassing all individual sequences of events.[20]

What she finally advocates, however, and what these studies could be a step toward, is an "alternate temporality" that merges calendrical, biographical, and historical time, one that is guided, in turn, by a narrative power capable of telling and untelling our temporal experience.

NOTES

1. Clifford Geertz, "Person, Time, and Conduct in Bali," in *The Interpretation of Cultures* (New York: Basic Books, 1973), 360–411.
2. Maurice Bloch, "The Past and the Present in the Present," *Man*, n.s., 12 (1977): 278–92.
3. Leopold E. A. Howe, "The Social Determination of Knowledge: Maurice Bloch and Balinese Time," *Man*, n.s., 16 (1981): 220–34.
4. R. H. Barnes, *Kedang; a Study of the Collective Thought of an Eastern Indonesian People* (Oxford: Clarendon Press, 1974).
5. Peter Munz, *The Shapes of Time: A New Look at the Philosophy of History* (Middletown: Wesleyan University Press, 1977), 35–36.
6. Johannes Fabian, *Time and the Other* (New York: Columbia University Press, 1983), 99.
7. R. W. Southern, "Aspects of the European Tradition of Historical Writing, the Sense of the Past," *Transactions of the Royal Historical Society*, 5th ser., 23 (1973): 127.
8. George Kubler, *The Shape of Time: Remarks on the History of Things* (New Haven: Yale University Press, 1962), 2.
9. Hayden White, *Metahistory* (Baltimore: Johns Hopkins University Press, 1973).
10. This account is indebted to Munz, especially pages 211–12.
11. Hayden White, "The Metaphysics of Narrativity: Time and Symbol in Ricoeur's Philosophy of History," in *The Content of the Form* (Baltimore: Johns Hopkins University Press, 1987), 169–84 [170].
12. Paul Ricoeur, *Time and Narrative* (Chicago: University of Chicago Press, 1984–88).
13. Jonathan Z. Smith, "A Slip in Time Saves Nine," in *Chronotypes*, ed. John Bender and David Wellbery (Stanford: Stanford University Press, 1991), 67–76 [71]. The quotation from Lévi-Strauss appears in his *The Savage Mind* (Chicago: University of Chicago Press, 1966), 95.
14. Richard Rorty, *Contingency, Irony, and Solidarity* (Cambridge: Cambridge University Press, 1989), 4.
15. James Fentress and Chris Wickham, *Social Memory* (Oxford: Blackwell, 1989), 26.
16. Edmund Leach, "Two Essays Concerning the Symbolic Representation of Time," in *Rethinking Anthropology* (London: Athlone Press, 1961), 125.
17. The quotation is from Trautmann's essay. See, too, his recent study "The Revolution in Ethnological Time," *Man*, n.s., 27 (1992): 379–97, to which I am indebted here.
18. Georges Charbonnier, *Entretiens avec Claude Lévi-Strauss* (Paris: Julliard, 1961), 34–36.

19. L. P. Hartley, *The Go-Between* (London: Hamish Hamilton, 1953), 9. Hartley's novel had gone through ten printings by the time Joseph Losey made his film of the same title in 1971. Soon afterward the opening sentence appeared in the *Oxford Book of Quotations,* which suggests that film, novel, and reference work contributed to the radial currency (and revision) of the opening sentence.

20. Janet Hoskins, *The Play of Time* (Berkeley: University of California Press, 1993), 381.

Part 1
Local Time

The Pasts of an Indian Village
Bernard S. Cohn

Events of the last two hundred years have led many scholars to divide the societies of the world into contrasting pairs such as developed-underdeveloped, advanced-backward, traditional-modern. Central to these concepts are not only ideas about the level of technological development but the very character of the societies and cultures that are compared and contrasted as well. Professor Edward Shils has recently remarked that the traditional society is one in which, among other things, there is a strong attachment to the past, by which behavior is determined and validated.[1] Solutions to problems old and new are determined by the past of the society, and modernization quickens when ties to the past begin to be cut and new criteria for determining and validating behavior are invoked.

Rural India would appear to be a typical traditional society; most behavior observed today has its roots deep in the past in customary behavior. The technology, the social structure, and the ideology of the Indian peasant seem to epitomize attachments to the past. It is the argument of this essay that, although the attachment to the past is undoubtedly there in the life of the Indian peasant, unless we ask what past or pasts does a given peasant refer to and unless we fully understand the potential complexities of the pasts of a traditional society, the dichotomy of traditional and modern may prove illusory.

Senapur, a large village in the Gangetic valley, has been described extensively in the anthropological literature.[2] It is a Thakur-dominated, agricultural village with over two thousand residents and twenty-three different caste groups. It lies in a heavily populated region twenty-five miles north of the city of Benares. It is not directly within the orbit of an urban-industrial center, but over the past fifty years members of the village have had a considerable degree of urban contact, and, although this contact and the general changes that have affected northern India have affected the village, the basic structure and culture of the village remains.

When the questions "What is the past of this village?" and "What is the past to which people refer to determine and validate behavior?" are asked, it becomes apparent that these questions cannot be answered in the singular but must be answered in the plural. There is not one past of the village but many; the following essay is an exploration of these pasts. I find it convenient to talk about two types of pasts: a traditional past, which grows out of the mythology and sacred traditions of north India, and a historic past, which is a set of ideas about the remembered experiences of a group of people in a local region. The traditional past usually refers to matters religious and cultural and the historic past to matters social and political.

In general terms, the varieties of pasts, traditional and historic, function in the same way for different groups (in this case, castes) in the village; but their content varies widely. The traditional past functions to validate a present social position and to provide a charter for the maintenance of that position or the attempt to improve it. It provides a much wider framework than do the local historic pasts. The traditional past relates particular groups to an extensive social network, in some cases stretching over most of north India. The historic past explains, supports, or provides a basis for action in the local social system. Analytically, the contents of the various pasts can be treated under these two rubrics, but the complexities of the differing contents are important in understanding the process of modernization in the society. They should not all be subsumed under one "attachment to the past." The student of the society must be able to include in his analysis the multiplicity of the pasts of the society and be prepared to understand how modernization affects not one past but all the pasts.

The Contents of the Pasts: Thakur

The traditional past of the Thakur explicitly goes back to the Ramayana and the Mahabharata. Rama and Sita, principals of the epic, are more than semi-divine folk heroes to the Thakurs of Senapur; they are their ancestors. The values and moral imperatives of the Ramayana are formally maintained for the Thakurs of the village by the annual presentation of the Ram Lila. The Ram Lila, which in Senapur lasts for six days, is the acting out by the Thakurs of the village of events in the Ramayana. The form is that the younger Thakurs of the village take the dramatic roles of the principal characters while the older Thakur males chant the verses that describe and comment on the acted-out portions of the story. As presented in the village, the Ram Lila is not a fixed traditional past, although the story is an old one and was originally reduced to writing over two thousand years ago by Valmiki. The version that is known in Senapur is the Ramacaritamanasa of Tulsi Das, and it dates from the sixteenth century. The Ram Lila is flexible enough to continue to incorporate the recent past. In some sections it may be used by some as a vehicle for comments on current behavior.

One of the first scenes acted out is the winning of Sita, daughter of Janaka, king of Videha, by Rama. Janaka has a contest to which all the leading Rajas are invited. The prize is the hand of Sita. The feat of strength and skill required is the drawing of Siva's bow. In the current Senapur version of this contest, the Raja of Manchester appears. He is dressed in a caricature of western dress, with an ill-fitting white suit and a large pith helmet. He looks much like a railway official and speaks in an English gibberish that sounds like announcements in an Indian railway station.

The maintenance of the legendary past of the Thakur is found not only in formal presentation of this past. The stories, incidents, and characters of the Ramayana are referred to frequently in daily life. Many of the names and nicknames of villagers come from this work. Events in the village are discussed on the basis of their similarities to events in the Ramayana, and the explicit moral judgments made by many Thakurs derive from the values of the traditional past. In addition, the traditional Thakur past includes the Thakur heroes of the eleventh and twelfth centuries, particularly Prithviraj, who battled the Muslim invaders of the time.

It might seem that this would be enough of a past for a "simple peasant people." But the Thakurs frequently invoke another past, which I label the historic past. This past relates to the period from the sixteenth century to the middle of the nineteenth century and focuses on known ancestors and events involving these ancestors. The Thakurs of Senapur (with the exception of two families out of fifty-six who are related affinally) are all descendants of Ganesh Rai, a Raghubanshi Rajput who in the sixteenth century migrated to the present Jaunpur district from what is today central Uttar Pradesh. With his two sons he established suzerainty over the existing population in a region of about seventy square miles, containing today a little over a hundred villages. In the Thakur version of the historic past, this population was led by the Soeris, a semiaboriginal people. Today there are well over twenty thousand descendants of Ganesh Rai in this local area (*taluka*). Ganesh Rai and the genealogical connections of the Thakurs in the village and lineage mates out of the village are a constant reference. The seemingly simple question of

"Whose land is this?" is almost invariably met with the reply "Ganesh Rai had two sons," and then the land is traced through fifteen or sixteen generations. All the lands in the village and the land throughout the *taluka* follow the descent line from Ganesh Rai. Every adult male Thakur can trace his descent from this ancestor. Two Dobhi Thakurs meeting for the first time will sit down and compare their formal relationship and appropriate behavior to each other. Not only is the formal kin history of the Thakurs discussed, but events in the lives of ancestors are talked about. Traditional enemies, such as the Raja of Benares, who in the eighteenth century tried to dislodge the hold of the Dobhi Thakurs on their land, continue to be castigated.

The historic past is not only a unifying factor but carries a divisive component as well. Feuding relationships in the village and in the *taluka* are constantly revived by reference to the historic past. Often I would be told a tale of one Thakur household burning out another in a dispute over landownership or following an assumed insult. I would suppose that the event had taken place within the last ten years, but after careful checking I would discover that it had taken place three generations ago. So far as the families are concerned, however, it did take place in the recent past.

The two pasts described for the Thakurs can also be found in almost every caste group in the village. Since the other twenty-two castes represented in the village all have at least mythological origins of their castes, one can talk about the twenty-three legendary pasts of the village. Until 1952 the other castes were the tenants and followers of the Thakurs, and their historic pasts are tied to the historic past of the Thakurs. I will illustrate these differing pasts with four groups: the Chamars, the Brahmans, the Telis, and the Muslims.

The Content of the Pasts: Chamar

The Thakurs are the top of the village and *taluka* stratification system. The Chamars—traditionally the agricultural workers of the Thakurs, whose ascribed occupation is skinning and tanning—are one of the lowest groups in the system. The Chamar traditional past has two elements. The first is seen in their origin myths, in which through trickery or accident they lost their previous high status, either as Thakur cultivators or as Brahmans, and were reduced to their present low status. There are several widely differing myths about this, the specifics of which are not important for my purposes here. The other element of the traditional past refers to a series of Chamar holy men of the fifteenth and sixteenth centuries. The most famous of these was Raidas, who achieved great spiritual power as a follower of the Ramuanja sect, which taught that emancipation could be attained through devotion (*bhakti*). Raidas and the Bhakti traditional past provide a way out of the present degrading position of the Chamars. I have suggested elsewhere that this part of the traditions of the Chamars has been brought to the rural Chamar largely through those Chamars who have become educated or who have had urban experience in recent years.[3] Thus the diffusion of this traditional past is a function of modern conditions, if not of modernization itself.

The historical past of the Chamars is a shallow one. At best Chamars can give their genealogies for only three ascending generations. I had to get information about Chamars' ancestors from their Thakur landlords. Some of the older Thakurs had knowledge going back farther than that of the Chamars themselves. There are three aspects to the Chamar historic past. There is a generalized feeling that things were much better fifty years ago when their Thakur landlords were richer than they are today because there were fewer of them and fewer Chamars. The landlords left their Chamar workers pretty much alone to carry on the agricultural operations in the village, while the Thakur was content to sit on the veranda of his men's house and smoke his *hukka* (water pipe). Today, on the other hand, the Thakur landlord is out in the field with the Chamar, directing his every move and in some cases working side by side with him. The Chamar's view of the past also includes the major events in the Thakur household to which he and his father had been attached. A Chamar describes with as much detail as would the landlord himself a wedding party or a fight in his landlord's house. Until very recently, part of the status of the Chamar within the Chamar caste in Senapur was determined by the status of the Thakur house to which he was attached. A rich house could afford to let more land to its tenants than could a poor one.

Until very recently, the Chamars of Senapur had no corporate historic past such as the Thakur have. One of the effects of recent social and economic changes in the society at large has been the beginning of a corporate past for the Chamars. In Senapur this results from a movement in 1948 and 1949 by the Chamars and other low-caste groups in which they contended for political power within the village. The dramatic events of this contest are now part of the historic past of the Chamars and provide a guide for current action.[4]

The Content of the Pasts: Brahman

Most scholars, when they think of Indian traditions, evoke certain aspects of what might loosely be called the Brahman tradition. Elements of this tradition may be characterized as follows: knowledge of and respect for the Vedas and Shastras, emphasis on spiritual activities, extensive ritual competence, concern for ritual purity, and otherworldliness. There are no resident Brahmans in Senapur today, although about thirty years ago there was a Brahman family. For most of the normal ritual services that Brahmans perform, the Thakurs and other castes of the village depend on a village of Brahmans about a mile away from Senapur. Theoretically, their legendary past should be very different from that of the Thakurs and should emphasize the themes noted above. In practice their legendary and historic pasts are much linked with the Thakurs' pasts, since they acknowledge ties to the Thakurs, who settled them in their village to provide themselves with ritual guides. The Brahmans around Senapur are much involved in the Ram Lila. Several of the younger Brahman men are leading exponents of a typical Thakur sport—wrestling. Brahmans derive most of their income not from their priestly functions but from their lands, as do their Thakur employers.

The Brahman tradition in Senapur is further complicated by the fact that some Thakurs in the village are ardent followers of the Arya Samaj, a Hindu reform movement of the late nineteenth century. The Arya Samaj emphasized the purification of Hinduism and attempted to base the revitalized Hinduism on its interpretations of the Vedas. The Arya Samaj was strong enough in the period before the First World War to make most of the Thakurs vegetarians and nonsmokers. Vegetarianism and nonsmoking are distinctly non-Thakur traits and are closely associated with the Brahmanic tradition. The dilapidation and eventual collapse of the village temple also date from the Arya Samaj period in the village. There is one interesting result of the Arya Samaj in Senapur: one meets individuals who seem to be bearers of the great Brahmanic tradition because they quote the Vedas and talk in the way that textbooks on Hinduism or Indian philosophy lead one to expect Brahmans would talk; however, these people are Samajists whose immediate tradition is a result of modernization.

The Brahmans' historic past, since they were brought into the region by the Thakurs, is tied to the Thakur historic past. They, too, refer to Ganesh Rai and his descendants and the fights of the eighteenth and nineteenth centuries. But, in addition, one type of historic past that Brahmans in surrounding villages recognize refers to their land grants, which were obtained as maintenance for their performing ritual functions, or special grants from particular Thakur households or lineages on particular auspicious occasions for the Thakurs concerned.

The Content of the Pasts: The Telis

The Telis are a low caste whose traditional occupation is oil pressing but who for many generations in Senapur and the surrounding region have been petty shopkeepers and traders as well. They are literate and because of their occupations probably have been literate for many generations. During the last one hundred years they have been trying to raise their social status. They have been successful to some degree: fifty years ago Brahmans and Thakurs would not take water from them, but now they do. Their social mobility is important in the Telis' structuring of their traditional and historic pasts. Today their traditional past includes major emphasis on the Brahmanic tradition. Like the Chamar traditional past, it includes an explanation for their low social status; however, the Teli fall from grace was not the result of trickery or accident but of the Telis' militant defense of Hinduism in the face of the Muslims.

The Telis have also been affected by the Arya Samaj. There are several Teli Samajists in the village. The Arya Samaj in Jaunpur district is generally looked upon now as being in the hands of Telis.

The historic past of the Telis centers around two periods. One began about seventy-five years ago and lasted until about 1910. This was a period of great prosperity derived from the manufacture and sale of sugar. The grandfathers of the present Telis in Senapur built and managed several sugar factories in Senapur. These were not large-scale industrial operations but small-scale local industries. When they were running they might employ fifteen or twenty people in their operation. In addition to manufacturing sugar from cane, some of the Telis carried on extensive middleman functions in the sugar trade. This period of prosperity is the "good old days" for the Telis. At that time they had large houses, many servants, good clothes, and good food. They could entertain the village and pay for religious rituals, and they had in their debt many of the Thakur landlords. Around 1910, with the spread of large-scale industrial sugar refineries in western Uttar Pradesh, the demand for the locally produced sugar diminished and Teli prosperity decreased. Some Teli families who had been wealthy during this period turned to urban pursuits in business and the professions. The Telis left in the village reverted to their roles as oil pressers and traders and once again came into the hands of their Thakur landlords.

The other aspect of the historical past is the Teli identification with Mahatma Gandhi and the Nationalist movement. Many of the leading figures in the Nationalist struggle in Jaunpur district were Thakurs, and in several cases they came from Dobhi *taluka*. But, except for personal identification with particular Thakur leaders, the Thakurs in Senapur appear to have been largely untouched intellectually by the Nationalist movement. The only group that continually refers to Gandhi and Gandhian symbols and ideas is the Telis. Gandhi is viewed as a validation of the Teli identification with the Brahmanic tradition. Some of Gandhi's social ideas fit in with Teli attempts to enhance their social status. The Teli identification may be the result of Teli urban ties; their occupation takes them to the local bazaars and to Benares and Jaunpur; they have many relatives who live in urban centers. This aspect of the Teli historic past has an urban as well as a local referent and support.

The Content of the Pasts: The Muslims

The Muslims of village Senapur, who form roughly five percent of the village population (fifty out of a little over two thousand people), have traditional and historic pasts radically differing from those of the other residents of the village. The pasts are connected to events and ideas that have their roots in the eighth-century Arabian Peninsula and in the rise and spread of Islam in the subsequent centuries. Unlike any other group in the village, they have a traditional past that ties them to peoples outside of India. Their religious leaders read and comment on the Koran and interpret Koranic law for them. Their rituals, life-cycle ceremonies, and festivals are markedly different from those of their Hindu neighbors. There are some village religious activities in which the rituals of the Hindus and Muslims are merged: Hindus worship at some of the Muslim shrines and graves in the area, and fifty years ago there was a sect, the Pancho Pir, which had Hindu and Muslim followers. Part of the effect of the Arya Samaj was to end Hindu participation in this syncretic religious sect.

The historic past of the Muslims is also different. The history of Senapur that the Muslim relates is significantly different from the one related by the Thakur. The Muslims believe that the village was wrested by the Thakurs not from the semiaboriginal Soeris but from four Muslim landlords who had subdued the Soeris previously. Although they do not claim direct kinship with these ancient landlords, it is still a matter of pride among them that Muslims held the village before the Thakurs did.

Modernization and the Attachments to Pasts

I have been documenting one point, which is that if we characterize India as a traditional society we must be prepared to analyze not one tradition and one past but many traditions and many pasts. Similarly, in thinking about modernization of such a traditional society, we must think in the plural—about effects on multiple segments, each attached to its own peculiar past, within one traditional society. In a changing social situation we expect to find a transitory attachment to the past, often in the irrational form of a nativistic movement. It is significant that the anthropological literature on this sort of attachment to or revitalization of the past has focused on the more spectacular messianic cults such as the Ghost Dance, the cargo cults, and the Mau Mau. But we have not yet analyzed the complex interaction of modernization and traditionalization, such as is found in the Arya Samaj, or the traditionalization of Chamar religious life as a result of literacy and urban experience. The apparent conflicting values, institutions, and behavior found in India seem to our minds rationally incompatible. We cannot build their coexistence into our theories of change; we can only describe them. Perhaps it is characteristic of members of modern societies to believe that consistency in social and intellectual life is a prerequisite to efficient functioning of either a social system or a theory.

The description of the multiple pasts of one Indian village does not invalidate categorization of societies into traditional and modern, but it may point to another dimension necessary in the categorization. All Americans share a past created by our educational system and media of mass communication. We can invoke this past and have it be meaningful across regional and class lines. Indians do not as yet share such a past. An appeal for action on the part of the central government, based on what is thought to be a universal identification with a traditional or historic past, is meaningless or leads to antagonistic reactions of major parts of the population. The Brahmanic past can be an anathema to non-Brahmans in Maharashtra or south India. Evoking the "First War of Indian Independence" (the Sepoy Mutiny of 1857) means nothing outside of the Punjab, Uttar Pradesh, Bihar, and a few parts of western India, except to the urban educated classes. The evocation of Mahatma Gandhi is sometimes viewed by Thakurs in north India as symbolizing not the new India but their being dominated by socially inferior merchant castes.

I would speculate that a society is modern when it does have a past, when this past is shared by the vast majority of the society, and when it can be used on a national basis to determine and validate behavior.

NOTES

1. Shils's views were presented at a seminar of the Committee on the Comparative Study of New Nations at the University of Chicago. These remarks stimulated me to

reexamine some field data that I had collected in a north Indian village in 1952–53 to try to determine what pasts were to be found in this village. My thinking about the problem was furthered by discussions of an earlier draft of the essay with Professors Milton Singer, Sylvia Thrupp, and Eric Wolf.

2. The basic description of the social, religious, and economic structure can be found in Morris E. Opler and Rudra Datt Singh, "The Division of Labor in an Indian Village," in Carleton Coon, *A Reader in General Anthropology* (New York, 1948), 464–96; and "Economic, Political and Social Change in a Village of North Central India," *Human Organization* 11 (1952): 5–12. See also Morris Opler, "Factors of Tradition and Change in a Local Election in Rural India," in Richard L. Park and Irene Tinker (eds.), *Leadership and Political Institutions in India* (Princeton, 1959), 137–50; Rudra Datt Singh, "The Unity of an Indian Village," *Journal of Asian Studies* 16 (1956): 10–19; Bernard S. Cohn, "The Changing Status of a Depressed Caste," in McKim Marriott (ed.), *Village India* (Chicago, 1955), 53–77; and Bernard S. Cohn, "Some Notes on Law and Social Change in North India," *Economic Development and Cultural Change* 9 (1959): 79–93.

3. "The Changing Traditions of a Low Caste," in Milton Singer (ed.), *Traditional India: Structure and Change* (Philadelphia, 1959), 207–15.

4. Details of this contest are in Cohn, "The Changing Status of a Depressed Caste"; see also note 2, above.

The White Man's Book: The Sense of Time, the Social Construction of Reality, and the Foundations of Nationhood in Dominica and the Faroe Islands

Jonathan Wylie

The Settings

Dominica is the northernmost of the Windward Islands in the Lesser Antilles. The capital is Roseau. The official language is English, but for most of the predominantly Negro population of about eighty thousand the language of everyday life is a French-based creole called *patois*.[1] Dominica was granted full independence from Britain on 3 November 1978. From November 1977 to December 1978 I lived there in a fishing village called Casse. Casse had about eight hundred inhabitants.

The Faroe Islands are an archipelago of seventeen inhabited islands in the North Atlantic, roughly halfway between Iceland and Norway. The capital is Tórshavn. The total population of about forty-seven thousand is of predominantly Norse stock. The first language is Faroese, a West-Scandinavian tongue, but until recently the official language was Danish. Since 1948 the Faroes have been an internally self-governing dependency of Denmark. From June 1971 to April 1972, and during several briefer stays since then, I lived in a fishing village called Álvabøur (pronounced AWL-va-bö-vur). Its population was 536.

Álvabøur and Casse are alike in some ways. Both are fishing villages huddled between hills and the sea. In both you find men gathered to chat where they can keep an eye on the seas from which they draw a living. But the better you know them, the more these villages seem as different as the cold, cod-rich reaches of the North Atlantic and the warm Caribbean, from which the day's small catch includes a bewildering variety of fish, alike only in possessing exquisitely evolved assortments of teeth, scales, spines, scutes, and even wings with which to harry and escape each other. In Álvabøur the weather is the first topic of conversation, and so the first conversations I felt at home with in Faroese concerned winds and tides, sunshine, rain, and degrees of drizzle. Trying to push beyond this topic, I quickly found that the Álvabingar would tell me next to nothing about the political factionalism I had hoped to study. They go out of their way to get along with one another; dissension is discussed privately in scandalized undertones. Instead of party strife, I heard dozens of times about the founding of the village in 1833–34, and found myself collecting genealogies of astonishing depth and accuracy. After this experience, Casse was a shock. Whereas Álvabingar disparage idleness as well as dissension, the daily round in Casse featured both heated public debates and long stretches of "cooling out," when people joked and chatted idly or just dozed in the stern of a beached canoe. People seldom talked about the weather except as a matter of immediate practical concern; I was able, with great difficulty, to collect only a few, unimpressive genealogies; and no one seemed to know or care much about the history of the village.

This essay concerns the sense of time and the social construction of reality in Casse and in Álvabøur. They could hardly be more different. Casse's past is shallow and unimportant; Álvabøur's is deep and a topic of general interest. In Casse reality is shiftingly construed, often through argument, as a matter of received opinion, or else it is founded distantly in the antithetical world of white men's ways and God's word. In Álvabøur the social order is construed in terms of such portions of reality as historical facts and the order of nature. I also want to suggest a corollary of these differences with the broader intention of comparing the cultural significance of nationhood in an Afro-Caribbean society and a Scandinavian one. What the Dominican press called "the move to independence" was profoundly ahistorical and culturally threatening; in the Faroes, gradual separation from Denmark has seemed an almost natural fulfillment of cultural development.

Casse, 22 May 1978

Baptiste and I were sitting in a bit of shade by the shore. He was mending a net, cutting out weak spots and patching new netting in. I was squatting on a rock I'd pulled up by way of a chair. From time to time we would look out at the tepid expanse of the Caribbean, here formed into a bay by a low spit reaching out to a headland called Labatwi.

There are a few ruins on Labatwi—a crumbling parapet and a stonework tank for storing rainwater. Labatwi belongs to the Queen, Baptiste said. There used to be forts there: look, you can still see the ruins "where the soldiers were." A couple of old cannon lie half-buried in the sand at the base of the headland, and sometimes he has found glass bottles "as thick as my hand."

Baptiste said it was the French who built the fort. Dominica lies between Guadeloupe and Martinique, he went on, and if the English had held Dominica it would have kept ships from passing between the French islands. But during the First World War—"1914, you know"—the British captured the fort from the French. And so they won the war. This is why Dominicans speak French, or anyway the "broken French" of patois, although Dominica is English. "It's a story."

Nodding sagely, I thought to myself that it is very nearly the sort of story we call myth, in which the doings of the gods are recounted to chart out the state of the world as people know it today. Baptiste's story explains why there are ruins on Labatwi, why Dominicans are "English" but speak "broken French," and, by extension, one reason why Dominica's independence, which was then being planned, almost in secrecy, by the British and Dominican governments, was widely feared: the French might take the island back. But Baptiste serves up this story as a factual account—as history. It is impossible history, of course. No large battle, let alone the decisive engagement of the First World War, could possibly have been fought on Labatwi, and in any event the British and French were allies in 1914. Exasperated, one wonders what actual events Baptiste's account might echo.

Left largely to the Carib Indians until the early eighteenth century, Dominica had for all practical purposes become a French colony by the time the Seven Years' War broke out in 1756. Although the island was nominally neutral, the British seized it in 1761, hoping to disrupt communications between Martinique and Guadeloupe. To defend it they erected a series of strongpoints, including the one on Labatwi, at strategic spots along the coast. In 1778, France tried to recoup her losses in the New World by allying herself with the United States. One of the first French actions was against Dominica.[2] Early on the morning of

> Monday, the 7th of September, in that year, a French armament, consisting of a forty-gun ship, three frigates, and about thirty sail of armed sloops and schooners, having on board two thousand regular troops, and a lawless banditti of volunteers, about half that number, appeared off the island,

under the command of the Marquis de Bouillé, governor of Martinico, and general of the French Windward and West Indian Islands.

The first frigate bombarded Roseau briefly before launching a diversionary attack in the north of the island. The second landed about eighty men, under cover of darkness at the mouth of a ravine several hundred yards east of Labatwi, who took the battery by storm. According to English sources,

> strange as it may seem, the case afterwards appeared to have been, that some of the French inhabitants had insinuated themselves into the fort a few nights before, and having intoxicated with liquor the few soldiers that were there on duty, had contrived to spike up the cannon!

Its lines of communications secured, the main French force descended on Roseau where the British capitulated after some sharp skirmishing on its outskirts. France ruled Dominica until the island was restored to Britain by treaty in 1783. Dominica remained a British colony until 1967 when it was granted the semi-independent status of Associated Statehood.

This cannot be the full history of the storm of Labatwi. One would like to know who lived in the two estate houses by the bay, and why the "inhabitants" had such easy access to the fort. Why was the fort apparently undermanned, and why did its garrison fail to notice that their cannon had been spiked "a few nights before" the attack? We might answer such questions by consulting French, British, and Dominican archives. But this would lead us away from Baptiste and his story and relieve us too quickly of our initial exasperation. Really, how is it possible to confuse 1914 with 1778?

We cannot dismiss Baptiste himself, a trustworthy and intelligent man who has seen something of the world. Unlike many villagers, he knows his birth date, 1919, though he must pause to work out how old that makes him now. He recalls working in Curaçao as a young man and, for "jig wages" of five shillings a day, for the American Army in Antigua during the Second World War. In 1955 he went to England, where he stayed until 1961, working for London Transport. He is quiet-spoken and methodical; it is easy to picture him tending the high-tension lines of London Transport with the same meticulous patience he exercises here on his battered nets, and with the same self-possessed air he carries about the village and into his "gardens" in the jungled hills above it, avoiding on his way home through the city the dread gangs of Teddy Boys who used to harass West Indians in London.

In short, Baptiste is a reliable informant; and, faced again and again with such obviously trustworthy people telling such obviously untrustworthy stories, our exasperation becomes general—a token of our reaction to the culture of this village. All too easily we may suspect that the people here cannot get anything right. They *will* muddle up everything—get dates wrong, wars wrong, armies wrong, speak languages by turns crazily grandiloquent and insouciantly debased, and put on shows of knowledge wonderful only for the ignorance they betray. But succumbing to culture shock is as dishonest as retreating into the archives. It shies clear of realizing that we must come to terms with two entwined cultures: our own and a local one in which the moorings of our scheme of things have been cut adrift to mark a different sense of the world. Obviously, "1914" means something peculiar to Baptiste; but it is not meaningless. If, then, we are to make sense of Baptiste's story as an artifact of his culture, we must first understand how the people of Casse conceive of the differences between their culture and ours.

In the local scheme of things, what we appreciate as cultural differences are construed in terms of race. (I hope it will be clear, incidentally, that the following ideas about racial characteristics are not my own. Indeed, one of the more unpleasant features of doing fieldwork in Casse was confronting beliefs I found repugnant. I trust, however, that I have represented them here as fairly as those I found merely strange.) In the local view, there are two main races: white and black.[3] It is said that black people—one's fellow villagers and Dominicans—are generally jealous, grasping, suspicious, ignorant, untrustworthy, and superstitious. White people are honest, knowledgeable, generous, trustworthy, and competent. Black people are always quarreling and trying to hold each other back. White people get along well with each other and help out white and black alike. Though it is often said, particularly by young people in conversation with whites, that white and black people just have different colored skins (but underneath everyone has the same red blood), the antithesis of black and white remains fundamental to local thought. It is much better, a man a bit older than Baptiste told me one day, to work for a white man than a black one. The white man knows he is superior, so he will help you. The black man will try to hold you back, so you don't become as good as he is. I began to protest, but my friend said no, don't object—he had worked for both black and white, and he knew. It is often said, similarly, that the islands ruled by white people—Martinique, Guadeloupe, St. Thomas—are better off than those ruled by blacks—Jamaica, Trinidad, Grenada. Dominicans feared independence for this reason as well as for the French threat: freed from British, white restraint, the black government would try to hold the people back.

Local culture is not wholly black, however. It is, rather, a continuum, embracing both a black pole, so to speak, and a white one. Since the latter offers a common ground with the world actually inhabited by white people, it is both a point of entry for features borrowed from the white world and, by the same token, a focus of local aspirations to respectability.[4] Villagers who have only a few, mostly abusive or peremptory words of English often try to "speak English" with their children rather than use patois, which is not considered to be a "real language."[5] As this example suggests, however, the coherence of the blacker reaches of local culture is inadmissible—or, more exactly, while black norms remain vital they are relatively inexplicit and are often radically at odds with the white ones people also espouse (Wylie 1989b). One result is that borrowed bits and pieces of the white world are fitted piecemeal to a black scheme of things. There they serve as earnests of respectability—labels, as it were, meaningful more for their source than for whatever is written on them. The results of this process are sometimes rather bizarre. My favorite linguistic example is something a young woman called out to me in English one day: "*Beke* ['white man'; also my nickname], I want you to remove my snap!" I realized that she was making a common request in an uncommonly elegant way. "Remove my snap" was a fancy way of saying "take out my snap." "Take out a snap" is a relexification of patois *tiwe yō pôtwe,* from the French expression *tirer un portrait,* 'take a likeness'. Immediately I made a date to take her picture. A similar logic has shaped Baptiste's tale. Since it concerns the white world, Baptiste modestly displayed his acquaintance with European history by placing the decisive engagement of the First World War on Labatwi and carefully dating it to "1914." But since the tale is, after all, about local, black culture, he dismissed it as just a "story."

Let us take up Baptiste's story again. Its historical pretension and its graftings from white culture suggest how radically, if inadmissibly, Baptiste's sense of time differs from our own.

The story's main borrowing is the First World War—"1914, you know." Shorn of its European significance, this event becomes a badge of whiteness, lending the story a respectable frame and its teller a respectable air, and signalling its distance from everyday life. The careful date thus paradoxically lends the story the timelessness of "once upon a time." It also subverts white world history to a black myth addressing a central local concern—the relationship between English and French. To this end, too, an English attack on a French fort (in "1914") has replaced a French attack on an English fort (in 1778).[6]

Another quasi-mythic aspect of the story is its distance from village life.

Its only characters are the Queen and the French and English armies. No villagers—not even any Dominicans—help to build the fort, defend it, or storm it. The "inhabitants" who plied the defenders with liquor and spiked their cannon have been forgotten. This reflects the powerlessness felt by the people of Casse today. After all, what do they have to do with the decision to grant Dominica independence? If the French invade Dominica, a villager announced one day, we'll ask the English for help. No, he was told by an educated type who had come up from Roseau to buy fish, independence means we can't ask the English for help anymore. Scandalized, the first man went off to proclaim round the village how perfidious were the government's plans for independence. No one paid much attention. What could they do about it? Besides, maybe the man from Roseau was kidding. Since they do not feel they have control over their fate, the foreseeable future does not stretch very far ahead for the people of Casse. Events are not so much planned—and then in secrecy lest a jealous neighbor spoil them—as anticipated, so that they have the mysterious inevitability of fashions. Independence, Carnival, Saint Peter's Day, the season for dolphin-fish—a feeling builds up that they are "going to happen anytime," announced over the radio or by the priest, or in spreading rumors and dreams. Dreams are said to be messages from God, and the future is in God's hands.

I found it difficult to understand how my friends and neighbors in Casse could get along between such a foreshortened past and such an unforeseen future. How could Baptiste possibly date the French expulsion from Dominica to only five years before his birth? But in fact this dating is not as remarkable as his speaking at all of events that took place "before I know myself" (*avā mō kônet kô mwē*). "Before I know myself" is the time before a person is about eight or nine years old. Not themselves remembering them, people tend to dismiss accounts of events "before I know myself" as old people's maunderings or as lies: "Some people just saying what they want." As a rule, the most distant period about which people will speak is their "grandparents' time." Although a few people can trace genealogies to an allegedly white great-grandparent, one's "grandparents' time" is the era to which clearly fantastical tales are dated and in which slavery, actually abolished here in 1834, is said to have existed. Even extraordinary natural events quickly sink beneath the surface of time. Did no one in Casse know about the eruption of Mt. Pélée, clearly visible across the channel on Martinique? When it blew up in 1902, the town of St. Pierre was destroyed and some forty thousand people were killed. I was never told about this spectacular—even, one would think, memorable—catastrophe. Nor did people volunteer much information about hurricanes, a perennial danger in this part of the world. One muggy, breathless day in September (as it happened, I now see in my journal, the two hundredth anniversary of the attack on Labatwi), I became alarmed by radio reports of brewing tropical storms. I asked around about hurricanes. A man in his forties told me that the last big one here had been in 1929—before he was born, he remarked, but he had heard people talking about it. An older man told me about hurricanes' great winds and waves. This was obviously news to his teen-aged son who was listening in. He thought a hurricane would be exciting, but the old man said rather sharply, "*pa māde pu wê-y*": Don't ask to see it.

How then do people pass knowledge along?[7] Didn't someone a bit older than Baptiste ever let on that in his or her day Dominica had been English too? But people talk mostly to people of their own age and from their own experiences. They may overhear their elders' reminiscences, but Casse has no personages or institutions dedicated to storing and passing on such lore. Nor can people trust accounts of events they did not witness, since so often "people just saying what they want." Personal histories are unreliable, while history in the sense of a collectively maintained description of the past scarcely exists.

One result is the continual reinvention of the past, almost from scratch. One day a smart and engaging young man called Kodak told me something he'd just learned from some friends of his, something he thought might interest me: Winston Churchill had once lived in the village, in a house that had been converted into a school shortly "before I know myself." He was an important man, Kodak said, a white man from England. Had I ever heard of him? Yes, I said, he was a famous man: What did he do? He won the war, Kodak answered. What war was that? I asked. A flicker of doubt crept into Kodak's eyes (questions like this disturb him; at school you are beaten for wrong answers), but he persevered: the Vietnam War. Several months later I discovered the basis of Kodak's story. The house had been owned by Mr. Winston, a well-to-do black merchant from town.

Kodak's tale, like Baptiste's, is a kind of myth without a mythology, a history without historiography. It is not embedded in a system of knowledge or confirmed (or confirmable) by specialists. Indeed, it seems that any form of systematic knowledge about local affairs is dangerous: *savā* means not only 'knowledgeable, clever' but also 'secretive, sly, evasive'. The indigenous, essentially religious expertise of "old niggers" (*vye neg*) versed in obeah, or of the *sukuyā* (night-flying sorceresses, able to change their skins and shapes, spoiling people's luck and sucking their blood), is obviously dangerous not only because its adepts may harm their fellow villagers but also, one suspects, because the very fact that it is organized knowledge lends its possessors untoward power. The few villagers—the schoolteachers, a Baptist preacher, scattered individuals—who do profess systematic ideas about the way the world works boast knowledge founded safely in the alien worlds of the English language, foreign textbooks, white people, or God's word. History of this book-learned sort is an occult science, like witchcraft, but safe and respectable because it has so little to do with everyday life.

Baptiste's story suggests a further difference between his culture and ours: a sense of reality not only very different from our own but subversive of it. For us, in principle if not always in practice, present reality and historical truth are impersonal, existing independently of what any individual thinks or knows about them. The order of things is "out there," theoretically discoverable by any rational investigator. From contemporary records a historian should be able to piece together a reasonable account of what happened on Labatwi in 1778; an anthropologist who masters his personal frustrations should be able to describe the culture of this village in abstract terms and with telling concrete examples. Moreover, we like to organize and institutionalize our knowledge, entrusting it to specialists whom we may consult if need be.

In Casse, however, in principle if not always in practice, reality and historical truth have no existence independent of what people say about them—and often "people just saying what they want." Moreover, since organized knowledge is dangerous, its explicit formulation is best left safely distant from local life, in white hands or in God's. Nature, for example, is accepted as "God's work," and people do not inquire deeply into its workings (Ringel and Wylie 1979). Otherwise, the truth is established from moment to moment as a matter of received opinion. Thus, discussions of apparently trivial facts often become complicated exercises in social relations. Whose reputation is greater? Who controls the place where people are discussing something?

Which is larger, Dominica or Guadeloupe? This was the question under consideration one morning by a group of fishermen. The majority opinion, and that of Franklin, one of the best fishermen in the village, near whose boats we were lounging about, favored Dominica. The minority opinion, and that of his younger brother, Bateson, who has spent a good deal of time in Guadeloupe, was undecided or tended to favor Guadeloupe. I was particularly friendly with Bateson, but in different ways we both depended on Franklin's patronage. At length Bateson asked my opinion—and didn't I have a book that could tell me? I said I thought Guadeloupe was bigger, but if they wanted

I could get the book. Yes. I fetched the almanac I had brought along to Dominica and read that Guadeloupe was nearly twice as big. But this was not to be. For one thing, the book said that Dominica's area was 305 square miles, whereas everyone had learned in school that it was 306 square miles. Where had the extra mile gone? "*Beke,*" Franklin concluded with a friendly but rather pitying smile, "*liv-u pa bõ*": White man, your book is no good.

By midafternoon Bateson had reached a compromise position. He told me privately that he thought Guadeloupe itself was smaller than Dominica but was larger when its outlying islands were taken into account. He was still wrong, according to the book, but I didn't bother to tell him so. Our friendship dictated a compromise fact, just as Franklin's reputation dictated one we must all accept so long as we were hanging about near his boats. Guadeloupe's size is as immaterial as the First World War.

These discussions were friendly. But they grow loud and angry when someone's reputation is seriously at stake. At any given moment an argument is likely to be raging somewhere in the village streets, as claims and counterclaims are stated, the support of "friends" is sought, and derogations of "enemies" are denied. Although such "debates" (*deba*) sometimes degenerate into mere quarrels (*gumẽ*) or fights (*batay*), they are far more stylized than they may seem, and serve as a way of establishing contested matters of fact. In the local legal process, for example, they regularly provide an effective means of publicly defining and investigating delicts, precipitating the formation of accused and aggrieved parties, and—most important—establishing the litigants' reputations. Just as Baptiste's "1914" is nonsensical in European terms, however, the legal process of *deba* and counter-*deba* usually runs afoul of the quite different procedures and precepts of Dominica's formal, basically British, legal system, by the lights of which litigants' reputations are immaterial and the "facts of the case" are "discovered" to have a fixed, historic sort of reality.[8]

In everyday life, reality—including its bygone portion, the past—has little fixity in Casse: it is social, shifting, fragmentary, a matter of shifting coalitions of convenience. Embroiled in the continual recreation of their home world, the people of Casse thus remain alienated from the white world to which they aspire. Yet what both white (with exasperated condescension) and black (with aspiring self-denigration) may call these villagers' ignorance (or worse) actually measures the inadmissible strength of black culture. It still gains its subversive triumphs. Depending on whom you ask, Winston Churchill lived in Casse, Dominica is larger than Guadeloupe, and the British captured Labatwi in 1914. As Franklin said, the white man's book is no good.

Álvabøur, 29 February 1972

Spring had not yet come, but the sun was returning to Álvabøur. During the winter it does not rise above the bluff south of the village. A few days earlier I had met Ólavur Hansen "down on the roost," a short stretch of road between the post office and Álvabøur's diminutive movie theater. In good weather old men like to lean against a railing there—like chickens on a roost, the village wags say—gossiping, spinning yarns, and keeping an eye on the tides as they look out over the harbor and the wide fjord to the long ridges and rearing sea cliffs of the next islands. According to his own account and that of the carefully kept village register, Ólavur was born on 29 June 1912. He calls himself a fisherman, but he doesn't get to sea much anymore.

I had asked Ólavur if I might visit him sometime to ask about his family history. He was surprised. In the Faroes one calls on people without invitation, without even knocking at the door. You walk in the front door, take off your shoes in the hallway, come into the kitchen, and sit down by the stove. The woman of the house makes tea, and in the slow way Faroese conversations have, punctuated by the several equivalents of "yep" and "I reckon so," the talk gradually works its way through a discussion of the weather—fair, or more often stormy—to the affairs of the day and whatever business you have in mind. But yes, Ólavur said, I could visit him.

Now we sat in the parlor while his wife listened to us through the door as she set out little cookies and the good china on the kitchen table. I was being treated rather formally. First we discussed the weather. It had been warm in the morning but had turned cold when a southerly wind came up.

Ólavur turned out not to be a particularly good genealogist. He could go back only to his great-grandparents' generation, and even then he made some mistakes. I don't know how far back my best informants could have gone. I arbitrarily concluded a genealogy when it stretched beyond Álvabøur or past its founding in 1833–34. However, Ólavur did tell me about the founding of the village.

I had already heard this story many times and was to hear it many times more. I had gotten into the habit of politely hearing it out with a smile I hope was not too fixed, noting to myself only small additions or variations of detail and any unusual conclusions. The story inevitably begins, "Yes, Álvabøur is a young village." It is the stock response to such questions as "What kind of a place is Álvabøur?" I heard it most often when people asked me what I was doing. "Studying the folk-life," I would say. "*Ja, ja.* Yes, Álvabøur is a young village. . . ." Álvabingar clearly feel that the character of their village is neatly contained in this phrase and the story that follows it. According to Ólavur, the first man to settle in Álvabøur was Mikkjal Joensen from Strandarvík. Ólavur concluded that the village Young People's Club he had belonged to broke up because "some people" denied this fact.

I found Ólavur's conclusion interesting. The "some people" were clearly the descendants of Jóhan Hendrik Davidsen, the other (and usual) candidate for Álvabøur's "first inhabitant." And I had not heard much about the Young People's Club, let alone its breakup. Álvabingar rightly insist that they are cooperative and egalitarian. Ólavur's story suggested that some Álvabingar might be more equal than the rest, that there might be latent divisions among them along family lines, and that the resulting tension was expressed in disagreements about who was Álvabøur's first settler.

Who *was* Álvabøur's first settler—Jóhan Hendrik or Mikkjal? Like many other minutiae of Faroese history, this question supports a small literature, from which we may draw a rather fuller account than those I was hearing.[9]

An old ballad relates that a man called Álvur lived here in Viking times. Although modern scholarship has revealed that the ballad is, in fact, a rendition of an Icelandic saga about an incident in Norway, the plot of which was applied to this island by a chance congruence of names,[10] archaeological evidence has confirmed that a Viking Age homestead stood on the legendary site of Álvur's house. The site may have been inhabited off and on until the early seventeenth century as well. Thereafter no one lived in Álvabøur, undoubtedly because it is a notorious *brimpláss*—a shore with heavy surf. A boat was kept there, however, for Álvabøur is the most convenient landing place on this side of the island. The boat and the pasturage round about belonged to people in Strandarvík, the nearest village. Since there were no shelters or boathouses in Álvabøur, travelers who came ashore there had to make the trek—a dangerous one in bad weather—across the hills to Strandarvík.

Álvabøur's modern history begins in 1815, when Emilius Løbner, the Danish commander of the small Tórshavn garrison, proposed that a few families be settled there to ferry travelers across the fjord and shelter them in foul weather. Nothing came of Løbner's proposal, or of a similar one by his successor. Finally it was resubmitted by Governor Frederik Tillisch and approved by the Danish government in 1832. (Between 1816 and 1852 the Faroes were virtually ruled by appointed Danish governors.) Tillisch envisaged a settlement large enough to man an eight-man boat for the compulsory ferry service (*skjútsur*) across the fjord. The people would live primarily by fishing. It might be noted that around this time the Faroes' Danish administrators were seeking to improve the quality of life in this backward province, in part by encouraging the growing population to rely more heavily on fishing. Álvabøur was thus one of a number of "new settlements" founded with official encouragement in the early nineteenth century (R. Joensen 1966).

Tillisch did not find it easy to recruit settlers. Both he and the Strandarvíkingar feared that poor or landless folk might find it easier to steal sheep than to fish; and Strandarvíkingar were as reluctant to settle in Álvabøur as they were to give up their land there to others. Tillisch eventually persuaded Mikkjal Joensen, the young heir to the leasehold whose lands included the site of the proposed settlement, to come to Álvabøur. The only other suitable settler he found was Jóhan Hendrik Davidsen, from Hestur. Tillisch met with both men in Álvabøur on 1 April 1833. Work on their house—or houses—began soon afterward. Tillisch provided state funds for materials and labor.

How many houses were built that summer is a vexed question. So is when they—or it—were finished. Oral sources suggest that only one of two houses was completed in the summer of 1833. One of Jóhan Hendrik's greatgranddaughters told me, for example, that his first son, Jóhan Pauli, was born on Hestur on 19 August 1833, "the same day that the rooftree of the first house in Álvabøur was raised."[11] The first house, she said, was Jóhan Hendrik's. She heard the story from Jóhan Pauli himself, who had heard it from his father. Documentary sources roughly confirm this date, for the carpenters hired by Tillisch received their last large payment on 16 August. They were also paid a small sum in late September, probably for work on outbuildings, although conceivably for additional work on a dwelling. The documentary evidence makes it virtually certain, however, that there was only one main structure—a long double house, half for Jóhan Hendrik and half for Mikkjal—and that both halves were at least roughly finished in mid-August.

The next question is when Jóhan Hendrik and Mikkjal moved in. Jóhan Hendrik surely began living in Álvabøur as soon as his house was habitable. He was joined by his wife and their infant son soon after the new year. Mikkjal's whereabouts are not so clear. He most likely commuted to Álvabøur from his home in Strandarvík from the late summer of 1833 through the winter of 1833–34. He took on a hired hand in Álvabøur in March of 1834, and, when a census was taken in the Faroes that August, Álvabøur's inhabitants were listed as Mikkjal Joensen ("unmarried, who lives there"), his hired hand, and Jóhan Hendrik, his wife, and their child. The governor complained in October, however, that Mikkjal had spent "almost the whole summer" in Strandarvík. It is quite possible that, as some people claim,

Mikkjal did not finally settle down in Álvabøur until as late as November 1835, when he was married.

Mikkjal was certainly not in Álvabøur on the night of 26 February 1834, when the house was badly damaged in a storm. It is said that Jóhan Hendrik, his wife, and little Jóhan Pauli took refuge in a boathouse overnight and the next day made their way to Strandarvík where her sister and brother lived. Documentary evidence shows that Jóhan Hendrik wanted to go back to Hestur. Tillisch persuaded him to stay in Álvabøur only by offering to defray the costs of rebuilding and to provide him with an oven. It was only now, some people say, that a duplex house was built. Documentary sources make it clear, however, that the existing structure was simply reroofed and had its stone walls strengthened.

Soon the village grew. A third couple arrived in 1836, children were born to all three families in 1837, a fourth family came in 1842, and a fifth in 1844. The new village had five households and thirty-three inhabitants in 1850. Mikkjal unfortunately was not among them. He had died in 1846 in an accident on the bird cliffs. His widow eventually returned to Strandarvík, and his brother-in-law, Peter Winther, was granted partial rights over most of Mikkjal's lands in Álvabøur in return for agreeing to move there. Jóhan Hendrik and his heirs were granted ownership of the rest by a series of agreements beginning in 1849. Peter Winther and his heirs remained small-time farmers; their lands were not extensive, and Álvabøur's only farmer has always fished as much as he farmed.

Álvabøur thus became willy-nilly pretty much what Tillisch had envisaged: the island's port of entry, whose inhabitants lived more by fishing than from the produce of a small infield. Since the holdings Álvabingar gradually accumulated by marriage, inheritance, purchase, and immigration from Strandarvík were never large, the tenor of village life has been set by the spirit of cooperative enterprise that fishing requires. Similarly, since there was never much land to build them on, Álvabøur's houses are crowded together. Strandarvíkingar say the place is *too* crowded—they say Álvabingar can (and do) look into each others' windows all the time. But Álvabingar claim they get along well. Certainly my wife and I found that our neighbors' knowledge of one another's and our affairs was as discrete as it was extensive. As one man told me, "Álvabøur was built from the sea, so we have always had to help each other out—you see, Álvabøur is a young village."

What elements of this history do Álvabingar know? How do their versions differ? What does it mean to them?

One man showed me a fire-blackened stone he had picked up at the archaeological dig on the legendary site of Álvur's house and told me how the ballad recounts its burning. This tale is widely known in the village, but it is not part of the story beginning "Álvabøur is a young village. . . ." The late-medieval homestead is not known at all. As in other Faroese oral histories, the Danish authorities have been edited out of the tale. In keeping the Faroese emphasis on natural conditions, however, Álvabøur's bad surf is almost always mentioned. So are the ferry route, Strandarvíkingar's fear that Álvabingar would steal their sheep, and the fact that this has always been a fishing village. The events of 1833–34 are recounted in detail. Sometimes people give the date. They always identify the main characters by name and village of origin, and often by kinship to themselves or other living Álvabingar: "So-and-so is Jóhan Hendrik's great-grandson, you know." The storm is always mentioned. So are the duplex house (you can still see some rubble from its walls beneath the movie theater) and a number of details about land tenure. In short, the story beginning "Álvabøur is a young village . . ." concerns the site's modern settlement and the natural conditions in which Álvabingar have found themselves. By contrasting Álvabingar and Strandarvíkingar (Strandarvík is a very old village), and by editing out Tillisch and the workmen he hired, the story also sets forth what Álvabingar consider to be their distinguishing qualities: egalitarianism, cooperativeness, and self-reliance.

The way all this is put together varies slightly. Ólavur and one other man told me the most radical variation. They claimed that Mikkjal came to Álvabøur and built a house, but it blew down in a storm and he returned to Strandarvík with his wife. He came back the next year. Then he and Jóhan Hendrik built a house together. Ólavur, who is proud to be a Winther on his mother's side, also gave me a genealogy making the present farmer a direct descendant of Mikkjal. Most of this is simply wrong. Mikkjal was in fact unmarried in 1834, it was not his house (or his house alone) that blew down, and the present farmer is not his direct descendant. What Ólavur had evidently done—or what whoever told him the story did—was to substitute Mikkjal for Jóhan Hendrik in the standard, "Davidsen" version of the tale: Jóhan Hendrik came here first, but his house blew down, and so forth.

In a sense, however, these variations matter less than does the persistence of the tale. At first hearing, it is clearly a kind of history: a matter-of-fact account of events happening in a specific place at a specific time. The actors are all human, and even the undocumented events have nothing inherently improbable about them. I have not checked the parish register for Hestur, but there is no a priori reason to doubt that Jóhan Pauli was born there on 19 August 1833 or that this was "the same day that the rooftree of the first house in Álvabøur was raised."

Still, this is history served up like myth. It explains why Álvabøur is the way it is, doing so in conventional phrases and in episodes that, as in Ólavur's version, remain constant even while their characters exchange roles. The story explains, usually at the end, why Álvabøur's houses are so tightly clustered, why this is a fishing village, why there is still a certain tension between Álvabingar and Strandarvíkingar, why Álvabingar are neighborly, egalitarian, and cooperative, and even why there is a pile of rubble beneath the movie theater. In a way, "Álvabøur is a young village" is the community's constitution, its charter of basic values. Not surprisingly, therefore, times of tension and transition have been marked by disagreement about who settled the village first.

Ólavur's conclusion is a case in point. When he was young, he said, there was a Young People's Club in Álvabøur. Among other things, the club sponsored dances. One time—I didn't ask when, but this must have been shortly before Álvabøur's centenary in 1933—the club decided to erect a monument commemorating Álvabøur's settlement. A bit of land and a stone were bought. The stone was to bear the first settler's name. "Some people" insisted that "Old Jóhan Hendrik" should have his name on the stone. Others, Ólavur presumably included, held out for Mikkjal Joensen. A compromise, that both names be put on the stone, proved unacceptable. The club foundered. The monument was never put up. (The stone was eventually erected as a memorial for men lost at sea and on the bird cliffs; ironically, therefore, it bears Mikkjal's name but not Jóhan Hendrik's.)

I found Ólavur's conclusion interesting because it suggested that the Davidsens enjoyed a rather precarious prominence as the first citizens of an egalitarian society. Genealogical and other data reveal why this was so.

The Davidsens form what Faroese call an *ætt,* or more loosely and colloquially a *fólk:* essentially a patrilineage, generally exogamous within three or four degrees of kinship, whose members are identified with an actual or ancestral homestead.[12] There are five or six major *ættir* in Álvabøur: the Davidsens, the descendants of Álvabøur's third settler, those of some intermarried sets of siblings who moved there later on, and so forth. *Ættir* may endure for centuries, of course, but neolocal marriage by all but one of a family's children attenuates the identity of collateral lines of descent after two or three generations. In other words, the increasingly distant cousins who do not live in the eponymous ancestral homestead become less and less clearly identified as members of the original, central lineage. Thus, while all of Old Jóhan Hendrik's resident patrilineal descendants are sometimes still spoken of as a whole, only those individuals who grew up or still live "In the Cage," as his house and its successor have been called, are known as the "Cage-People" (*Búrafólk*). The upshot of all this is that local society retains a long-term identity as a set of *ættir* and homesteads but is integrated by their attenuation and, more immediately, by intermarriage between them.

The "Cage-People" had been slow to be drawn into the village kinship network by marriage. None of Old Jóhan Hendrik's three boys married village girls, and his daughter moved to Strandarvík. Of his eighteen village-born grandchildren, only two took Álvabøur spouses (six took spouses from elsewhere, four married elsewhere, two remained single, and four died young). Moreover, by the 1930s most of his descendants no longer lived or had grown up "In the Cage." In other words, the Davidsens were rather tenuously enmeshed in the village kinship network to begin with, and their *ætt* was reaching a natural point of attenuation after three generations. It remained, however, a focal point for the identity of the village as a whole—the more so as Davidsens continued to hold a disproportionate number of the positions mediating Álvabøur's relations with the outside world. Old Jóhan Hendrik himself had been Álvabøur's first postmaster and merchant (the original "Cage" was a cubicle inside his house that served as a shop) and was in charge of collecting the tithe and assigning men to ferry duty. These last positions were abolished around the turn of the century, but the "Cage-People" have provided all of Álvabøur's postmasters, at least one merchant and one member of the village council at any given moment, several lay officials in the church, and, when I lived in Álvabøur, the proprietor of the movie theater and the manager of one of the village's two banks.

More to our present point is the fact that Old Jóhan Hendrik's grandson and namesake was Álvabøur's leading citizen for many years. Born in 1869, this younger Jóhan Hendrik was married in 1896 to the daughter of the sheriff in Strandarvík, an office previously held by her grandfather, and, from 1911 until 1950, by three of her brothers in turn. Like his father-in-law, this Jóhan Hendrik was an early leader of the nationalist movement and an adherent of the Self-Rule Party (*Sjálvstýrisflokkur*). He represented the district in the Faroese parliament in Tórshavn from 1897 to 1908 and again from 1916 until 1932. (He was succeeded by his grand-nephew and foster grandson from 1932 until 1936.) He was also Álvabøur's schoolmaster from 1912 until 1939. Thanks partly to his efforts and to his political and matrimonial alliance with the sheriff's influential lineage in Strandarvík, Álvabøur was a remarkably lively, prosperous place in the 1920s. The Faroes' first major public-works projects outside Tórshavn were the construction, largely by local labor, of Álvabøur's harborworks in 1922–26, followed by the building of the Faroes' first real road, between Álvabøur and Strandarvík. The schoolmaster was also a leading spirit behind the construction of a combination dance hall and playhouse, in 1922, where locally written skits as well as published plays were produced, often under his direction. This was the first proper playhouse in the Faroes.

The late 1920s and early 1930s, however, were troubled, transitional times in Álvabøur and the country as a whole. Given Álvabingar's reluctance to discuss such things, it is hard to be sure in detail about troubles in the village, but it is clear that their main cause was economic. As fish prices continued to fall, the depression that had beset the Faroes after the First World War kept deepening. The depression was perhaps particularly severe in Álvabøur, a fishing village without public-works projects now that the harborworks and the road were finished. In national politics, a two-party system was giving way to a three- and then four-party one as economic issues overtook nationalist ones. The Self-Rule Party lost its near monopoly of the vote in Álvabøur. It had captured about 90 percent of Álvabøur's vote until 1924, but this figure fell to 77 percent in 1928 and 76 percent in 1932, the year the schoolmaster retired from politics. It received just 59 percent in 1936.[13]

The shift in Álvabøur's voting patterns is clearly related to the demise of the Young People's Club around the same time. Both involved failures of ideological unanimity, and both took the form of a reaction against the Davidsens' position as Álvabøur's first citizens—directly against the younger Jóhan Hendrik in the case of political preferences, and displaced onto Old Jóhan Hendrik in the case of the memorial stone. Neither development, however, entailed the sort of open confrontation Álvabingar find intolerable. Differences of political opinion were expressed in the privacy of the voting booth or through delegates in Tórshavn, while the club apparently just dissolved when it threatened to become a forum for debate.

The years around 1950 were another troubled time. Again the cause was partly economic. Here, as elsewhere in the Faroes, the fishery had emerged from the Second World War scarcely competitive and desperately in need of modernization. Political alignments were shifting again with the resurgence of nationalist sentiment. The Self-Rule Party enjoyed a brief revival in Álvabøur but was rapidly eclipsed by the new, nationalist Republican Party (*Tjóðveldisflokkur*). The schoolmaster had long since retired (he died in 1954), but Álvabøur's leading citizen, albeit at a more modest level, was now his son Niels Davidsen, who succeeded him as schoolmaster in 1940, became deacon of Álvabøur's church in 1941, and was elected foreman of the village council in 1950. Like his father, Niels was a local leader of the Self-Rule Party. He went on in the early 1960s to achieve a certain prominence in national politics. Several of his younger cousins supported the Republicans, however, and one—the heir to the postmastership—was drawn to the conservative Union Party.

Village life was more seriously disrupted by religious disarray than by political factionalism, however. Until 1948, when missionaries of an evangelical sect, the Assembly of God, arrived in the village, all Álvabingar belonged to the established Lutheran church.[14] (Álvabøur has its own church building but is served by the minister in Strandarvík.) The missionaries condemned as sinful not only such current fashions as curling women's hair but also such traditional customs as infant baptism, drinking and dancing, and the more recent practice of staging plays. The sect collected about thirty adherents, mostly people who were economically and socially marginal. The sect's public proselytizing, confrontational style ("Are you saved?"), emotional services (which included speaking in tongues), and the fact that children took to singing its catchy hymns while playing outdoors violated many local norms. Many Álvabingar found all of this offensive, but the evangelical ideology enjoyed the advantages of being explicitly formulated and based in an unimpeachably sacred text. How could anyone refute it?

People had mixed reactions to the sect. Attendance at Lutheran services swelled, but the minister, a Dane whose long tenure—he had come to Strandarvík in 1930 but never learned to speak Faroese—had done little to endear him to Álvabingar, especially among the nationalist-minded young, disappointed his parishioners by deciding that the best tactic was to say nothing about the Assembly of God. The evangelical arm of the Lutheran church sent a missionary to Álvabøur who, although personally popular, did little more than polarize opinion further. Prominent Álvabingar—Davidsens of various political persuasions foremost among them—opposed the sect openly, but their outspoken antidisestablishmentarianism was not wholly popular either. For a time, lest they offend neighbors who belonged to the sect or sympathized with it, many Álvabingar quietly eschewed "sinful" behavior. Plays were no longer produced, dances were sparsely attended if they were held at all, and a woman who was a Davidsen by both birth and marriage told me that she was one of the few village women who defiantly continued to have their hair done. This informal sort of neighborly solidarity was, of course, well in keeping with Álvabingar's reluctance to disagree openly, but it also hastened the demise of such traditional means of promoting community unity as village dances and at least tacitly acknowledged that a nonlocal, pseudohistorical charter drawn from the Bible was as valid as the "constitution" embodied in the story of Álvabøur's founding.

Álvabøur's founding was apparently much discussed around this time, as if to make local history more explicit and give it a textual basis of its own. The discussion even reached a national audience in the summer and fall of 1950 as professional historians and local laymen debated points of detail in letters-to-the-editor columns of the Tórshavn newspapers. What did the documents say? What did someone's grandfather remember? Álvabingar found to their dismay that even in Tórshavn they had acquired a reputation for contentiousness. How unfair, I was told twenty years later: after all, "this is a young village."

The religious furor died down within a few years. The fishery revived, and few people found themselves economically predisposed to declare that Jesus would solve their problems. Álvabingar found that getting along together involved more than seeming to live sinlessly. But their memories are long, the sect still has a handful of adherents, and I found Álvabingar as reluctant to discuss religion as local politics.

Agreeing about the past is an important part of getting along in Álvabøur. One may claim, of course, that Álvabøur's founding was not the real point at issue when the Young People's Club broke up in the early 1930s—that Álvabingar were really worried about the depression, the decline of the Self-Rule Party, and the schoolmaster's prominence in village society, and that these concerns were shaped by the evolution of Old Jóhan Hendrik's *ætt*. One may likewise say that Álvabingar were really vexed in the years around 1950 by the Assembly of God, resurgent nationalism, and an unprofitable fishery. But it is also true that in good times as well as troubled ones, and in the normal course of events as well as at moments of crisis, Álvabingar know their place in the world by knowing about their past. Like the discussions of the weather that inevitably precede getting down to other business, agreement about Álvabøur's founding confirms a reality apart from Álvabingar's immediate social relations and frames their understanding of them. Mild disagreement gives Álvabingar a way to express divergent understandings of local society without opposing each other directly. Serious disagreement is, if possible, avoided. In short, while from one point of view reality is as much a social construct in Álvabøur as it is anywhere else, from another point of view (which corresponds to Álvabingar's own) society itself is construed in terms of such external realities as historical fact and the order of nature.

Using the past in this way requires the constant reaffirmation of knowledge about it. Indeed, Faroese have long engaged in rituals of collective recollection. Until the late nineteenth century, legends (*sagnir*) and other oral literature were recited at evening gatherings called *kvøldsetur,* when families and hired hands spent the long winter evenings entertaining each other while they carded, spun, knitted, and did other indoor tasks.[15] Legends sometimes take on semisupernatural overtones, but they abound in genealogical, geographical, and other circumstantial details that can often be checked against documentary evidence. Their accuracy is often remarkable. So is their depth of time. Most date from after the late sixteenth century, but, opening at random the great collection made by the Faroese philologist and folklorist Jakob Jakobsen in the 1890s, I find a little tale beginning, "Before the Black Plague came and devastated the Faroes . . ." The bubonic plague swept the Faroes in about 1350. Even earlier times are recalled in the heroic ballads (*kvæðir*) that are danced to at festival times. Many concern medieval Norse heroes, but some tell of Charlemagne, for example. Others are closer to myth or romantic epic and have little historical pretension. *Kvøldsetur* have long since disappeared, and balladry is no longer such a living tradition. But modern times have brought new literary and historiographic forms—novels, plays, local and national histories—and many informal oral genres remain vital: the generally satirical, lighter ballads (*tættir*); recollections of notable personalities, witticisms, mishaps, and natural events; homestead histories; genealogies; anecdotes about the day's fishing; the yarns men spin down on the roost; and, of course, such stories as the one beginning "Álvabøur is a young village . . ." For Álvabingar, as for other Faroese, the present is deeply rooted in the past.

Nationhood in Time

Time is deep in Álvabøur; in Casse it is shallow. History is a matter of popular concern in Álvabøur; in Casse it scarcely exists. In Álvabøur reality is held to exist independent of the social order while local society is constituted in natural and historical terms. Reality is an immediately social construction in Casse while local society itself is seen self-denigratingly in terms of white ways and God's word.

These observations raise a few more general questions. What groups or classes have taken the lead in defining the worlds in which Faroese and Dominicans find themselves? How has social change affected them? How have Faroese and Dominicans envisaged *future* social orders, especially as they have looked ahead to attaining political independence?

In Álvabøur, and, we may suggest, throughout Scandinavia, people take bearings for their collective identity from things that, while local, stand outside the immediate social scene. Society is understood in terms of, for example, the past and the order of nature. Just as Álvabingar ritually establish a common ground in weather talk before embarking on a serious conversation, "the farmers who share a valley [in western Norway] feel that their deepest common concern is the weather."[16] Indeed, whole nations define their unity and their place among other nations in ecological terms. Thus, Iceland pioneered the fifty-mile fishing limit, "an issue," the Prime Minister told the Nordic Council, "on which the whole nation is united," so that she might "assume her role in the international division of labor, which has been allotted to Iceland by nature" (quoted in Jónsson 1972:7). It was on similar grounds that Norway, though less unanimously, rejected membership in the Common Market in 1972, and that in 1974 the Faroese began successful negotiations for a special accommodation with Denmark and the European Economic Community.

History similarly provides a model for society. What book about Iceland, what tourist brochure, would be complete without an account of the Viking settlement and the sagas? The coherence of modern life depends upon them. Gylfi Gíslason (1973:91), noting that technological change and increasing contact with other nations are as desirable as they are inevitable, maintains that "this does not mean that Iceland may or should relax her efforts to preserve and promote her distinctive culture. Icelandic culture has always been and always will be the main argument in support of the nation's right to independence." Gíslason's comment suggests something else about how Scandinavians assert collective identities from the village level to national and even international ones. The distinctiveness of a village or a nation must be established in formal demonstrations of cultural worthiness. As he goes on, "the closer Iceland approaches other nations, the greater will be her duty to stand loyal guard over her culture, both for her own sake and that of the world."

As nationhood is construed culturally and culture is in large part construed historically, it is not surprising to find that the Faroese movement for greater autonomy from Denmark has at each stage involved formal demonstrations that the Faroese are respectable partners in the Scandinavian cultural heritage.[17] Faroese separatism began with appreciations of the ancient roots of Faroese folkways. Both among the nascent middle class at home and among the Copenhagen intelligentsia, an important focus of attention was "folkish" (Danish *folkelig*) ways and local survivals of once widespread literary forms, especially the "medieval" ballads. People were also interested in the Faroes' language, for liberal opinion in Denmark had adopted the view that language was "the primary distinguishing feature of a people who, it was claimed, should be granted the right to political self-determination" (Oakley 1972:172–73). In a sense, the ideal language was the medieval Icelandic of the sagas, which was sometimes confused with the Common Scandinavian mother tongue. The first decisive step in the reformulation of Faroese culture was the publication in Copenhagen in 1846, by the Faroese theology student V. U. Hammershaimb, of an orthography that made clear the ennobling affinity of Faroese to Icelandic.[18] By the end of the century, the Faroese found themselves able to demonstrate their worthiness to Copenhagen and to themselves by writing their own histories and systematically collecting and publishing their own folklore. They went on to build up an impressive modern literature, at first mostly lyric evocations of the Faroes' natural beauty, followed by plays, novels, and so forth. Political separatism followed a pace behind, dependent on the successful demonstration of cultural distinction. The first major institutional expression of Faroese political separatism was the Self-Rule Party, which was founded in 1906. In 1948—belatedly, almost all Faroese felt, and in too small measure, since something close to a majority now felt that the Faroes were entitled to complete independence—separatist goals were realized in the Home Rule Law by which the Danish government granted the Faroes internal self-government on, precisely, the cultural-political, historical, and natural grounds wherein Scandinavians recognize their distinctiveness, or, as the preamble to the Home Rule Law put it, "in recognition of the special position which the Faroe Islands occupy within the realm in national, historical and geographical respects."

We should note two things about all this: a persistent and widespread distinction between everyday and official levels of thought and action, and the manner in which fundamental values are carried through from one level to the other. Egalitarianism is particularly highly valued in Álvabøur as elsewhere in Scandinavia. As Barnes observed in a Norwegian village, "the relationships that are most highly valued are . . . to be found in the shifting middle ground of social intercourse between approximate equals."[19] Moreover, people insist that they get along well. This creates a dilemma, however; for, as Gullestad (1984:29) has aptly remarked of Norwegians, because "equality is codified as similarity, as being and doing the same, . . . handling all kinds of differences becomes problematical." It is, of course, particularly problematical in political life, which by definition involves inequality and disagreement. As the dissolution of the Álvabøur Young People's Club suggests, "some people's" standing was precarious precisely because their social prominence and their descent from Old Jóhan Hendrik set them above other Álvabingar, or in a "cage" apart from them. Simply avoiding dissension won out in this case: the club just dissolved when its affairs became politicized. If an institution is to survive as a political institution, however, it must somehow preserve the fundamental equation between agreement and equality even as it provides a forum where rival leaders may express disagreement. In general, this trick is carried off by removing both leaders and disagreements to neutral ground where they may meet as equals.

Anciently, "neutral ground" was not just a metaphor. All of Scandinavia's courts or parliaments (Old Norse *þing*), among them the Faroes' Løgting and six district courts, traditionally met at central but uninhabited or sparsely inhabited spots. Park (1972:9; cp. Hollos 1976) notes that in Helgoland, in Norway, "the principle that major decisions for a community are made when its leaders congregate elsewhere has a long history . . . and has probably played a formative role in the intellectual culture of rural communities." I have suggested that the natural world and the past are also neutral grounds of a sort—places, as it were, where people agree or at least agree to differ.

Language is a neutral ground as well. Thus, in rural Norway the standard language, Bokmål, which "carries connotations of differences in rank which are unacceptable in the realm of informal local relations," is used between speakers of different regional dialects to demonstrate their shared identity at a higher level" (Blom and Gumperz 1972:433). Written Faroese is similarly "neutral." It was explicitly developed not only to demonstrate an affinity to the language of the sagas (and hence a status equal to that of Scandinavia's other written languages) but also to avoid lending primacy to any of the Faroes' several phonetic dialects (Hammershaimb 1891:1, lv). Language, however, is not just a neutral ground in its own right but also a primary means of gaining access to other neutral grounds—the medium in which laws are stated, histories recounted, the weather discussed, natural beauties evoked, and the language itself glorified.[20] As an early nationalist poem put it:

> Hear the surf crash on the shore!
> Hear the storm sough in the mountains!
> > That's the Faroes' language.
> Hear the waterfall play on the rock face!
> Hear the echo on the cliff![21]
> > That's the Faroes' language.

> Hear the dance go in the room!
> Hear the long ballad about heroes!
>   That's the Faroes' language.
> Hear the old man by the hearth
> Gather children around old tales!
>   That's the Faroes' language.

In short, from the most informal, everyday levels of Scandinavian society to its most official, formal ones, one moves through tiers of equals. At each level, works of language are universally available means of gaining access to the enduring moral resources of nature, the past, and language itself.

Herein, however, lies a further dilemma, which social change has made more acute. In the *kvøldseta* or "down on roost," one person might know more stories or tell them better, but at least anyone might have a go at it. In Faroese ballad dancing, similarly, one man (or woman) sings the verses while everyone joins in on the refrain, and when one ballad is finished someone else starts up another one. But, just as socioeconomic change in the late nineteenth century produced a new, native middle class whose members asserted the goals of Faroese separatism, so the movement they led entailed establishing institutions and levels of culture to which access is *not* universal. As Faroese national culture has been defined by publishing ballads and legends, schoolbooks and scholarly histories, poetry and newspapers and novels, the formal articulation of Faroese culture has increasingly become a matter for specialists. Many people can (and do) buy the village histories (*bygdarsøgur*) whose litanies of genealogical and geographical detail mimic the norms of oral historiography. People may even read them—certainly they put them on their bookshelves. Yet, with the demise of *kvøldsetur,* the decline of ballad dancing, the fixing of legends and ballads in written form, the advent of schools and scholarship, and a growing divergence between "good" and colloquial Faroese, people have to some extent lost immediate access to their past. "History" is something children learn about in school, along with foreign languages and "Faroese," arithmetic and natural history. "Folklore" is taught at the university level in Tórshavn and abroad. Few young people care to learn the ballads' endless verses, which are chiefly preserved by dance clubs.

In other words, Faroese culture is changing in a rather complicated way. The fundamental distinction between everyday and formal levels of culture, which was once roughly aligned with the distinction between Faroese and Danish, has been preserved along with the primary value of egalitarianism, with the use of nature, language, and the past as moral resources, and with the role of language as a means of gaining access to them. But relations between the everyday and the formal have become increasingly complex with the standardization and institutionalization of Faroese high culture. A certain tension between folk and specialist cultural expertise was evidently felt as early as the late nineteenth century. David Margolin has argued convincingly that the "Ashlad" folktales popular in the 1890s both reflected and reflected upon

> the creation of, as it were, a class of Øskudólgurs [Ash-lads 'men unlike other men' who by magical means overcome poor, anomalous beginnings to marry the princess and inherit the kingdom] to assume . . . the role of defining Faroese culture from above—folklorists, poets, journalists, bibliophiles, teachers, historians, shippers, politicians, bankers, artists, and a Faroe-born priesthood. (Wylie and Margolin 1981:65)

Ólavur's version of "Álvabøur is a young village . . ." may be read as a sort of Ashlad tale. He was saying clearly, if rather elliptically, that "some people" may be schoolteachers and politicians, well married and prominent in national affairs, but that their social standing doesn't make their version of the story correct or their primacy in local society a settled thing. His nonstandard account of Álvabøur's founding and his conclusion about the demise of the Young People's Club express the enduring, if somewhat beleaguered and occasionally subversive, vitality of everyday culture in the Faroes. Ólavur himself thus resembles Ashlad's brothers, for Ashlad "ruled the kingdom both long and well, [but] his brothers always tried to make trouble for him" (Wylie and Margolin 1981:51).

Dominican independence was not the result of a popular movement, much less one glorifying local culture or the local past. The British wanted to be rid of their West Indian Associated States. Dominicans wanted more foreign aid than the British could provide and the dignity and trappings of statehood.

As independence neared, the central issue was starkly political: who would control the government? The opposition Freedom Party, which favored independence in principle but opposed it under a government as venal as that of the then premier, Patrick John, maintained that the Association Act of 1967 required that a referendum be held before independence was granted. Apparently acting in collusion with the British government, the ruling Labour Party contrived to avoid one. A constitution was drawn up semisecretly in London, probably by British civil servants nominally in consultation with representatives of the Dominican government. Finally, on 13 July 1978, "after some fourteen hours of nonstop debate, the Dominica House of Assembly adopted a resolution calling on the British Government to grant independence to Dominica on November 3rd [of that] year" (Anon. 1978). The debate lasted so long partly because the independence resolution had to be approved in an ostensibly democratic manner so that Britain could announce on schedule that it had acceded to Dominica's request for independence. During the debate, the government defeated a motion to discuss the constitution clause by clause, and at 1:45 in the morning the leader of the opposition, Mary Eugenia Charles, fainted from fatigue as she was speaking. It was said that she had not seen the constitution's final draft until the night before, and had been up since then studying it.

Dominicans were profoundly ambivalent about independence. On the one hand, it seemed an event of almost eschatological proportions—a time when the unworthiness of local culture might be shuffled off, inaugurating an almost white way of life blessed with unity, progress, prosperity, godliness, a strong family life, and all the other things so notably absent from life as it was actually lived from day to day. "ITS HERE," the *New Chronicle*'s banner headline proclaimed on 4 November 1978. Its editorial, "A Nation under God," began with a pastiche of Jefferson quoted (more or less) from the preamble to the new constitution.

> A new nation is born, the Commonwealth of Dominica, a nation under God. We affirm by our constitution that Dominica is founded upon principles that acknowledge the supremacy of God, faith in fundamental rights and freedoms, the position of the family in a society of free men and free institutions, the dignity of the human person, and the equal and alienable [*sic*] rights with which all members of the human family are endowed by their Creator. What does this mean as we enter nationhood and become politically independent?
>
> It means first of all, that love of God should at all times be our principal concern and our guide to action. We must not allow the affirmative [*sic*] to remain just another expression in the constitution. It must become for us, a way of life. Love of God must mean love of neighbor. . . . Our neighbor must be all the people of Dominica.[22]

Similarly, at the Freedom Party's convention earlier in the year, a speaker had been enthusiastically applauded for comparing Miss Charles to Jesus. Miss

Charles, a woman of the highest moral integrity who seemed almost grotesquely out of place amid the self-serving partisanship of Dominican politics, seemed embarrassed by this impromptu apotheosis. But its logic was impeccable. Being black in pigmentation but white in behavior, she embodied a nearly supernatural transcendence of black and white.

On the other hand, independence seemed an ominous prospect. In patois, *ēdepādā* means not only politically independent, but also, in a more basic sense, personally free to act as one wishes, hence willful, irresponsible, "just doing what you want"—or, in short, free to construct reality and to determine the future as one chooses, perhaps even to dabble in the occult arts. To be *ēdepādā* is individually desirable, but an *ēdepādā* person is socially dangerous. So is an *ēdepādā* government, at least a black *ēdepādā* government. (There were several stories about Patrick John's occult expertise. It was said, for example, that having "invoked" his mother [called her from the grave] in order to win the 1975 elections he needed his sister's help to get her back again.) A man in Casse suggested the menacing possibilities of Dominica's becoming *ēdepādā* when he went about drunk on independence morning singing

Ēdepādā,
Ēdepādā guvelmā
Nu ka plêwe

[Independent,
Independent government,
We are crying.]

As things turned out, Patrick John's *ēdepādā guvelmā* performed pretty much as many people feared, becoming embroiled in a series of increasingly grotesque scandals until it fell in 1980. The Freedom Party then gained a parliamentary majority, and Miss Charles duly became prime minister. The Reagan administration patronized her for her "conservatism" and welcomed her support for the invasion of Grenada in 1983. She is not, however, particularly "conservative" in the Reaganite sense. (Grenada's nominally "leftist" government must, in her eyes, have seemed dangerous chiefly because it resembled the Dominican Labour government under Patrick John.) She is better understood as heir to the traditions of the old mulatto elite, which controlled the House of Assembly for most of the late nineteenth century, and especially of that elite's successors, a well-educated, "liberal-minded middle-class group of professional men and farmers," including Miss Charles's father who dominated Dominican politics between the First and Second World Wars (Honychurch 1975:93). Her popularity in Casse—for Casse was a Freedom Party stronghold—must also be understood in these terms.

Awkwardly straddling Dominica's white and black worlds, the class into which Miss Charles was born maintained a distinctive brand of local culture until, roughly speaking, the late 1950s. Its members "still danced lancers and flirtations at their subscription parties" and "took the two days [of Carnival] to 'run masked' and escape from their respectability" (Honychurch 1975:94–95). Unlike the "professional men and farmers" who founded the Faroese separatist movement, however, this native Dominican elite did not attempt to make local folkways or the vernacular respectable. Indeed, its respectability depended on keeping a careful distance from the local, the vernacular, and, in a word, the black. A number of factors—emigration, agricultural decline, the rise of labor unions, wider suffrage—led to this elite's loss of economic and political influence after the Second World War. By the time I lived in Dominica, it had been succeeded, in a sense, by civil servants and other members of the professional middle class, mostly in Roseau. Such people supported a dance troupe and a theater company, for example, and someone had written a patois Mass that was occasionally said round the island, but otherwise there was little evidence of any concerted attempt to create a respectable register of local black culture.

Casse had also had a landed elite—four or five well-to-do families that mediated relations between Casse and the outside world both politically and, through their estates and fishing and smuggling operations, economically.[23] They likewise reconciled, to some extent, the white and black branches of local culture. On the one hand, for example, the heads of these families were properly married, regular churchgoers; on the other hand,

> at least some of them upheld an ideal of male sexual prowess by fathering illegitimate children, and the two wealthiest men in the village were widely believed to be *lugawu:* men who have contracted with the Devil in order to win riches at others' expense. (Wylie 1993:373)

Casse's landed elite also apparently preserved some sense of the community's continuity with the past. In general, people in Casse can and do give detailed accounts of how landholdings have been inherited over several generations. Baptiste, for example, can tell you in detail how he acquired his land in the hills, although he is baffled if you ask what his father was doing in "1914." Probably because their estates were larger and older, the more substantial landholding families even claimed some knowledge of historical facts without any obvious practical importance, although the stories I collected from these families' surviving members scarcely added up to a coherent history of the village. Thus, one man told me that there used to be a church and a graveyard in Casse. Parish records and a manuscript diocesan history (Anon. n.d.:197) confirmed that a chapel and graveyard had existed in Casse, probably passing out of use in the mid-1840s. But I was also told that the "head" (headland) at Labatwi takes its English name from a British captain and priest who was beheaded there by the French (his head is allegedly "somewhere in Roseau"). Someone else told me that the place got its name from a Britisher, allegedly my informant's great-grandfather, who was beheaded there. In fact the headland is named for a British officer who, so far as I can determine, served briefly in Dominica in the eighteenth century, left no descendants there, and ended his career peacefully in England. The same man said that "people" used to live on Labatwi. Thinking this unlikely, I asked Baptiste about it. He said no, only British soldiers had ever lived there, during "the Boer War." (I suppose this must have been after they captured it during the First World War.)

There was not much left of Casse's landed elite by the time I lived there—a pair of aged sisters entrenched in an embittered, dilapidated gentility, a poor widow, a few younger men who worked abroad, the owners of a couple of decaying estates who lived chiefly on their savings. It had fallen victim to the abysmal state of the market for limes, the rise of artisanal fishing and small-scale interisland marketing ("huckstering"), easier communications with Roseau, the virtual disappearance of smuggling, and, perhaps most important, emigration and people's reliance on remittances from abroad instead of on the patronage of the local well-to-do. The same factors also apparently undermined the integrity of the more black reaches of local culture, weakening both traditional customs of cooperative endeavor and the social constraints on folk-religious beliefs in witchcraft. A cultural result of Casse's more diffuse economic integration with the rest of the world was a contentious, intensely ingrown dissociation from it. Politically, however, a result was not factionalism but unanimity: nearly universal support for the Freedom Party, and especially for Miss Charles, who, it was often said, was so rich she would not have to steal. In other words, she represented at the national level an idealized version of what was missing at the local level.

Conclusion

Time is deep in Álvabøur; in Casse it is shallow. If Álvabøur's memory were as short as that of Casse, men now middle aged would scarcely know that their fathers built the harborworks and the road to Strandarvík "before I know myself." If Casse's memory were as long as that of Álvabøur, people would remember emancipation as well as Álvabingar recall the founding of their village in the same year; the heroes of the last Maroon War (1812–15) as well as Faroese recall the loss at sea of their hero Nólsoyar Páll (1808); the Middle Passage and the establishment of the West Indian plantation economy as well as Faroese recall the bitter, hungry years when they were ruled by the Danish courtier Christoffer von Gabel and his son Frederik (1662–1708); the affairs of African states as well as Faroese recall the death of the Viking warlord Olaf Trygvason and the coming of Christianity to the Faroes (ca. 1000); and perhaps even a southern branch of the Moorish expansion, which to the north, as Faroese ballads tell, doomed Roland at Roncevaux in 778—exactly a thousand years before a more fortunate French force stormed Labatwi. Or Dominicans might have picked up French or English history, assuming a European or Euro-American past in the same way that a Polish-American, say, might appropriate Valley Forge, the Pilgrim fathers, and the Magna Carta.

History, I have argued, is a matter of popular concern in Álvabøur because the past is conceived of as a part of reality independent of the current social order, in terms of which local society is constituted. History scarcely exists in Casse because reality is an immediately social construction, while local society itself is explicitly (and self-denigratingly) seen in terms of white ways and God's word. More generally, Faroese have approached independence through demonstrations of their native culture's worthiness, expressed in part historiographically. Dominicans visualized independence as dangerous in black terms but desirable in terms of aspirations to godliness and white norms. There is irony in this: Faroese have sought to achieve culturally an independence they have not achieved politically; Dominicans were granted politically an independence they did not seek to deserve in terms of local culture.

Behind this irony lie several more complex contrasts. Both Dominica and the Faroes were colonial societies, many of whose institutions were controlled from abroad—from London and Copenhagen. The Faroese, however, were in a sense a species of Dane who maintained a distinctive brand of Scandinavian rural culture while partaking of European civilization through their Danish connection. Social and economic changes in the nineteenth century fostered the growth of a native middle class that found itself in an increasingly awkward position between the local and the cosmopolitan, the Faroese and the Danish, the traditional and the modern, the everyday and the official. In part by exploiting their Danish connection, the members of this middle class founded a separatist movement, which undertook, with remarkable success, to clear up their own and the Faroes' increasingly ambiguous status by creating a national culture out of a selective version of the folk one. The very fact that Faroese national culture was a high culture, however, tended to replicate—internally, now—the distinction between the everyday and the official.

Like its Faroese counterpart, the native Dominican elite has found itself balanced between the cosmopolitan and folk branches of local culture. Its situation has been rather more clear-cut, however, and its orientation toward the white end of things more marked. Perhaps independence will help to complicate its situation enough to inspire its members to embark upon a course like the Faroese one—that is, to establish a national culture in which folkways are given a respectable form. On the other hand, the difference between white and black is far greater than the difference between Danish and Faroese, and correspondingly more difficult to bridge. It is possible to imagine a patois literature, but hardly, for example, the legalization of obeah.

In other words, both the creation of a Faroese national culture and Dominica's actual attainment of independence have in some measure perpetuated a dislocation between everyday and official levels of culture. It is a long way from Ólavur's parlor to Baptiste's shady bit of shore, but the circumstances in which the two men find themselves are not, after all, incomparable. From their point of view—and also, one might add, from the point of view of a historically minded ethnographer surveying the literature about their societies —it might be said that there are too many books about the Faroes and too few about Dominica.

NOTES

The fieldwork in the Faroe Islands on which this essay is based was supported in part by the Comparative International Program, Department of Social Relations, Harvard University. Further research was supported by a Fulbright research fellowship at the Sosialantropologisk Institutt at the University of Bergen and by the Danish Marshall Fund of America. Fieldwork in Dominica was carried out under the aegis of the Island Resources Foundation, St. Thomas, United States Virgin Islands. The American Council of Learned Societies supported some preliminary work on comparisons between the Faroes and Dominica.

I am grateful to Ms. Gail Ringel for the use of her field notes from our stay on Dominica, and to the Center for International Studies at Cornell University for sponsoring a lecture based on a draft of this essay, a version of which appeared in *Comparative Studies in Society and History* (Wylie 1982).

Casse, Álvabøur, and Strandarvík are pseudonyms, as are the names of living people in both villages. I have, however, disguised only the family names of deceased Álvabingar. Unless otherwise noted, the ethnographic present in this essay is 1971–72 for the Faroes and 1977–78 for Dominica.

1. Patois—the Dominican variety of Lesser Antillean French Creole—has no standard written form. Except for the name *patois* itself, which is pronounced as in French, I render it here according to the phonemic system proposed by Douglas Taylor (1977:198–204).

2. The following quotations in the text are from Edwards (1807:1, 435–36).

On the general background of the attack, see Dusseigneur (1963). The clearest picture of the attack itself is given by a map, a photocopy of which (from the original in the Bibliothèque Nationale, Paris) may be found in the Map Room of the New York Public Library, entitled "Plan de la Conquête de l'Isle de la Dominique par le M^r le Mis de Bouillé en 1778." This map may be part of a fuller map of Dominica, which de Bouillé sent to his superiors in Paris but which failed to reach them (de Bouillé 1778). The attack is illustrated, in some respects rather fancifully, by Godefroy (n.d.). Edwards's account, like most in English and some in French (e.g., Daney 1815:4, 127), relies on that of Thomas Atwood ([1971] 1791:108ff.). Most French accounts rely on documentary sources and on the Marquis de Bouillé's correspondence and memoirs (Barrière [1859:46ff.]; Guerin [1851:44–45]; Lacour-Gayet [1905:181–83]; Lerville [1972]). For a somewhat different French view, see Loir (1894:19ff.). For official lists of the personnel on the French side, see Roussel (1779:161–62, 175–76; see also Dahlman [1877:11]). For a quasi-official British military view, see Beatson (1804:4, 384ff.). For British personnel, see, e.g., Porter (1889:1, 205) and, of course, the relevant *Army Lists*.

The British fortification at Labatwi was probably supposed to be manned, at least in part, by the local "gunners, matrosses, and negroes" required to "do duty at certain signals and alarm posts and batteries" by a law of late 1777 or early 1778 (Laws of Dominica 1818:xiv). This law was evidently widely disregarded (Andrews 1786:3, 177). Further particulars about the state of Dominica's defenses await discovery in British archives; see, e.g., Ragatz (1923:11).

The British often named strongpoints after officers. Thus Scotts Head, at the southern tip of the island, is named for Lieutenant Colonel George Scott, the military governor of Dominica from 1765 until 1768 and an officer so unmemorable that as early as 1800 Europeans were calling it Scotsman's Head (McKinnen 1804:46; A Resident 1828:33).

3. There are also several minor races, including Caribs, Portuguese, and East Indians. Black people themselves range in color from "black" to "red-skin" to "clear-

skin." The patois term *was* is in some senses synonymous with "family" (*famiy*) in the latter's sense of people related by blood.

4. I am using the terms *respectability* and *reputation* in the senses proposed by Wilson (1973:2ff.) to mean dialectically related principles of social stratification based respectively on nonlocal and local values and pursuits. Wilson argues persuasively that this dialectic is fundamental in West Indian cultures. In a less extreme form (because not aligned with alleged racial characteristics) it is probably common to all provincial cultures. I suggest below that in Dominica, at least, an elite, formed of what might be called "people of substance," rather precariously mediates respectability and reputation. A preacher, for instance, is respectable, a good fisherman is reputable, and a well-to-do landowner is a man of substance.

5. Patois is everyone's first language in Casse. Older children are beaten for using patois in school, however, and some people claim that God does not understand it. People's competence in English varies immensely. Most people's English is probably best described as a nascent English creole influenced by patois as well as by standard English and by existing West Indian English creoles (Amastae 1979).

6. Only Baptiste told me the story of the siege of Labatwi. It is evidently not commonly known in the village, although most people are probably aware that Dominica used to be French. The enmity between French and English is sometimes attributed to the latter's having executed Joan of Arc for working obeah. Probably popularized by French priests in the late nineteenth century, the story of Joan of Arc has been perpetuated by several movies that have toured the island.

7. I am speaking here of knowledge with no obvious practical importance. Practical knowledge is learned mostly by rote and imitation, and it quickly fades if its applications disappear. Thus, the few constellations people once knew were forgotten when night smuggling ceased.

8. For an extended discussion of the legal system(s) in Casse, and of "debates" generally, see Wylie (1989b).

9. The definitive work (Clementsen 1983) was written after I lived in Álvabøur. For an earlier treatment, also relying mostly on documentary sources, see Bjørk (1968). The discussion by Hjalt (1953:95ff.), which relies more heavily on oral traditions in Strandarvík, must be used with caution since his dating is off by at least a year. I am grateful to Niklas Jacobsen for letting me copy his file of newspaper clippings (J. M. Poulsen 1950; Øssursson 1950; Nolsøe 1950) upon which the following account is partly based. I have also used scattered material from, e.g., Clementsen (1981), Johannesen (1976), Waag (1967), and Øssursson (1963).

10. J. H. Poulsen (1963:46–58). "Álvabøur" itself more likely means 'elves' infield' than 'Álvur's infield'.

11. Literally, "that the first house (or 'the first building': *tað fyrsta hús*) in Álvabøur was raised (*reist*)." Completion of a house is customarily dated to the raising of the rooftree. She spoke of "the house" in the singular, but the whole issue is complicated by the fact that in Faroese a single dwelling or homestead is generally referred to

in the plural. Thus, *tey fyrstu húsini* might mean either 'the first houses' or 'the first house' as well as 'the first buildings'.

12. For a discussion of the Faroese kinship system, see Wylie (1974).

13. Elsewhere in the Faroes, the Self-Rule Party drew only about 35 to 40 percent of the total vote in 1920-36. In general it was their conservative rival, the Union party (Sambandsflokkur) that lost votes to the new Social Democratic Party (Javnaðarflokkur). In Álvabøur, however, the Social Democrats took votes away from the Self-Rule Party.

14. Non-Lutheran, mostly Protestant, denominations are commonly called "sects" (*sektir*) in the Faroes. Both the Protestant sects, the first of which arrived in the Faroes in the late nineteenth century, and their evangelical rivals within the Lutheran church have condemned many elements of traditional Faroese culture (J. P. Joensen 1980:174-75; Hansen 1984).

15. Jakobsen (1898-1901) is the standard collection of Faroese legends and folktales. Many others are recorded in, for example, the numerous "village histories" (*bygdarsøgur*) published since the 1930s. For a history of the Faroes as seen through the legends, see Jakobsen (1904). For a sustained analysis of Faroese legendary historiography and "village histories," see Wylie (1987).

16. Park (1972:9). Park says this betokens "a positive resistance to community." I would argue that it exhibits exactly the sort of agreement on which a sense of community is based. For analyses of views of nature elsewhere in Scandinavia, see, e.g., Frykman and Löfgren (1987), and Pálsson (1991).

17. Space permits only a cursory discussion of Faroese cultural and political history since the early nineteenth century. For fuller accounts in English, see Wylie and Margolin (1981:73-94), West (1972), Jackson (1991), and Wylie (1987). Most histories of the Faroes have stressed the "folkish" foundations of Faroese identity. More recent scholarship stresses the uses to which surviving elements of the peasant culture were put by an emergent middle class and other untraditional social types. For an incisive overview in English, see J. P. Joensen (1989).

18. Hitherto there had been no satisfactory system for writing Faroese, which was in any case rarely written and then usually quasi-phonetically in awkward accordance with Danish conventions. For accounts in English of the development and symbolic significance of Hammershaimb's morphophonemic and etymological orthography, see Wylie and Margolin (1981:82-94), and Wylie (1987:89ff.).

19. Barnes (1954:54). Most descriptions of Scandinavian communities stress the high valuation of egalitarianism. Thus, Tomasson (1980:195) notes that "there are few things Icelanders believe more about themselves than that theirs is a country where there is equality among interacting individuals and equality of opportunity"; Hollos (1976:242) notes that "egalitarianism was highly valued" in a mountain community in eastern Norway; Yngvesson (1978:81) notes that in a Swedish fishing community "the ideal to which community members aspired in their relations with one another seemed to be patterned on a 'team model' [in which] the equality of all full members of the

community was stressed"; and Anderson and Anderson (1964:111) note that "Class stratification is ideologically unacceptable to modern Danes." I have elsewhere (Wylie 1989a) reached similar conclusions about Scandinavian culture by a rather different route, beginning with Barnes's seminal article. Very similar conclusions about everyday (or informal) and official (or formal) levels of culture in Scandinavia have been reached in different ways by Gullestad (1989a, 1989b), and Pálsson (1991).

20. The glossolalia practiced by the Assembly of God is interesting in this light, as a nonstandard (even humanly incomprehensible!) means of communication with the source of the moral order.

21. Literally, 'hear the dwarf speak on the cliff'. 'Dwarfs' language' (*dvørgamál*) is one term for "echo" in Faroese.

22. Such exhortations and expectations have been common in the West Indies. Manning (1979:19) reports, for example, that the Bermudan Progressive Labour Party, advocating independence from Britain in the elections of 1976, appropriated the language, meeting style, and personnel of evangelical movements and campaigned to strengthen "the type of family prescribed by the church—the nuclear, monogamous unit based on marriage" so strikingly different from "the black family." Progressive Labour's campaign

> was likened to a crusade against evil waged by a people whom God has chosen to remake and inherit Bermuda. Party leader Lois Browne . . . struck the theme repeatedly in the campaign:
>
> *"God doesn't mean for oppression to win. So ultimately we will win. . . .*
>
> *"We have faith, strength. Even if we don't win, we're going to go on. It's inevitable. We know we're going up and the others are coming down. We will claim victory in 1980, or 1984, or whenever. It is God's work to take us there. . . .*
>
> *"The party wants to build up idealism and restore it to our lives and politics. Our members are quality people. They are made in the image of God, and will represent you."* (original italics)

23. The following remarks draw on snippets of information I picked up in Casse, on personal communications from Jon Amastae and Jonathan Wouk, and on Wouk (1965). For a somewhat fuller account of Casse's social history since the Second World War, see Wylie (1993).

REFERENCES

A Resident. 1828. *Sketches and Recollections of the West Indies.* London: Smith, Elder.

Amastae, Jon. 1979. "Dominican English Creole Phonology: An Initial Sketch." *Anthropological Linguistics* 21:182–204.

Anderson, Robert, and Barbara G. Anderson. 1964. *The Vanishing Village: A Danish Maritime Community.* Seattle: University of Washington Press.

Andrews, John. 1786. *History of the War with America, France, Spain and Holland; Commencing in 1775 and Ending in 1783.* 4 vols. London: John Fielding.

Anonymous. 1978. "Independence Resolution Adopted, November 3rd, Independence Day. 16 Yes, 8 No." *New Chronicle* (Roseau) 71:28, 1 (15 July).

———. N.d. *History of the Diocese of Roseau.* Book 2: *Beginning and Development of the Various Parishes of Dominica.* Diocesan Archives, Roseau. Manuscript.

Atwood, Thomas. [1971] 1791. *The History of the Island of Dominica.* London: J. Johnson. Facsimile ed., 1971. London: Cass.

Barnes, J. A. 1954. "Class and Committees in a Norwegian Island Parish." *Human Relations* 7:1, 39–58.

Barrière, M. Fs., ed. 1859. *Mémoires du Marquis de Bouillé.* Paris: Didot Frères, Fils.

Beatson, Robert. 1804. *Naval and Military Memoirs of Great Britain from 1727 to 1783.* 6 vols. London, Edinburgh, and Aberdeen: Longman, Hurst, Rees and Orme, and J. and J. Richardson; A. Constable; and A. Brown.

Bjørk, E. A. 1968. "Elsta søga Skopunar." *Úrval* (Tórshavn) 3:1, 3–37.

Blom, Jan-Petter, and John J. Gumperz. 1972. "Social Meaning in Linguistic Structure: Code-Switching in Norway." In *Directions in Sociolinguistics: The Ethnography of Communication,* John J. Gumperz and Dell Hymes, eds., 407–34. New York: Holt, Rinehart and Winston.

de Bouillé, François-Claude-Amour, Marquis. 1778. Untitled letter, Archives Nationales, Série Colonies C[8A], Martinique (Correspondance à l'Arrivée), article 77, folio 90.

Clementsen, Ólavur. 1981. *Søga og skemt av Sandi.* Tórshavn: Klovin.

———. 1983. *Skopun 150 Ár.* Tórshavn: Klovin.

Dahlman, G. Déhon. 1877. *Historique du 12e Régiment d'Infanterie de Ligne.* Paris: Ch. Tanera, and Librairie pour l'Art Militaire, les Sciences et les Arts.

Daney, Sidney. 1846. *Histoire de la Martinique, depuis la colonisation jusqu'en 1815.* 6 vols. Fort-Royal: E. Ruelle.

Dusseigneur, F. 1963. "Les Antilles et la Guerre d'Indépendance." *Revue Historique de l'Armée* 19:1, 79–89.

Edwards, Bryan. 1807. *History, Civil and Commercial, of the British Colonies in the West Indies.* 4th ed. ("with considerable additions"). 3 vols. London: John Stockdale.

Frykman, Jonas, and Orvar Löfgren. 1987. *Culture Builders: A Historical Anthropology of Middle-Class Life.* Trans. Alan Crozier. New Brunswick, N.J., and London: Rutgers University Press.

Gíslason, Gylfi. 1973. *The Problem of Being an Icelander: Past, Present and Future.* Trans. Pétur Kidson Karlsson. Reykjavík: Almenna bókafélagið.

Godefroy, François. N.d. [1783?]. *Recueil d'estampes representant les différents événemens de la Guerre qui a procuré l'Indépendance aux Etats Unis de l'Amérique.* Paris: M. Pouce, and M. Godefroy.

Guerin, Leon. 1851. *Histoire maritime de France*. . . . Paris: Dufour et Mulat.
Gullestad, Marianne. 1984. *Kitchen-Table Society: A Case Study of the Family Life and Friendships of Young Working-Class Mothers in Urban Norway*. Oslo, Bergen, Stavanger, and Tromsø: Universitetsforlaget.
———. 1989a. *Kultur og hverdagsliv: På sporet at det moderne Norge*. Oslo: Universitetsforlaget/Det Blå Bibliotek.
———. 1989b. "Small Facts and Large Issues: The Anthropology of Contemporary Scandinavian Society." *Annual Review of Anthropology* 18:71–93.
Hammershaimb, Venceslaus Ulricus. 1891. *Færøsk anthologi*. 2 vols. Copenhagen: S. L. Møller (Møller & Thomsen). Facsimile ed., 1969. Tórshavn: Offset-Prent/Emil Thomsen.
Hansen, Gerhard. 1984. *Vækkelsesbevægelsernes møde med Færingernes enhedskultur*. Annales Societatis Scientiarum Færoensis, Supplementum X. Tórshavn: Føroya Fróðskaparfelag.
Hjalt, Edward. 1953. *Sands søga*. Tórshavn: Varðin.
Hollos, Marida. 1976. "Conflict and Social Change in a Norwegian Mountain Community." *Anthropological Quarterly* 49:4, 239–57.
Honychurch, Lennox. 1975. *The Dominica Story: A History of the Island*. Barbados: Letchworth.
Jackson, Anthony. 1991. *The Faroes: The Faraway Islands*. London: Robert Hale.
Jakobsen, Jakob, ed. 1898–1901. *Færøske folkesagn og æventyr*. 2 vols. Copenhagen: Samfund til udgivelse af gammel norsk litterarur. Republished in 3 vols., 1964–72. Tórshavn: H. N. Jacobsens Bókahandil.
———. 1904. *Færøsk sagnhistorie, med en indledende oversigt over øernes almindelige historie og litteratur*. Tórshavn and Copenhagen: H. N. Jacobsens Forlag, and Vilhelm Priors Hofboghandel.
Joensen, Jóan Pauli. 1980. *Färöisk folkkultur*. Lund: LiberLäromedel.
———. 1989. "Socio-Economic Transformation and the Growth of Faroese National Identity." *North Atlantic Studies* 1:1, 14–20.
Joensen, Robert. 1966. "Hvussu gouml er bygdin?" *Varðin* (Tórshavn) 38:1–2, 26–31.
Johannesen, Marius, ed. 1976. *Lærarafólk í Føroyum, 1870–1976*. 2d ed., expanded. Tórshavn: Føroya Lærarafelag.
Jónsson, Hannes. 1972. *Iceland and the Law of the Sea*. Reykjavík: Government of Iceland.
Lacour-Gayet, G. 1905. *La Marine Militaire de la France sous le règne de Louis XVI*. Paris: Honoré Champion.
Laws of Dominica. 1818. *The Laws of the Colony of Dominica, commencing from its earliest establishment to the close of the year 1818; and tables of the several acts, with preface and index*. Roseau: Wm. F. Steward.
Lerville, Edmond. 1972. "Rivalté Marine-Armée de Terre à la Martinique en 1779 vue à travers les documents chiffrés." *Revue Historique de l'Armée* 28:3, 33–52.

Loir, Maurice. 1894. *Jean Gaspart Vence, corsaire et amiral (1747–1808)*. Paris: Baudoin.
Manning, Frank. 1979. "Religion and Politics in Bermuda: Revivalist Politics and the Language of Power." *Caribbean Review* 8:4, 18–22, 42–43.
McKinnen, Daniel. 1804. *A Tour through the British West Indies in the Years 1802 and 1803, Giving a Particular Account of the Bahama Islands*. London: J. White.
Nolsøe, Páll. 1950. "Tá Skopun varð endurbygd: Hvat skjölini siga frá nýbúsetingini." Interview, *14 September* (Tórshavn), 2 October.
Oakley, Stewart. 1972. *A Short History of Denmark*. New York and Washington, D.C.: Praeger.
Øssursson, Janus. 1950. "Meira um Skopun." Letter to the editor, *14 September* (Tórshavn), 21 August.
———. 1963. *Føroya Biskupa-, Prósta- og Prestatal*. Tórshavn: Mentunargrunnur Føroya Løgtings.
Pálsson, Gísli. 1991. *Coastal Economies, Cultural Accounts: Human Ecology and Icelandic Discourse*. Manchester and New York: Manchester University Press.
Park, George K. 1972. "Regional Versions of Norwegian Culture: A Trial Formulation." *Ethnology* 11:1, 3–24.
Porter, Whitworth. 1889. *History of the Corps of Royal Engineers*. London: Longman, Greene.
Poulsen, J. M. 1950. "Hvör bygdi Skopun?" Letter to the editor, *14 September* (Tórshavn), 24 July.
Poulsen, Jóhan Hendrik Winther. 1963. "Um Finnbogarímu færeysku." *Skírnir* (Reykjavík) 137:46–58.
Ragatz, Lowell Joseph, comp. 1923. *A Guide to the Official Correspondence of the Governors of the British West India Colonies with the Secretary of State, 1763–1833*. London: Bryan Edwards.
Ringel, Gail, and Jonathan Wylie. 1979. "God's Work: Perceptions of the Environment in Dominica." In *Perceptions of the Environment: A Selection of Interpretative Essays*, Yves Renard, ed., 39–50. Caribbean Environment, Environmental Studies, no. 1. Barbados: Caribbean Conservation Association/Association Caraïbe pour l'Environment/Asociación para la Conservación del Caribe.
Roussel, M. de. 1779. *État militaire de France pour l'année 1779*. Paris: Godefroy.
Taylor, Douglas. 1977. *Languages of the West Indies*. Baltimore and London: Johns Hopkins University Press.
Tomasson, Richard F. 1980. *Iceland: The First New Nation*. Minneapolis: University of Minnesota Press.
Waag, Einar. 1967. *Val og Valtøl 1906–1966*. Klakksvík: Privately published.
West, John. 1972. *Faroe: The Emergence of a Nation*. London and New York: C. Hurst, and Paul S. Eriksson.
Wilson, Peter J. 1973. *Crab Antics: The Social Anthropology of English Speaking Negro Societies of the Caribbean*. New Haven and London: Yale University Press.

Wouk, Jonathan. 1965. "Witchcraft and Sorcery in a Dominican Fishing Village." Senior honors thesis, Harvard College.

Wylie, Jonathan. 1974. "I'm a Stranger Too: A Study of the Familiar Society of the Faroe Islands." Ph.D. diss., Harvard University.

———. 1982. "The Sense of Time, the Social Construction of Reality, and the foundations of Nationhood in Dominica and the Faroe Islands." *Comparative Studies in Society and History* 24: 3, 438–66.

———. 1987. *The Faroe Islands: Interpretations of History*. Lexington: University Press of Kentucky.

———. 1989a. "The Christmas Meeting in Context: The Construction of Faroese Identity and the Structure of Scandinavian Culture." *North Atlantic Studies* 1: 1, 5–13.

———. 1989b. "The Law of the Streets, the Law of the Courts, and the Law of the Sea in a Dominican Fishing Village." In *A Sea of Small Boats*, John Cordell, ed., 152–76. Cultural Survival Report, no. 26. Cambridge, Mass.: Cultural Survival.

———. 1993. "Too Much of a Good Thing: Crises of Glut in the Faroe Islands and Dominica." *Comparative Studies in Society and History* 35: 2, 352–89.

Wylie, Jonathan, and David Margolin. 1981. *The Ring of Dancers: Images of Faroese Culture*. Philadelphia: University of Pennsylvania Press.

Yngvesson, Barbara. 1978. "Leadership and Consensus: Decision-Making in an Egalitarian Community." *Ethnos* 43: 1–2, 73–90.

Time and Memory: Two Villages in Calabria

Maria Minicuci

I am going to describe the idea of history and the measurement of time in two Italian villages less than two miles away from each other. Their history goes back several centuries and is documented by written records of which the inhabitants are totally unaware, or almost. What is more important is that they do not consider this their history at all. It has no interest for them, it does not concern them.[1] The history they recognize as their own occupies a defined space and a time frame made up of different durations, which are measured by internal village standards.

Oral traditional and local customs are responsible for most, and the most significant, social relations and orient assessment and choice of action. These features might encourage us to label the two villages "traditional societies." But Leach, speaking of tribal societies, discourages the use of the term *traditional* because it conveys the sense of a static society, one that has long been unchanged (Leach 1989). Firth, among others, disagrees and considers it "a convenient but admittedly slippery term. In its ordinary dictionary meaning, tradition refers to a statement, belief, or practice transmitted (especially orally) from generation to generation," and he does not believe it precludes change (Firth 1989:49).

Leach's warning seems appropriate to the villages I will be discussing. It is not just ethnologists who offer a view of traditional society as a kind of solid homogeneous block that may be struck by events and/or processes that shatter or crush it. Informants also offer ethnologists news of a traditional society immersed in one unchanging course of time, a society in which things have "always been done this way." I want to consider whether this image is a true one—whether a single period or even whole eras are actually perceived as a uniform continuum that denies historical change. Is time considered a function of a particular type of discourse, and if so, what kind of discourse is it?

True, there is systematic recourse to tradition, and the past is presented with the features I have just mentioned, but this does not mean that tradition is understood as something permanent, constant repetition of the same thing, nor does it mean that there is only one way to calculate time. On the contrary, there is an awareness that changes have taken place and new elements assimilated, but what matters is the final result of the adaptation process, which makes it possible to create the reassuring image of a stable whole that can face the uncertain and alarming aspects of outside historical developments that affect them but over which they have no control. There are two essential elements that make this image essential for them: the "historical marginality" (Pizzorno 1967) in which these poor, oppressed communities have lived for centuries, and a consequential way of understanding tradition in these historical conditions.

The two communities on which I focus are in one of the most underdeveloped regions of Italy, Calabria, a region in which the population has been halved by emigration. (According to official estimates, four hundred thousand Calabrians emigrated to Argentina between the two wars.) Until the middle of the twentieth century the two villages were dominated by large landowners and devastated by frequent earthquakes; they turned in on themselves to find the means of survival and to reproduce for the future. Since the only real capital of the two communities is people, they have used the institution of marriage and a network of exchange relations to organize strategies of adaptation, resistance, and reproduction, and in order to do so they have made choices and set rules to make sure the strategies are respected. From this perspective, tradition is not only a set of orally transmitted beliefs, customs, knowledge, and practices, validated by the authority of the ancestors who have handed them down. More importantly, tradition is a set of norms considered responsible for establishing and maintaining balance in a larger situation of widespread precariousness.[2]

This equilibrium has in the course of time constantly readjusted to internal and external events (e.g., demographic change, emigration, war, and specific individual choices), but the profound fracture occurred after the middle of the present century as the result of a complex of historical contingencies that also affected these communities, which had in the meantime become more open and less marginal and had gradually been led to abandon the underlying logic of their own social organization. This abandonment marks a break between the present and the past and tends to site even the recent past in some remote era that envelops a host of times past. But, while there may be a host of times past, one of them is dominant: the past that memory is able to reconstruct.[3]

I will be considering two villages in southern Italy. One is fairly large (about one thousand inhabitants), and the other is quite small (about two hundred people). Moreover, the smaller community is part of another municipality. While the two communities are in the same geographical area and have similar economies, they have different characteristics.

There are two reasons for comparing them. One is to disprove the commonplace notion that traditional culture is essentially uniform over vast areas, in which differences between single communities are attributed solely to the pace of modernization. Too often, at least in studies of European societies, and especially Italy, the analysis of a single village or area has been considered as representative of the conditions of wider zones, for example, Banfield's approach to southern Italy (Banfield 1959). Or whole regions, such as Calabria, have been studied on the grounds of their comparable development process, degree of backwardness, and cultural values, as if geographic and productive space were experienced and used the same way by all the inhabitants and as if historical events were experienced, interpreted, and remodeled in the same way by various local cultures.

The second reason, which follows from the first, is to see whether there are significant differences between two seemingly homogeneous communities, which are similar in their structure and which are closely related through a policy of marriage-exchange they have followed for more than two centuries.[4] The two communities are marriage partners, exchanging men, women, and culture. In this uninterrupted flow of ideas and people, how have differences been structured and maintained?

Comparing the two communities requires us to distinguish the basic elements each has chosen as a means of organizing itself, following parallel courses that sometimes intersect and sometimes diverge, but nevertheless hewing out their own separate histories. Their histories are expressed in time, but they do not measure time in the same way nor appropriate it to the same degree. When I say "appropriate," I mean to sense time as one's own, to know and control a time that serves as a link with that otherwise unrelated infinity in which one is situated. What makes it possible to appropriate time and to construct one's own history is memory. So I will also consider the functions that memory serves—how it is used and to what ends—in two villages that are not versions of a vaster reality nor islands in a nation. They are not static "traditional societies," nor have they ever been. They have created their own history and oriented it according to parameters of their own, but it is based on an armature that is common to both communities—the armature of kinship.

The Villages

In describing the two communities, Fitili and Zaccanopoli, I will give only the background essential to our present purpose and refer the reader to the bibliography for the main historical sources. There are no specific monographs on these villages, but there are a great many studies of Calabria, the part of Calabria in which the two villages are located, and the city of Tropea, which has exercised political and administrative control over Fitili and Zaccanopoli in different ways at different times.[5] Both communities are now part of the province of Catanzaro.

Fitili is about 200 meters above sea level and Zaccanopoli 430 meters. Both communities are near the Tyrrhenian coast. Fitili is administratively part of the municipality of Parghelia. Zaccanopoli was part of Parghelia until 1918 when it became a municipality itself. Until the middle of the present century, the inhabitants of Zaccanopoli were shepherds mainly, but except for a few people who owned small flocks they did not own the sheep they tended. Large flocks, between four and five hundred head on average, were the property of noble and middle-class families in Tropea and other nearby communities who rented them to the people of Zaccanopoli. At the same time the inhabitants cultivated their own very small parcels of land and the land of the masters. Sources going back to the sixteenth century mention Zaccanopoli as a sheep-rearing community, a producer of excellent cheese and also of wheat, wine, oil, linen, silk, legumes, cereals, and fruit of all kinds, but in the course of time—as shown by the land registry of 1747 and notarial acts of the eighteenth, nineteenth, and twentieth centuries—it became solely a producer of cheese, legumes, and cereals, and increasingly less wine. Today there are only four shepherds left. They own flocks of no more than one hundred head each, and the people grow grain, corn, and beans.

The social make-up has been diversifying substantially since 1960 with more widespread schooling, work in the tertiary sector, and entry into the professions, but most of the population, especially those over fifty, are still peasants. Nowadays they cultivate their own land, but a few rent land from locals who have emigrated, the occasional absentee landlord, and the Church. What remains to be said is that Zaccanopoli has never had a major landowner in loco, except for the Church, and it has experienced two kinds of emigration: the first, which began early in this century (although there had been sporadic earlier departures), was to Argentina; the second, later movement was to the industrial areas of northern Italy. Together these migrations have halved the community (Minicuci 1989).

The same sources mention Fitili as a producer of excellent oil, linen and silk, wine, and principally fruit of different kinds. The inhabitants worked the land, kept vineyards, and fished. Over the centuries they also appear in notarial acts renting herds, together with the inhabitants of Zaccanopoli, and, also on their own, leasing cereal-growing lands, but the main produce remained fruit, vegetables, and olives, while the vine began to disappear. Few inhabitants owned their own land; for the most part, they farmed for the Church and the nobles and burghers of Tropea and Parghelia.

Unlike Zaccanopoli, Fitili had a resident landowner. He lived in a nearby village but spent long periods of time in Fitili, where his palace cellars were used to store the produce of the countryside cultivated by the inhabitants. Fitili underwent the same migratory phenomena as Zaccanopoli and to the same destinations. In addition, people left Fitili for nearby communities, including Zaccanopoli and Parghelia, and in this way Fitili also lost more than half its population. More than two-thirds of the population left between the beginning of the century and the present. Only a few of the men who stayed behind are still peasants, and they produce fruit and vegetables primarily. Most of the men of working age have blue- or white-collar jobs or work as seamen, while the women generally cultivate kitchen gardens. In Fitili and Zaccanopoli alike, people of retirement age have one or more state pensions.

Both communities are closely built with houses adjoining one another, the only separation being streets and squares. The land the inhabitants cultivate is outside the village, some of it even in the territory of other municipalities. The same road connects both communities to the Poro plateau in one direction and the sea in the other. Zaccanopoli is about three miles from the sea and Fitili just over a mile.

I now turn to some features of the social organization.[6] Both communities practiced intense village and kinship endogamy as far back as the sources go, that is, from the middle of the eighteenth century until the 1960s, and to some degree they still do. There is a high percentage of consanguineous marriage within the degrees prohibited by the Church. Marriage dispensations show a rate of 11 and 12 percent within the third degree of consanguinity, but this figure is only approximate. Genealogies based on the records give an even higher rate and indicate a more complex situation that other elements in the kinship system help to clarify.

First of all, the most common consanguineous marriages are of the first and second degree. Zaccanopoli has an explicitly declared preference for marriage between first cousins. In Fitili this choice is considered a necessity for economic reasons and because the village is so small. Both communities exclude patrilateral parallel cousins as possible spouses. This kind of marriage is prohibited in Zaccanopoli and strongly discouraged in Fitili for two different reasons: in Zaccanopoli because blood is transmitted by the males and the children of brothers would have the same blood; and in Fitili because they have the same surname.

Kinship terminology distinguishes these cousins from the others and associates them with brothers and sisters. They are referred to as *giusti* ("proper") cousins and addressed as *frateu* (males) and *surrea* (females). (*Frati* and *soru* are the terms for "brother" and "sister.") The usage is still current in Zaccanopoli where the younger generations extend the terms to all first cousins, although they occasionally specify that the actual "proper" cousins are the patrilateral parallel cousins. The usage has been lost in Fitili, however, and only a few of the older people know the terms *frateu* and *surrea*. Some distinction does nevertheless survive: patrilateral parallel cousins are considered *più intimi* ("closer") or *proprio intimi* ("really close").

What is considered the best marriage, especially in Fitili, is a union with a matrilateral parallel cousin. In addition to exchanges in the area of blood relatives, marriage with affines is very common.

In addition, sororate is practiced, and there are two particular kinds of marriage exchange, one called *dubbro* or *dubbrera* ("duplicate" pair) and the other called *affruntato* ("opposite" pair). The former is the union of two pairs of siblings of different sex, and the latter two pairs of siblings of the same sex. Often these couples are blood relatives. Another form of marriage involves two siblings taking two spouses who are each other's cousin.

I will not go into the nature of these kinship systems, since it is not pertinent to the present theme, but these few elements suggest the frequency of consanguineous exchange and marriage with affines—practices that create dense networks of alliances as they are repeated over the course of time.

The other important element, in addition to the choice of a spouse, is the transmission of property (houses, land, and money). The declared rule is that goods must be divided equally among all the children, but there are many variations in practice, and goods are allocated differently for males and females. In both communities the house goes to male children, all of whom have rights of ownership. In practice, only one son actually inherits the paternal home if there is more than one son; the others receive goods in compensation from the inheriting brother or from the parents. What matters, and all agree on this, is that the house be handed down in the male line.

The two communities behave differently in respect of the rest of the property and in respect to the women. In both communities all the daughters receive dowries of equal value. In Zaccanopoli a woman's dowry is, by definition, land. All women receive land upon marriage, and all receive equal shares. Only if some remains after the women have been endowed may males receive land. There are many variations of this norm (Minicuci 1980, 1989), but what is particularly significant is that women have prior and often exclusive claim to the land. A mother's dotal land is passed on to her daughters; if males receive land it is a father's land or land that parents acquired after marriage. At the time of my study, money was not passed on because, as soon as there was enough, it was invested in purchasing more land. Finally, women received land on marriage, while the men received the house and any other eventual goods as inheritance.

In Fitili, however, few people owned land, and it was divided into equal parts for transmission to males and females. Since males also received the house and since property had to be divided equally among all the children, women received more land than men did. But in actual fact the Fitilesi have avoided this by compensating the women with money and leaving the land to the men. This strategy worked if women married locally and especially among relatives, but it was impossible if they married men from Zaccanopoli, who expected land as a condition of marriage. In any case, it was up to the parents, insofar as they were able, to provide all their children with the minimum required to establish a new family nucleus, and males and females alike received their share on marriage.

These elements, which have many points in common, are the grounds on which each of the two communities constructs its own history, its own perception of the past.

History and the Past

It might be helpful to describe the way I worked on the theme of memory, the occasions, settings, and purposes of talk about the past. In a general way, the past was always present whatever our topic of conversation, except for exclusively contingent situations, and it was usually present in conversation among the villagers themselves. I began my work in Zaccanopoli by compiling files on each family and then constructing genealogies. I visited all the houses in Zaccanopoli, and after twice renting separate lodgings I took up residence in the home of two elderly women. When introducing themselves and their families, and hence the village, the inhabitants constantly referred to the past, either to describe it or to contrast it with the present. All their conversation about family, land, work, and the social life of the village unfolded along the dual track of before and after, and not just in conversation with me. Around the fireside in the evening there was talk of many things but always in connection with local people or others somehow connected with them. Reference to the past was frequent. For example, talk about repairs to a house would lead to consideration of what the village used to be like, recollections of the old days and past inhabitants. Similarly, peasants would gather at the barber shop at the end of the day, old and young alike, to discuss local politics or recent events, and without any encouragement from me the conversation would shift from today to yesterday.

As to memory and its functioning, it was activated by the construction of genealogies, not just at my explicit request but normally when the conversation turned to others. Other circumstances also stirred memory processes, for example, the return of an immigrant, an engagement, or a wake, but there are no "official" occasions or rituals in which genealogical narrative is used for any particular purpose. The different circumstances in which it is used—to identify someone, to trace past connections in order to establish the genealogy of a plot of land, or to recall the nature of relations between the families being discussed—set the course, duration, and direction of a story but do not give rise to separate kinds of representation.

I worked with older people, men and women alike, but I discussed genealogy primarily with the women because they have more extensive memories. They are the ones who "arrange" marriages and whose responsibility it is to trace all the links that might revive old alliances or establish new ones. Genealogical memory, like the memory of local history, is not the exclusive preserve of one sex, one social category, or particular individuals. Everyone can, and does, when necessary, draw up his own genealogy or that of others, although the younger people are less adept at it.

If you ask the inhabitants to recite the history of their communities, all of them, even the young and "educated," say that they are unable to do so, and the older people are stunned by such a bizarre request. The most frequent responses are: "How would I know?" "The people who remembered the old days are dead now." "All I know is what my folks told me. I don't know what happened before. I wasn't born yet." But there are positive responses as well: "What I know is that we have always lived on this land." "Zaccanopoli has always been a village of shepherds." "We've always been peasants in Fitili." "Of course we all come from here. My grandparents were born here and so were their parents." "What kind of history do you expect villages like ours to have? Toil, toil and poverty, I guess it's always been like that here, because my father used to say that things were worse in his father's day." These responses already indicate some important features: a way of understanding history as something that is reconstructed on the basis of direct experience or memory handed down; and a sense of time circumscribed within a duration that memory establishes on the basis of generations going back to a point beyond which there is "always."

I should like to take these elements apart to see how they work. But I would also point out that the image created by these responses should not be taken as representative of the whole reality, which is far more complex, and that our knowledge of history should not be used as a yardstick to measure theirs. Knowing as I did the history of the whole area, I implicitly expected to find traces of that history in some form in *their* history, at least for more recent times. What I found instead was silence. It was immediately clear that their perspective was different from mine. The challenges then were to understand why events in which they had taken part, and events that they had learned of in their few years of schooling, had not been stored in memory and organized as knowledge, however fragmentary; and to learn, instead, what references were used to construct a local history. Local history is not recounted as a whole. It begins with individual experience, the history of single families, episodes witnessed, and things learned by oral transmission. The style of narration depends on the perspective of the speaker and shifts quite smoothly from the present to the past. It is not a unitary history of events unfurling one after another in the course of time. It is, instead, a myriad of histories, fragments that cohere because they belong to the same shared space and time. There are three referents on which this type of history is built: space, time (especially the past), and the organization of the community.

"All I know is that we have always lived on this land." "Of course we come from this village. My grandparents were born here and so were their parents." Living in the same place for a long time, as long a time as memory can reconstruct, provides the first essential element of identity. It establishes the categories "we" and "others."[7]

The inhabitants almost never describe themselves as Calabrians or Italians. Regional or national identity is invoked only by those who have traveled and come in contact with people from elsewhere. It is chiefly the men who describe themselves as Calabrians, when speaking of having emigrated for work, of their military service, or of the war, but they do so only under specific circumstances. Otherwise being Calabrian means simply that their community is *geographically* located in a region called Calabria. The same is true of being Italian, but in this case there is also an awareness of being part of a nation that is felt to be remote, hostile, and oppressive.

The people consider themselves Fitilesi and Zaccanopolesi—they are different from other people, even from Calabrians and Italians. Their identity is not part of a wider regional or national identity, except when more distant identities force them to be. *We* refers to people in the same village, and that is the term most often used. Then there are the others, the foreigners and strangers. Foreigners are people from distant cities who speak a language the inhabitants do not understand, and strangers are everybody else not born in the village and those who live outside it, even if they live only a mile or two away.

Nor is it enough to have been born and raised in the village to be considered part of it. One's forebears, as far back as genealogical memory goes, have to have been born there, too. People whose parents or grandparents came from even a neighboring community, say Alafito, which is little more than a mile from both Fitili and Zaccanopoli, are still called strangers. People who have moved away continue to be considered locals, but descendants born elsewhere are not. Conversely, people born in Argentina or Milan, say, are considered Argentinian or Milanese, even if their forebears are local. These people are less foreign than others, and they are associated with the locals when they come to visit, but their different origin is always specified.

Two different temporal dimensions distinguish one's own ethnic identity from that of others. One's own requires deep roots in the same territory. But it is not sufficient to be a descendant of people who shared the same past—as shown by the attitude toward those who move away; a person's present history also has to be sited within the same kind of collective history. That the two elements are indivisible is demonstrated by the term *stranger,* which is applied even to people who have been in the community for some generations. These people share the same present, but they do not share the same past.

The other element is time itself. It would be misleading to say that the villagers had a cyclical or linear conception of time. Time is broken into segments of different duration and may be presented as cyclical or linear depending on the speaker's point of view and the use that is made of it.[8]

"In the old days," "today and yesterday," "then and now," "in olden times," and "in modern times" are phrases that recur constantly when an inhabitant describes his or her individual experience and that of the community, and the two are constantly compared. "The old days" refers to a time "before," time that is enveloped in a kind of fog—it is a time without confines, a time with no identifiable beginning or end. It is all the time that has passed away, and nothing is known about any earlier time. This side of "olden days" there is measurable time divided into past and present. This past is indicated by the expression "used to be" and more often by "in my father's day," "in my grandfather's day," and so on. This past time can be measured and dated approximately in reference to the sequence of the generations. Genealogical memory gives a sense of duration, which is linked to the earliest generations that can be remembered and to those still alive.

But the two villages have a different attitude toward the past. In Zaccanopoli the past is spoken of as a time when things were worse, yet in some ways better, because there was more of a community, there was less individualism. People helped each other more, they were together more, and they spent a lot of time on the land. The land is the locus of memory par excellence; it is where the best things happened. When the community past is described, the toil and poverty evoked in other circumstances are not mentioned, while positive features are appreciated. This is the time of their shared history, all of which evolves around the two axes of village and land in a continuous interaction in which memory picks out particular events and moments.

Toil and poverty, on the other hand, are evoked when the inhabitants recite their own biographies. But this talk of the past does not evoke an individual past, for that is only a single life, an experience of brief duration, the testimony of a single experience. It is rather the sum of all times past in its shared aspects that constitutes the past of Zaccanopoli. This is clearly evident in the manner of narration. In speaking of one's own life, there is constant shifting back and forth between present and past tense. Time is divided into segments that correspond to the actual duration of experience and is perfectly datable. But two past tenses are used in speaking of the past in general: the past perfect for specific episodes that can be approximately dated and the past imperfect for everything referring to "those times" ("they used to do," "there used to be"). The narrative style is different, more like literature than history. Each time the past is described, it acquires new variations and takes the form of a story. There is no dating of this time; the only term used to specify it is *before* or *once upon a time,* and the adverb *always* is very common.[9]

That's the way they did it because "that is how it was always done," which means that one's ancestors did it that way because that is the way their ancestors always did it. Ancestors are the guarantors of the norms and still take part in the life of the village. They come back to visit the houses and the village in spirit form, "the spirits of the dead." And they return in dreams to reiterate the validity of the norms and make sure they are respected (Minicuci 1989). They do not survive merely in memory; they also have independent lives of their own: "they come when they want to," and they come from the beyond. The beyond is imagined as a place where there is "no end of time," a faithful mirror of the village on earth. All the images of the beyond depict a place (usually a green meadow) where all the Zaccanopolesi live together in continuation of the relations they had in this life. They are sitting there at rest, side by side, because "there, in that world there is no end of time," a time that joins the "before" of those already dead with the "after" of those still alive.[10]

The present is still populated by these presences, and this keeps the past from ever becoming completely past and allows them to flow smoothly into the present.

In addition to individuals and families, there is another reference point in speaking of time and even dating it: the cycle of the seasons and work on the land. This is also true of Fitili. According to the informants, "in those days" (i.e., in the past), time was calculated by the stars (the rising of the sun, the waning of the moon). There was planting time, threshing time, and harvest time. "The land," a peasant woman explained, "has its own time. First there is the ploughing, then the planting, etc., and we lived on the land, and so that's how our time passed too."[11]

This is a different kind of discourse, and it refers to a different past, which pours into the family and individual past whenever the reference is to work and the relationship with nature. These two kinds of past are not mutually exclusive, nor are they antithetical, even though they use different standards of measurement. They coexist in different positions, and the hierarchy of these positions depends on the kind of discourse involved.

Generally speaking, the people of Zaccanopoli consider the past to be a time that is known, subject to control, and certain, while the present is felt to be elusive and drawn out. It is still connected with the past, though a clear dividing line cannot be established; indeed, the boundaries are constantly being redrawn.

In Fitili it seems that the only past is that of individuals, their conjugal families, and families of orientation. There seems to be no memory whatsoever of village events except for the unusual, say, a murder. It is as if nothing had ever happened in the course of time. As the inhabitants tell it, life is a ceaseless story of toil, from early childhood, that begins at dawn with the departure for the fields and finishes at sundown with the return to the village. This is how an old peasant summed up the past: "In my day, and in my father's day and my grandfather's day, you were born and you died and all you did was work. That's our history. The years went by, it was always the same, nothing ever happened."[12]

But things, of course, did happen, things they knew about by means of personal experience or report. The people of Fitili can call to mind various kinds of episodes when pressed to do so, but it is always with reluctance. They did not do so spontaneously, nor do they consider such events worth mentioning, unlike the people of Zaccanopoli who have countless anecdotes to tell.

But the Fitilesi do not remember, or rather they do not care to remember, village events. What they do remember, and in detail, are working conditions, relations with their masters, gains (more often losses), times of famine, epidemics, and bad harvests, everything that characterizes the past as a time of toil, fatigue, poverty, and subordination. This is the collective past that people, especially family members, use as a reference to date single circumstances and the whole duration of past time. That was all before. This is the present, and now there is more prosperity and more security. There is no gradual transition from past to present, just an abrupt break. What marks the shift from "before" to "after" is the end of certain production relations that more than one informant described as "slavery."

The only sense of continuity between past and present is in the fact that people consider themselves, and introduce themselves, not as individuals but as families descended from other families that have "always" lived there.

Genealogical memory is the device that makes it possible to reconstruct family history and use it to provide reference points for measuring temporal duration and situating oneself in it and in relation to others.

Memory

Neither in Fitili nor in Zaccanopoli does genealogical memory constitute a separate type of memory, distinct from individual, domestic, collective, or historical memory; it is the ensemble of all these memories organized on and around genealogies, which serve as guides and supports in placing people and events in time. What is specific to it is the idiom it uses for discourse. What is almost totally absent is what Collard calls "social memory": "it is social because it derives from a social milieu not in this case one derived . . . from genealogy, the life cycle, or the history of the village . . . but from the social effects of history itself" (1989:90). Not even Fitili has a social memory in Collard's sense. Granted, in Fitili the past is experienced and assessed in terms of social relations of production, but it is experienced as the past only if it is anchored to the genealogical dimension.

Like the Zaccanopolesi, when the Fitilesi speak of their own history and compare it with that of others, they constantly speak of masters, also called "lords," and call the period when they dominated local life "the time of the masters." How did this time start, and out of what historical situation did it arise? The answer to the first question is everywhere the same. There have always been masters, and what *always* means here is mainly their fathers' time, their grandfathers' time, and as far back as genealogical memory goes.

But it also means "since eternity" for different and contrasting reasons. For many people, especially the old women, God created the rich and the poor. The rich have always gotten richer and the poor poorer because each person has his own destiny, and you cannot change destiny. They have a great many proverbs to support this indisputable truth. There are people like this in Zaccanopoli and Fitili, but so are there everywhere. Such a view is common to southern folk culture and was affirmed by the local Church.[13] Priests explained that people had to bear up and accept things as they were, because this was the will of God, who loved the poor more than He loved the rich. "What they didn't explain to us," one old woman remarked, "was why He created the rich in the first place."

Others think the division necessary because there always has to be someone to command and someone to obey. This is how a shepherd explained it.

> There have always been rich and poor in the world, the same way there have always been men and women and there has always been the family. In a family there is always one person who commands and thinks for all. But the trouble with the rich is that the richer they get the more wicked they get. And that is because of envy, which has always existed in the world. One person wanted to be richer than another, and so they had wars, and they got us into them. Instead of thinking of us, they only thought about themselves. That's the way it is nowadays with the government. They steal everything and leave nothing for us. But what can you do about it? We're born that way.

When referring to an origin myth, discourse shifts from historical time to mythical time, but as soon as it is evoked it is immediately brought back into historical time with the words "and that is how we were born." The statement that the myth is absolutely true is directly followed by an explanation of who "they," the poor, are, who their masters were, and how they were treated. Memory is reactivated through the narration of their history and that of their forebears.

The other version is similar in substance, but it makes man rather than God responsible. We are all born equal, but man, like the animals, has a natural instinct to dominate his neighbor, and everyone tries to outdo the next man. So there are always winners and losers. This simpler explanation sites the difference in nature and, unlike the other, is not static. And, precisely because the condition is natural, it is not subject to doubt. But here it is the acts of men that orient the course of events. It is a constant struggle in which one person loses and another wins. You win because you are stronger, luckier, or more wicked. It is usually wickedness that is ascribed to the masters, a condition usually illustrated by local events.

This second explanation, which is more common in Fitili, would seemingly make it easier to identify at least the local historical events responsible for the period of the masters in these communities. But this is not quite the case. The following kinds of explanations are offered. In Fitili they say

> We of the village have always been peasants and there have always been masters, like I told you. They got rich on the labor of the peasants, and if the harvest was bad that year they took your land away, even your house, everything, and that way they got even richer. And the Church did the same thing, it has always owned land.

In Zaccanopoli you hear

> The old folks used to say that Zaccanopoli was founded by shepherd families from Aramoni.[14] Who knows if it's true? What's certain is that we have always been shepherd families; even the name of the village says so. A *zaccanu* is the corral where you keep baby lambs apart from their mothers, and so we Zaccanopolesi are shepherds. This is beautiful, fertile land, and they took it all, and the priests helped them. The priests are the worst thieves. And we tried to keep our own small pickings, but you couldn't buy any more land even if you lost what you had. When they lent you money for a dowry, or just to get along, you had to pay back twice the amount and that way they kept you in slavery forever.

These two accounts are full of features to which we shall return, but they still do not tell us anything about how and when the masters arrived, or who they were. Further questioning obtains vague and sometimes tautological answers: "They became the masters because they were richer," "because they came from rich families," "because they have always dominated the whole area, they had money and castles." Question: "When?" Reply: "In the old days." Question: "What is a feudal holding?" Answer: "Hundreds and hundreds of hectares." Question: "How did they get them?" Answer: "How do I know? I wasn't there. I guess they inherited them from their families."

This could continue ad infinitum without revealing any specific historical circumstances responsible for the current situation. And it is the same if you ask about periods in history. If you ask the inhabitants about something with which everyone is presumably familiar, say, about the unification of Italy, very few people will know when it occurred, and their answers will be vague. If you ask what Italy was like before and what has happened since, few will have anything more than vague memories of school days. They remember the name Garibaldi only "because we have a street named for him; they say he came to Calabria." People who have gone to school know that Italy has had many foreign conquerors, but the usual comments are: "How do you expect me to remember those things?" "The old people could have told you." "If my grandfather was still alive he could tell you. Maybe some of the old folks still remember."

It may seem surprising that, while local historians always mention the remote origins and particular history of an area, the inhabitants have no memory or awareness of it. And this is not just the case of the illiterate or semi-literate people among whom the present study was conducted, but also of more acculturated villagers. Although the village population now includes students, white-collar workers, and professionals, their numbers were slight at the time the present work was conducted in Zaccanopoli: ten in all, excluding secondary-school children. The same was true in Fitili. Both villages lacked the figure of the local intellectual who plays such an important informational role in rural areas. Neither village had a newsstand, much less a library or bookstore. Before television, and before a local secondary school was opened in Zaccanopoli, education was limited to elementary school and missionary sermons. Beyond all this, the absence of historical memory must be related to the way space is represented and experienced.[15] The geographical knowledge of the older female population is limited to the places they know directly. For the oldest of them, this does not extend beyond the province or at most the region. The men are somewhat better informed, having been elsewhere for work or for military service. But even Argentina and northern Italy, where so many villagers have moved, are hard to place for many of the older people, especially the women: northern Italy is "up there," and Argentina is across the sea.

Working on the idea of space in Zaccanopoli, asking people to make drawings and maps, and then registering their definitions of space, their ways of measuring it and representing it, showed that there were two main terms of reference—the village in which they lived, and the land on which they worked—on which the opposites inside/outside, above/below, and known/unknown were based. As to anything elsewhere, they knew it existed, sometimes vaguely but sometimes identifiable, yet always described in terms of near or far from their point of observation. Space that was known, regulated, and experienced comprised the pathways trodden by them or their relatives, always identifiable, however, in relation to persons. Significant in this regard is the account I had from an elderly woman contemplating her passage into the next world. She named the people she expected to meet there, but they did not include people who had died in Argentina: they died so far away in another land. Living and thinking about one's life in a space established by everyday experience enforces the sense that history is something bounded in that space.

The other kind of memory that is missing is the memory of the folk heritage. Even the old people have forgotten the folktales and songs, especially in Fitili. The older people in Zaccanopoli remember a time when stories were still told. Some beliefs have survived, a few customs, and many proverbs, which are usually quoted to confirm a rule or sanction a deviation (Minicuci 1989). What is particularly interesting is the way some folktales have been transformed into local anecdotes. For example, one woman told of a witch who rubbed herself with magic unguent every night so that she could fly to a witches' sabbath. That was in her grandmother's day, and the witch was the wife of a shepherd who worked the land next to that of her grandmother's cousin. The witches' sabbath was held by a walnut tree in an olive grove, so it must have been on the property of X or Y because in her grandmother's day they didn't have any, and so on. And at this point the woman reconstructed the transfer of ownership and the kinship relations between the people who had occupied the land then and those who have it now. But more than folktales or stories of magic, this sort of thing is done with ordinary accounts of possible events that do not require supernatural powers or presences.

To take another example, it was said that bandits used to hide out in a cave at Alafito where they lived on the sheep they stole during the night. The Fitili version was that the bandits stole from the gardens at night.

These stories have to be linked to local history to be remembered, but to establish a link the stories have to have some connection with the inhabitants or with their dwelling place. In any case, the older people all agree that in their day, that is to say, when they were children, a lot of tales and fables were told, but there is no way of knowing what fables they were and what function they performed because nobody remembers them anymore. There is, however, one particular kind of narrative bound to a social category and to certain particular persons: the stories of spirits. There is a very widespread belief in spirits, though it takes many forms. There are two kinds of spirits. The spirits that come into homes and walk through the village are called the souls of the dead. The spirits that walk the land outside the village may be souls of the dead, but they are the souls of people who met a violent end or corporeal representations of supernatural powers. Only shepherds can see these manifestations of supernatural powers, and not all of them, only a select category of shepherds who have received "the call," and they have the task of interpreting the spirits' messages, controlling them, and dominating them. I will not expand on this theme except to note that this is the only area in which there are "official" custodians of memory, entrusted with knowledge of a sacred nature that must remain largely secret, and that the function of these custodians is not so much to preserve and transmit knowledge to a select few as it is to take action to control and tame space, define its limits, protect the village, and open new pathways (Minicuci 1992).

The story of the witch exemplifies something I said before: genealogical memory guides the narration of events and thus performs one of its most important functions, to create history. Let us see what other functions and purposes genealogical memory serves.

All the inhabitants of the two villages know each other and can place each other in terms of family or "membership group" (I will explain what I mean by this term later). No one is ever identified solely by given name and surname (partly because of the number of people with the same name) because the community recognizes individuals always as members of a family and a wider group. The question is not "Who are you?" but "To whom do you belong?" or "Where do you come from?" The answers are "I am one of the B's" (the nickname of the group to which he belongs) and "I come from the C's" (the family surname). One answer is usually sufficient, the nickname, because the information that follows, whether asked for or volunteered, will refer to the person's parents and, if necessary, specify their origin as well. Strangers from nearby villages are asked the same kind of question to determine whether their families are known directly or by repute. Anyone about whom no family information is available is decidedly an outsider. "Nobody knows who he is because we don't know who he belongs to" is the most common explanation of why marriage to an outsider is discouraged. Genealogical information is indispensable for identifying a person and for placing people in relation to each other and to the outside world.

Genealogical memory is exercised every day because conversation about others always involves mention of kinship links, the ones that best identify them, even at the simplest level. The more information that is needed, the farther the genealogy is explored. It would be incorrect, however, to think that a genealogy is reconstructed in the sense of an organic representation of a set of kinship relations. Indeed, the people do not usually draw up a whole genealogy. Rather, they retrieve bonds of consanguinity or affinity, moving in some directions and not others, in particular circumstances and for specific purposes such as engagements, marriages, funerals, and political alliances. People are perfectly capable of reconstructing their genealogies, and they did so at my request. But they followed their own routes and would not be forced into such rigid schemes as the chronological sequence of generations or lines of ascent and descent.

Their memory does not operate in linear fashion but in bits and pieces. Let us consider the differences between the two villages.

The inhabitants of Zaccanopoli have a very vast genealogical memory; with older people and especially the women, it may embrace between one hundred and two hundred people.[16] The older people are the archivists of memory. They organize the facts and information about themselves and others and transmit the data as circumstance may demand, but they also transmit this information for pedagogical reasons in ordinary conversation with younger people, the stated purpose being to keep things from being forgotten: "if you lose your family, you have lost everything," or "without our kin we are nothing."

Memory is more extensive on the collateral plane than back in time. Starting from ego, the ascendant line does not reach farther than the great-grandparents and almost never extends to brothers and sisters. The paternal great-grandparents are usually more familiar than the maternal. Almost every relative of the grandparents' generation is known, but there is more detail about the female lines. Starting from his own generation, ego knows all his collateral and descendant relatives.

Memory is shorter in Fitili. The people do not like to give their own genealogy even when asked. Only reluctantly do they provide the names of their parents, uncles and aunts, paternal and maternal grandparents, and first cousins. Often they have only incomplete information about second cousins. Information is fuller and more exact about the male lines, although it usually extends to no more than twenty or thirty people in all. When speaking of other things—a job, for example, a neighbor, a departure for Argentina, or local political events—a person may remember connections he had not thought of before, and he can place them precisely in relation to himself. But this never includes a very large number of people. He tends to create lines rather than pursue relationships, and he can do this for his own line and that of others as well.

In Zaccanopoli, instead, the starting point is relationships, and each genealogy is gradually enriched with segments of other genealogies. Indeed, every married person provides the identity of the spouse and often part of the spouse's genealogy. For example, mention of one's father's uncle is followed by the explanation that he was married to X. She was the daughter of Y and Z, so she belonged to such-and-such a family, a member of which had married a blood relative or affine of a blood relative in one's own genealogy (and thus retrieving the genealogical link). At this point, the previous line is abandoned, and a transition is made to the genealogy just mentioned, meanwhile explaining the earlier and later alliances that established the relationship between the last two people mentioned. When all of these genealogies have been accounted for, you return to your own. In this process not only is there transition from one genealogy to another but there is passage up, down, and across the generations. Genealogy is an ongoing process.[17]

Memory is bolstered by the three elements of the naming system: given name, surname, and nickname. Given names are passed onto alternate generations first in the male line and then in the female. Thus, a firstborn son is named for his paternal grandfather, and a firstborn daughter is named for her paternal grandmother. The second son is named for his maternal grandfather, and the second daughter is named for her maternal grandmother. There are variations, but this is the rule, and it is generally respected (Minicuci 1983).

Given names that have been forgotten can be deduced from the current bearers since it is usually the names of the older generations that are forgotten. The surname is not very helpful as a tool of memory because it is shared by so many people in both villages, but it is used in Fitili to distinguish blood relatives in the male line from those in the female line, which are very often forgotten. The nickname is the real axis of memory. Everybody has a nickname and often more than one. In Zaccanopoli a distinction is made between the "family" nickname and the "personal" nickname. In Fitili the distinction is not verbalized, but it is equally present. The nickname is transmitted along the male line and the female line. In Zaccanopoli a great many nicknames originate in the nickname of a woman, and this is frequently the case in Fitili where matronymics are also very common (for example, Gerardo di Isabella, Laura di Paola di Isabella, and so on). A nickname originating from that of a woman may be transmitted for several generations to her descendants in the male line and the female line; it may be transmitted for one generation in both lines and then become male; it may be transmitted chiefly along the female line; or it may be extended to a husband as well, if he is a stranger in the sense I have described. If the nickname originates from the nickname of a man, it may proceed in comparable fashion. There are innumerable combinations, but they all entail constant segmentation, even when they are transmitted along the same line, as each new nickname comes to the fore and prevails. The addition of a second family nickname reflects a power relationship between two lines with one prevailing over the other, while the appearance of a new nickname marks a point of segmentation.

Since the nickname is not an automatic addition to the surname, it has to be sought out. One must trace its transmission and so reconstruct the alliances that were made and segmentations that have occurred. At the same time, when one knows how a nickname has been transmitted, it is possible to deduce missing surnames.

In Fitili the surname serves to place individuals in a line of descent of blood relatives who recognize each other as such because they have the same surname. The Zaccanopolesi, instead, do not consider the surname important in itself. It is the nickname that helps them retrieve the remotest lines of consanguinity, the female lines, which are obscured by the way surnames are transmitted. It is the nickname that makes it possible to discover previous alliances with a view toward new ones. The final result is that no genealogy is separable from those of others; a genealogy is built out of marriages rather than lines of ascent and descent.

So it comes as no surprise that the Zaccanopolesi, except for a few young people, showed no interest in the genealogies I constructed covering a period of more than two centuries. The image that my work offered was static, without time or movement. This was strikingly clear when I showed my genealogies to informants with long memories. They viewed my genealogies as something that had nothing to do with them. They could not identify themselves in that ordered set of names and dates.

As Merzario observed, the mere legal fact means nothing in itself,

> it is only a thread in a complex pattern, where each item of information is related, adds to or subtracts from some other item in a process of collective integration. This is how the oral register must be structured in order to resist the erosion of time: it is only the written record that can have the simple, elementary quality of a list, an ordered series of names, one by one, laid out according to a preestablished standard. In the oral world, dates are constantly listed comparatively: a death, a birth or a marriage

interact with each other, they are dated and redated by each other, and since the flow of demographic events is unceasing, so too is the work of oral codification from below. (1983:51, 1013).

Moreover, people could never be located in the pattern of relationships without the nickname, except for the youngest generations. As one person remarked, "If we don't know who they belong to, how can we know how and why they are relatives?"

Genealogical memory comes into play in the same circumstances in both villages, but the purposes it serves in Zaccanopoli and Fitili are not entirely the same. It is extensive in Zaccanopoli and limited in Fitili, and in both communities it helps to construct a kinship reseau. The question, however, is what kinship is recognized as such?

Fitili, like Zaccanopoli, practices intense kinship endogamy in the same forms of exchange as its neighbors, but it has a more restricted kinship area, sets limits to affinity, and accentuates patrilaterality. Only ego's closest blood kin are called relatives. In his own generation, ego considers only first cousins as genuine kin, and his patrilateral parallel cousins are closest. Second cousins, who are also referred to as *procugini* (great-cousins), are considered distant relatives, and many people do not consider third cousins to be relatives at all.

An informant explains, "Proper cousins are first cousins; second and third cousins are distant relatives, but only first cousins are close relatives." One woman answered the question "Whom do you call cousins?" by saying "The really close cousins, for example, children of father and mother." Another woman said:

> Real cousins are father's and mother's, and they are cousins, like when there are children and they are cousins. There was a time when we considered a man's children close cousins but not a woman's; like when she was a cousin of mine doesn't mean anything.

The obscure remark "like when she was a cousin of mine doesn't mean anything" is explained by other witnesses. For example, when one man mentioned a daughter of his father's sister, that is, a cross-cousin, he described her as a second cousin. And another man explained, "The sons of two brothers are true and proper cousins because they have the same surname. When the surname changes, they are second cousins."

These remarks help us to see several important things. One, the sex variable shows that the male is dominant in establishing the degree of kinship. When one or both of a pair of parental siblings is a woman (cross-cousins, matrilateral cousins), the cousins are reduced one degree and are farther from ego, while the children of two brothers are drawn into the category of ego's brothers and sisters and are called children of father and mother and not children of brother and sister. Another thing we see is that proper cousins are such because they have the same surname, that is to say, there is no female line interrupting the continuity of descent in the male line.

In the descending line, ego considers as kin his own children and those of his siblings and first cousins, but he "forgets" the descendants of his great-cousins. He thus effects an additional break, cutting off a consanguineous line as soon as it begins. I use the word *forget* because ego no longer mentions them as kin but does include them in the wider area of distant relatives so that he can retrieve them should the need arise, for political alliance, for example. In the ascendant line, ego stops at his parents' siblings, and forgets those of his parents' parents. In this case, *forget* means not remembering them at all. Moreover, he singles out his own filiation and goes deeper into it.

The genealogical story unfolds first in terms of blood relatives in the male line, registering and placing everyone who has the same surname, on the basis of which the kinship relationship, acknowledged as such, can be retrieved. Then comes the turn of the affines, and, as in Zaccanopoli, ego looks for any eventual kinship relationship with one of his blood relatives but over a shorter time span. Here the family is the model: his own family, that of his parents, and that of his children, including first cousins in each generation. The people of Fitili call this ensemble descending from a shared ancestor *family*. If they mean to refer solely to the descendant generation they use the term *heirs*.

The Zaccanopolesi use a different term, *casata,* to designate all the descendants of the same ancestor in the male line, and all the descendants of all the lines that memory can trace are considered kinfolk. The term *family* is used exclusively for the conjugal family and the family of origin. The kinship area they recognize is very extensive, and in everyday usage the term *kin* may be extended far enough to include affines as well, though specifying, if necessary, that they are not relatives. A distinction is also made between close and distant relatives but across a wider spectrum than in Fitili. Patrilateral parallel cousins are associated with siblings and considered the closest relatives outside the filial line, but cousins in the female line are not distanced. Second cousins are considered cousins, while third cousins are considered "great-cousins." Conversely, they recognize the siblings of grandparents, and they have a separate term to distinguish them from the siblings of the parents. The parents' siblings are called *zio* and *zia*, "uncle" and "aunt"; the grandparents' siblings are called *ziano* and *ziana*, as it were "unclish" and "auntish." Male and female lines are not separated a priori, for potentially all might be included.

The principle of consanguinity is based on the male line, where the term *casata* applies, but it does not exclude the female line. The *casata* is distinguished by having the same surname, which is transmitted only by the males, but since so many families have the same surname this is not sufficient in itself to identify individuals and their *casatas*. Consider, for example, that in the village there are about 350 people whose surname is Mazzeo.

This is where the nickname comes in, but it, too, may be subject to segmentation when it becomes extended too far for purposes of identification, and, as we have seen, the nickname is often transmitted by the females. The set of people defined by a nickname is what ego recognizes as the group to which he belongs. Given the way, or rather the ways, in which the nickname is applied and transmitted, plus the fact that it sometimes includes affines as well, it does not constitute groups with linear progression but associations of blood relatives sometimes in one line, sometimes in the other, and sometimes in both. The Zaccanopolesi do not have a term to define these groups, they simply call them by name.[18]

The same expansive standard used for consanguinity is also applied to affines and to another kind of social relation outside the field of kinship: *suppessaraggio*. In his own generation only ego's spouse's siblings and the spouses of his own siblings are considered brothers-in-law and sisters-in-law, but in everyday conversation the term is extended to the spouses of a wife's or husband's blood relatives (though not to the blood relatives of the spouses of ego's own siblings). This is the only kind of relationship in which an affine term is extended outside. *Suppessaraggio* is the link established between two parental couples when their children marry, a kind of "co-inlawship." On marriage the bride's parents and the groom's become each others' *suppessari*. In ordinary usage, the Zaccanopolesi also apply the term to the blood relatives of the two members of both couples, provided they are of the same or an older generation. In Fitili the terms *brother-* and *sister-in-law* and the term *suppessaro* or *suppessara* are limited to those rightly so called.

These are the different features of the two communities, and they deserve separate consideration partly because of the light they shed on the way the community is organized in the two villages. But this is impossible in the course of a brief essay, and I will limit myself to those elements that seem important and significant in terms of memory and the past.

Conclusion

Memory, as I hope I have demonstrated, is the main thread of all conversation about oneself and others, so the question arises: Why does it work differently in the two villages?

As already mentioned, the two villages are marriage partners that follow the same norms in choosing a spouse. They exchange spouses among and between themselves, but property and the way it is transmitted are not exactly the same. The inhabitants of both villages make their living primarily as hired labor. They tend other people's flocks and rent land from the owners of those flocks, and often they are only day laborers. The sources tell us that in the past, too, they owned only small parcels of land. But the kinds of land and the use that is made of it differ in the two communities. In Zaccanopoli mostly cereal grains are grown, while the farmers of Fitili cultivate fruit trees and olive groves. The difference is significant in itself, but the two villages both rent the same kind of land in their own and other villages as well. The economic base is similar but not identical. The two communities possess or have access to the same kind of capital, but they use it differently.

The economy of Zaccanopoli is centered on two poles: sheep herding and agriculture. A many-faceted strategy is employed to root the village in the land it works and inhabits. The land is handed down to the women, and this constant fragmentation systematically precludes accumulation of land: the people of Zaccanopoli have no masters in loco, and they do not want any. Village endogamy prevents the land from falling into the hands of strangers, and the bond of kinship means that the land is never wholly lost but only moves from hand to hand and from sign to sign. It provides the basic minimum for survival for each new group that is formed and assures its reproduction.

In order to work, though, the system must create a solid network of relationships and monitor them. In this framework, genealogical memory does not start out from an ego in search of an ancestor, but from a marriage in search of other marriages. In so doing it makes use of the history of dowries, and it also reconstructs that history. It moves from existing alliances in order to establish new ones. Genealogy is a constantly moving process that does not fix a group of blood relations in motionless time but, as it were, accompanies the group as it forms and reforms with each generation. And, as the group moves through successive generations, it sets its norms so that the movement does not result in disorder. It adopts the ancestors as guarantors of these norms, the people who came from afar to this territory, as families and as shepherds—two other components of their identity—the people whose choices made it possible for them to live today and forever in the same place.

But, so that the forebears can continue to be present and operative in village life—returning to redefine the community, via their ritual itineraries on established occasions and days but also "whenever they choose"—they are categorized as "living ancestors," even before they die, not only by virtue of the genealogical link but also through a relationship arising from a marriage: *suppessaraggio*. The *suppessari* are the guarantors of the new reproductive nucleus. Once this function has been performed, they are still parents, but they begin to shift in the direction of earlier generations with which the extension of the term associates them, and the center moves in the direction of the next generation and the generations that will follow.

Since this is the case, the past is no mere repetition of the same events over and over again. It is the sum of all the acts of individuals who are associated in groups of different kinds that are not established once and for all by virtue of a single principle. They vary in time and according to the references on which they are established and the different functions they perform, the most important being the pursuit of equilibrium, whatever possible fluctuations there may be. Equilibrium is not a given. It is something striven for and won by playing ever-new hands at different tables, passing cards from hand to hand so that all can take part but always respecting the rules of the game. It is the stability of the rules and the acknowledgment of their validity that give the idea of the stability behind the remarks "the way they used to" and "it's always been that way." But, when the deck is changed and new rules have to be found or old ones adapted, then the present feels different and the future seems uncertain. This is happening now to the Zaccanopolesi because the deck really is changing. It does not mean that the cards never changed over the course of time but that the players were different. The Zaccanopolesi of today have inherited the winnings and the losses of the past, but, since they are part of a wider horizon now, they are playing with a great many more partners.

Pursing the metaphor of card playing, we might say that the inhabitants of Fitili use approximately the same deck but that the object of the game is different. Fitili is no poorer than Zaccanopoli, but the presence in loco of one landowner and the extreme proximity of another make it more subordinate. In addition, Fitili is located midway between hills and sea. The Fitilesi do not identify with the land; they are bound to it, but they are also projected toward the uncertain, open space of the sea. Identity centers on two opposite poles, the land and the sea, both of them sources of uncertainty, which are met by a strategy of resistance that is only seemingly passive. Fitili builds its strength on family. It follows the same reproduction rules as does Zaccanopoli but manipulates some of them in a different direction (e.g., the transmission of property) so that solidarity can be maintained without any group's existence becoming dependent on that of others. Fitili is trying to follow a path toward the outside world, and it is succeeding. It is now richer and more "modern" than Zaccanopoli. Fitili feels dominated and harbors seeds of rebellion that lead not to open revolt but to refusal to link its destiny to the land. The family is fundamental, and a family will enter into alliances with other families, but it usually draws away after a brief period so that its history will not be wholly identified with that of the others. It does not implement a strategy of conservation but a strategy of resistance (rooting itself in tradition in the sense described at the beginning of this essay) and a strategy of growth. To do so it has to be based on small solid groups. Memory establishes these groups, making drastic choices between the close and the distant. Once selected, family members know that they belong to an entity distinct from the other entities with which they must deal but with which they must not be confused.

The same is true of marriage partners in Zaccanopoli, women for the most part. Indeed, it is the men of Fitili who come to Zaccanopoli in search of wives, rather than the other way around. They share the same memory of the conditions and circumstances of work, partly because the people who intermarry are almost always the children of people who work adjacent plots of land or of shepherds who tend the same flocks, but they always maintain a reference to their own past as something totally contained within their single histories of origin, which are never superimposed or mixed. There is a keener sense of difference in the way kinship is considered. Since a man from Fitili is an outsider, he is drawn to his wife's family, which is considered stronger by virtue of the fact that it is native to Zaccanopoli. The husband works his wife's land and shares in the relations she has with her kinfolk. It is recognized, and the men are quick to point it out, that there are more kinfolk in Zaccanopoli and more extensive relations. A wife from Fitili is absorbed (as an affine) into the kinship of her husband from Zaccanopoli, but she does not become a pole of aggregation because she does not have her own family and kinfolk behind her. And a woman enters more deeply into the system of Zaccanopoli than a man does. For example, she may use kinship terms in the way they are used in Zaccanopoli, though always specifying "as you say here," when referring to her own origin, and when giving her husband's genealogy she will include anyone he considers a relative.

The place of residence determines certain practices, but there is no fundamental conflict because the underlying logic is the same; and distinctive features remain intact because differences remain intact, since someone from Fitili can never be considered Zaccanopolesi and vice versa. The descendants of a mixed couple usually take their reference points from the village where they live, the place they were born, and this is where their past and history reside. This is clearly sensed in Zaccanopoli but not in Fitili because at the present time there is no Zaccanopolesi now in residence. There were only a few in the past, and they lived outside the village on land that was adjacent to plots that belonged to residents of Fitili. The greatest number of exchanges with Zaccanopoli in the course of three centuries, and over the whole marriage area, are not enough, however, to have resulted in any fusion between the two village histories.[19] It might be expected that one system would exert greater influence over another, but that has not been the case. The tenacious attachment to one's own space as a significant element of one's ethnic identity has helped to maintain the differences. Similarities have, when needed, dimmed the differences and made encounter and repeated encounter possible together with exchanges on other levels that are useful for the reproduction and continuation of their respective social organizations.

Such being the case, it is no surprise that memory is short and that the events of a shared past are not remembered. What they have in common is the experience of domination; this is the past that concerns them. And it is perceived in opposition to the present, which seems to offer richer possibilities and promise a better future. All the inhabitants are convinced of this. And when we consider that the Fitilese prefer the version that attributes to man rather than to God the origins of class distinction, the division of humanity into rich and poor, it is clear to see that faith in the action of individuals and families may offer prospects for a better future.

In conclusion, it is clear that both villages make use of time remembered, but they scan it differently, they calculate its speed differently, and thus they represent it in different modes and at different levels. The modes and levels are incorporated in a specific history. Far from seeing themselves as communities without history, in contrast to what we are told about villages that have many features in common with Zaccanopoli and Fitili—I am thinking, among many others, of Minot in France (Zonabend 1980)—these two communities make their own history, try to steer its course, and use past experience to assess events and decide on action for the present and future. They face situations; they are not submissive. But the people of Fitili and Zaccanopoli know, consider, and assess the past differently, and they act differently, one village opposing the past and the other seeking to transform it.

NOTES

The research for this paper was partially supported by funds (40 percent) from MURST, 1991.

1. The present study concerns the situation that prevailed in the two villages until the end of the 1970s. I dealt chiefly with the older adult population on the topics treated in the present essay, since they were particularly interested in, and willing to talk about, the primary research subject, kinship structures. Nevertheless, the younger peasant generation, men and women alike, shared the same body of knowledge and were equally involved in the course of the research. The younger people, however, preferred talking about topics that were of more immediate interest to them, such as new job prospects, changes in farming, and, above all, the advantages and disadvantages of living in a village. This was the case in Zaccanopoli, and it was also true in Fitili, although that village is now inhabited chiefly by old people.

2. I do not use the term *norm* in the sense of a binding rule or regulation but rather in the sense of a principle *générateur et unificateur des pratiques [qui] est le produit des structures que ces pratiques tendent à reproduire* (Bourdieu 1972:1106).

3. For the nonrelevance of the question "What is the past?" and the need to consider more than one kind of past time, cf. Cohn (1961, and in this volume) and the Indian village he studied. Considerations of different kinds of past time include Pocock (1962).

For the importance of memory in constructing the past and identity, see *Il senso del passato* [The sense of the past], a collection of essays sharing an "awareness of the dynamic nature of what we call memory, its importance in defining and redefining identity and planning the future and, at the same time, the inextricable interweave that arises in real life between personal memory, group memory, and the memory of a whole community" (Jedlowski and Rampazi 1991:91).

4. This is by virtue of the fact that marriage partners are chosen. My first work on family and kinship was conducted in the years 1975–85, starting in Zaccanopoli and ending in Buenos Aires, with people who had emigrated to Argentina. I reconstructed the genealogies of the natives, including people who had emigrated to Argentina, back to 1755, the first year for which parish registry records were available. Other sources I consulted (land registry, notary acts, judicial proceedings, ecclesiastic documents, etc.) go back even farther. Having developed hypotheses about the kinship system and wanting to verify them, I investigated the entire marriage area of Zaccanopoli, certain

that studying only the village itself would not be sufficient for a consideration of complex kinship structures. Subsequently I extended the research to Fitili. Registry documents for that village date from 1634, but, since I was reconstructing the pattern of marriage exchange between the two villages, I used the documents beginning in 1755. Research in Fitili is still in progress.

5. There is a vast literature on Calabria, and I have consulted many sources, but the bibliography is limited to those I used most frequently for information on the specific geographical area with which I was concerned.

6. I consider only those features of social organization that, at least in these villages, seem to be inseparably related to the way the past is assessed and decisions taken in the present. Davis says that "the social relations of the people who make history determine in some part the meanings that they attach to events" (Davis 1989:104), and I might add that the social relations of these villages allow history to become such in the awareness of the people. Events are not assimilated without persons.

7. Strathern (1982:249) remarks that "because people live at one place, the place (and not region, street, household, family name *et alia*) should be a salient source of identification. He also notes that assuming "the village is a matter of membership within a community group" leads to an equating of village and community. For these villages the equation is valid. The village is a space that is culturally defined by its inhabitants and by their relations, primarily relations of exchange that are repeated in the area and used to control it. These relationships are inscribed within an area comprising the village, its inhabitants, their property, and the actions they perform on this property, and they give a sense of belonging to a place that is also a community, the spatial reference of which sums up its essence.

8. Cf. Farriss (1987, and in this volume) a propos of the coexistence of linear and cyclical concepts of time, the past as a general norm, and the need to distinguish the hierarchies between the two concepts. On the same topic, but from a different perspective, see also Bloch (1977, 1989) and Fabian's critique (1983).

9. The two tenses refer to two different orders of discourse, not to two concepts of time. Both tenses apply to historical time with durations that experience and memory are able to trace and calculate, however vaguely, as well as other durations that defy calculation but of which the salient features have been preserved and transmitted in memory and are evoked for ritual normative purposes depending on the context in which they are communicated. In this regard, see Bloch (1977, 1989). *Always* implies continuity not just repetitiveness. Narratives describing the way things have always been are not part of a fixed corpus but rather take the form of what Hugh Jones calls "stories about old people" (1989), which can be varied, enriched, and used in different ways and for the particular purposes of the narrator. Narrator and narration are central to the elaboration of the past, or rather of those pasts that share the same normative value (cf. Benjamin 1976) in assuring continuity (cf. Pocock 1962) and hence establish their own diversity (cf. Cohen 1982).

10. These beliefs, like those that follow, refer to the time of dead people in the sense of the time in which they now live. For example, as soon as the soul leaves the body, it must cross a bridge as thin as a strand of hair to reach that other world. At the end of the journey, the soul arrives in a place where there is all time in the world, which might mean a place where time does not exist. But this infinite time is actually articulated in what the people refer to as the times of the dead, which need to be localized in space as well. On the night between 1 and 2 November, the souls of the dead come back, walk around the whole village, and go into the houses for something to drink. They rise from their tombs at midnight and walk around the vicinity of the cemetery to capture any passersby. And when they come at midnight for someone who is dying, they announce their presence by knocking on the door. These and other beliefs, which are part of other hermeneutic frameworks (cf. Lombardi Satriani and Meligrana, 1982), show that the dead, too, live within a time scheme that is subdivided into periods that are calculated with reference to the living and their history, periods that are sited in defined spaces in which the dead perform such actions as drinking, walking, knocking, and so on. The time of the dead is a bridge between past and present, making the past vivid by giving it a spatial context while projecting the present from its limited and circumscribed duration into the dimension of the infinite.

11. This ecological time consists of cycles that recur regularly in the course of the year, but they are not conceived of as always equal: they do not all have the same duration and quality. Meteorological events as well as changes in agricultural techniques and relations may accelerate or decelerate the phases of the cycles, despite their linkage to the seasons, and lead to fracture and discontinuity. The reference to work and the social condition that derives from it channels the rhythms of historical time into "natural" time (cf., in this regard, Heald 1991).

12. As Bourdieu observes (1980:181), with regard to marriage transactions, "tout se passe comme si la ritualisation des interactions avait pour effet paradoxal de donner toute son efficacité sociale au temps, jamais aussi agissant que dans ces moments où il ne se passe rien, sinon du temps."

13. In Calabrian culture the theme of acceptance of a subaltern position due to fate and/or the will of God is accompanied by the theme of implicit or explicit rebellion. See Lombardi Satriani (1968, 1974), who theorized the constitutional ambivalence of folk culture.

14. This is the version local historians give of the origin of the three villages Zambrone, Zaccanopoli, and Zungri (Corso 1918; Valente 1973). They say they learned it from their forebears, and as evidence they cite the apotropaic mask on a fountain, which is reputed to have come from Aramoni. This mask is the only relic shown to outsiders as proof of the village's antiquity.

15. For the idea of lived-in space, see Fremont (1981); and for the importance of representations, see Quaini's preface to the Italian edition of Fremont (1981).

16. The genealogy of more than one member of a family covering more than one period would run to well over two hundred people.

17. Davis (1989:108) describes the model he found among the Zuwaya in Libya, a model he calls "generative," and says that genealogy is "a set of rules for deriving the present state of political and matrimonial alliances." For the villages of Zaccanopoli and Fitili, I prefer the term *practices,* in the sense given by Bourdieu (1972), rather than *rules.* Here, too, it is genealogy that orients subsequent behavior, but at the same time identity is defined and membership is established case by case, and generations move farther apart or closer together, as if, to borrow another expression from Davis, they are "playing with time."

18. We might perhaps call them *quasi-groups* in Mayer's sense of the term (1966). This definition is unsatisfactory if it is applied solely in reference to an identity established by consanguinity and alliance, but it is more pertinent when extended to the ways they operate.

19. The parish and municipal archives of Zaccanopoli indicate that between the eighteenth and twentieth centuries marriages contracted with outsiders accounted for about one-fourth of a total of 2,410 marriages. There was repeated marital exchange among seven villages in the course of time, of which Fitili is not the most important numerically. Others supply a large number of spouses, particularly in the last half of the twentieth century. Fitili's significance lies instead in the continuity of its participation in the marriage exchange over the course of two centuries, a participation that reached a peak in the nineteenth century.

BIBLIOGRAPHY

Banfield, E. C. 1958. *The Moral Basis of a Backward Society.* Glencoe, Ill.: The Free Press.

Barrio, G. 1571. *De antiquitate et situ Calabriae.* Rome. Facsimile reprint, Cosenza: Brenner, 1979.

Benjamin, W. 1976 (It. ed.). "Il Narratore. Considerazioni sull'opera di Nicola Leskov," in *Angelus Novus.* Torino: Einaudi, 235–60.

Bloch, M. 1977. "The Past and Present in the Present." *Man* 12: 278–292.

Bloch, M. 1989. *Ritual, History and Power. Selected Papers in Anthropology.* London: Altholone Press, 1–18.

Bourdieu, P. 1972. "Les stratégies matrimoniales dans le système de reproduction." *Annales ESC* 4–5: 1005–127.

Bourdieu, P. 1980. *Le sens pratique.* Paris: Minuit.

Cohen, A. P. 1982. "A Sense of Time, a Sense of Place: The Meanings of Close Social Association in Whalsay, Shetland," in *Belonging: Identity and Social Organisation in British Rural Cultures,* ed. A. P. Cohen. Manchester: Manchester University Press, 21–49.

Cohn, B. S. 1961. "The Pasts of an Indian Village." *Comparative Studies in Society and History* 3: 198–220.

Collard, A. 1989. "Investigating 'Social Memory' in a Greek Contest," in *History and Ethnicity,* ed. E. Tonkin et al., 89–103.

Corso, D. 1918. "Tradizioni sulla terra di Aramoni in Calabria," in *Archivio storico della Calabria,* anno VI, 1/2/3/4: 289, 97.

Davis, J. 1989. "The Social Relations of Production on History," in *History and Ethnicity,* ed. E. Tonkin et al., 104–20.

Davis, J. 1991. *Times and Identities. An Inaugural Lecture Delivered before the University of Oxford on 1 May 1991.* Oxford: Clarendon Press.

Fabian, J. 1983. *Time and Others: How Anthropology Makes Its Objects.* New York: Columbia University Press.

Farriss, N. M. 1987. "Remembering the Future, Anticipating the Past: History, Time and Cosmology Among the Maya of Yucatan." *Comparative Studies in Society and History* 3: 566–93.

Frémont, A. 1981, 2° (It. ed.). *La regione. Uno spazio per vivere.* Ed. M. Milanesi, preface M. Quaini. Milan: Angeli.

Fiore, G. 1743. *Della Calabria illustrata,* tomi 2, tomo I 1691, tomo II. Napoli. Ristampa anastatica, 1974. Bologna: Forni.

Firth, R. 1989. "Fiction and Fact in Ethnography," in *History and Ethnicity,* ed. E. Tonkin et al., 48–52.

Galanti, G. 1792. *Giornale di viaggio in Calabria.* Edizione critica a cura di A. Placanica, 1982. Naples: Società Editrice Napoletana.

Galasso, G. 1975. *Economia e società nella Calabria del Cinquecento.* Milan: Feltrinelli.

Heald, S. 1991. "Tobacco, Time and the Household Economy in the Two Kenyan Societies: The Teso and the Kuria." *Comparative Studies in Society and History* 1: 130–57.

Hugh-Jones, S. 1989. "Varibi and the White Men: History and Myth in Northwest Amazonia," in *History and Ethnicity,* ed. E. Tonkin et al., 53–70.

Jedlowski, P. and M. Rampazi. 1991. *Il senso del passato. Per una sociologia della memoria.* Milan: Angeli.

Leach, E. 1988. "Tribal Ethnography: Past, Present, Future," in *History and Ethnicity,* ed. E. Tonkin et al., 34–47.

Lombardi Satriani, L. M. 1968. *Contenuti ambivalenti del folklore calabrese: ribellione e accettazione nella realtà subalterna.* Messina: Peloritana.

Lombardi Satriani, L. M. 1974. *Antropologia culturale e analisi della cultura subalterna.* Rimini: Guaraldi.

Lombardi Satriani, L. M. 1977. *Folklore e profitto. Tecniche di distruzione di una cultura.* Rimini: Guaraldi.

Lombardi Satriani, L. M. and M. Meligrana. 1989, 2°. *Il ponte di San Giacomo. L'ideologia della morte nella società contadina del Sud.* Palermo: Sellerio.

Marafioti, G. 1601. *Croniche et antichità di Calabria,* Padova. Ristampa anastatica, 1975. Bologna: Forni.

Mayer, A. C. 1966. "The Significance of Quasi-groups in the Study of Complex Societies," in *The Social Anthropology of Complex Societies*. Ed. M. Banton, ASA Monographs 4. London: Tavistock, 97–122.

Merzario, R. 1982. "La buona memoria. Il ricordo familiare attraverso la parola e il gesto." *Quaderni Storici* 51: 1001–25.

Minicuci, M. 1983. "Il sistema di denominazione in un paese dell'Italia meridionale." *L'Uomo*, vol. 7, 1/2: 205–17.

Minicuci, M. 1983. "Il disordine ordinato. L'organizzazione dello spazio in una comunità rurale calabrese." *Storia della città* 24: 93–118.

Minicuci, M. 1981. *Le strategie matrimoniali in una comunità calabrese. Saggi demoantropologici*. Soveria Mannelli: Rubbettino.

Minicuci, M. 1989. *Qui e altrove. Famiglie di Calabria e di Argentina*. Milan: Angeli.

Minicuci, M. 1991. "Il pastore e il meraviglioso," in *Per il decennale della biblioteca di Soriano Calabro 1981–1991*. Soriano: Jason, 107–23.

Padula, V. 1977. *Calabria prima e dopo l'Unità*. A cura di A. Marinari. 2 vols. Bari: Laterza.

Pizzorno, A. 1967. "Familismo amorale e marginalità storica ovvero perchè non c'è niente da fare a Montegrano." *Quaderni di sociologia* 3: 247–61.

Pocock, J. G. A. 1962. "The Origins of Study of the Past: A Comparative Approach." *Comparative Studies in Society and History* 2: 209–46.

Sergio, F. M. 1720. *Cronologica collectanea. De civitate Tropea eiusque territorio. Libri tres*. Stampa anastatica, 1988. Naples: Edizioni Athena.

Strathern, M. 1982. "The Place of Kinship: Kin, Class and Village Status in Elmdon, Essex," in *Belonging: Identity and Social Organisation in British Rural Cultures*, ed. A. P. Cohen. Manchester: Manchester University Press.

Tonkin, E., M. McDonald, and M. Chapman, eds., 1989. *History and Ethnicity*, ASA Monographs, 27. London: Routledge.

Valente, G. 1973. *Dizionario dei luoghi della Calabria*. 2 vols. Chiaravalle Centrale: Frama's.

Zonabend, F. 1980. *La memoire longue. Temps et histoires au village*. Paris: Puf.

Part 2
Big Time

Remembering the Future, Anticipating the Past: History, Time, and Cosmology among the Maya of Yucatan

Nancy M. Farriss

This essay is about concepts of time and the past among the Maya Indians of Yucatan in southeastern Mexico. It explores how these concepts fit into the Maya's general view of the way the world works and how they relate to certain dynamics of Maya history—as we define history—during their pre-Hispanic and colonial past. One inspiration has been the often baffling written records the Maya have left, from which we try to quarry historical facts without always inquiring what the records meant to the people who produced them. The other is the reminder, provided by recent historical work from anthropologists, that people do not record their past so much as construct it, with an eye to the present, and at the same time use that past in molding the present.[1]

Past as Prologue and Past as Prophecy

The shape people impose on the past depends on how they think about time and its movement. In good structuralist fashion we tend to cast our discussion of this thinking in terms of a binary opposition between two alternative conceptions, cyclical and linear.[2] According to a cyclical conception, time is a perpetual repetition, corresponding to the diurnal and seasonal rhythms of the natural world, and the past therefore is infinitely repeatable. In a linear conception, time advances along a path as an irreversible chain of events. In the one view past is prophecy; in the other it is prologue.

There also seem to be two main explanations for these concepts and the contrast between them. The first, though more recent, links the contrast between linear and cyclical concepts to the presence or absence of literacy.[3]

In what might be called a technological or Marshall McLuhanesque model of cognitive systems, stress is laid on the means of communication, which influences the content of what is conveyed. Writing and reading are necessarily unidirectional. One proceeds along a designated line—up or down or side to side, as the case may be. In nonliterate modes of communication, especially in visual imagery, information is conveyed as a totality of impressions without a starting or ending point; and the same applies to a certain extent to the "speech performances" of oral communication. Pattern or structure is emphasized in nonliterate modes, while in writing it is process.

The idea that writing has therefore "restructured human consciousness" (Ong 1982:78, 141–49) is an intriguing one. But the evidence on ways of thinking about time is equivocal at best. Most societies that have developed systems of writing—Ancient Egypt, India, China, and Mesoamerica among them—have also viewed time and the past in a predominantly cyclical fashion.[4] The literacy model has had to be more narrowly defined to deal with these exceptions. What counts, it seems, is not writing itself but the alphabetic script, first developed by the ancient Semites, and the widespread literacy that its easy mastery encourages (Goody 1968:20–24; Goody and Watt 1968:34–55; Ong 1982:85–93). Still, it is unclear why writing, in whatever form, should not always produce the same effect on those who master it. One would expect the learned minorities in China, Mesoamerica, and other societies with nonalphabetic scripts to conceive of time and the past in a linear fashion even if the untutored masses did not.

The correlation between literacy and linear thought seems to hold up in only a single case: that of western civilization with its roots in the Semitic and Hellenic worlds. Western civilization has frequently been assigned the role of anomaly in world history. Equally familiar is the notion that what is most distinctive in western intellectual development derives ultimately from linear modes of thought: from explanation based on cause and effect rather than on how parts fit into a whole. But even if we accept that secularization, the Scientific Revolution, the Enlightenment, and all the other major transformations leading up to modern western culture can be traced to a chain-of-events type of reasoning that we define as historical thought, it does not follow that alphabetic writing provided the impetus.[5]

Another, more venerable, explanation has been offered for the apparent dichotomy between western and nonwestern thought, one that emphasizes the same contrast between cyclical and linear perceptions of time and the past but focuses on the *content* of thought rather than on how it is communicated. It is a teleological model that locates the wellspring of western linear thought in the historicist theodicy of the Judaic—and ultimately the Judaeo-Christian—tradition. This theodicy is based on unique, unrepeatable events, starting with the Yahweh-Israel covenant—events that occurred in time and permanently altered not only the structure of the cosmos but also the manner of its movement.

The much-affirmed and much-analyzed historical content of the Old Testament needs no recounting here. What bears emphasis is the difference in consciousness, especially consciousness of the past, between the Old Testament and the sacred texts of neighboring groups. "At the heart of the religion of ancient Israel is a break with the entire universe, a vehement repudiation of both the Egyptian and Mesopotamian versions of cosmic order" (Berger 1969:115).

The idea that a historicist theodicy underlies the uniqueness of western thought has a long pedigree ultimately traceable to St. Augustine and other church fathers. But it has received much impetus from Mircea Eliade's explorations of cyclical perceptions of time and his concept of the "myth of the eternal return" as a universal mental construct from which only western historicism has deviated (Eliade 1954). According to this construct, the cosmic order, and the link between humanity and the sacred, must be continually and literally reestablished through the ritual enactment of the events by which they were originally created. Only in the Judaeo-Christian tradition does the idea exist that things were set in motion—or humanity redeemed—once and for all time.

The two explanatory models outlined here have much in common. Not only do they define cyclical thought as the norm and linear thought the deviation that needs explaining. They also trace the crucial innovation, whether teleological or technological, to the same source—the ancient Semites. (Someone better equipped than I might look into the coincidental emergence of alphabetic script and historicist theodicy and their roughly parallel development in time and place.) And they also share some of the same limitations. One of them is the failure to account for Islam in either schema. A more general though related problem is that both models seem to set up a false antithesis between linear and cyclical modes of thought. While not actually defined as mutually exclusive, the two modes are presented in terms of an either/or opposition that is not productive at this stage of our understanding. There is ample evidence, including in Eliade's study, that linear and cyclical conceptions can coincide within the same cognitive system and often, perhaps usually, do. What we should be asking, then, is how the two fit together. This does not mean that the two models have nothing further to contribute to the inquiry. If in the Maya case the issue of literacy does not get us very far, teleology does. But before exploring how the Maya related linear and cyclical thought, in theory and in practice, let me outline the evidence for their dual conception of time and the past.

Maya Perceptions of Time and the Past

Most of our information on the perceptions of time and the past held by the pre-Hispanic and colonial Maya comes from their own writings. Although some oral traditions have been incorporated into their postconquest documents and also recorded by the Spanish, it should be emphasized that we are dealing with a literate culture and one, moreover, that long preceded the arrival of Europeans. The Maya and several other groups in Mesoamerica present an unusual, if not unique, example of a literate tradition that retained unbroken continuity through a shift from one system of writing to another, totally unrelated one.[6] While Maya hieroglyphic writing died out, the Maya language has survived in written as well as spoken form, transcribed into the alphabetic script introduced by the Spanish.[7] Although the rate of literacy seems to have actually declined after the conquest, despite the new script's relative simplicity, the Maya elite continued to produce their own records and their own unfiltered versions of the past.[8]

From their earliest recorded history, the Maya displayed an intense interest, bordering on obsession, in measuring and recording the passage of time.[9] Evidence for a cyclical conception is especially abundant. Cycles of many different dimensions were calculated, based on the movement of different astral bodies and on more esoteric criteria, and all were enmeshed in an exceedingly complex system of computation.

Among the best-known representations of cyclical time, and one of particular interest here because of its clear association with the recording of historical events, is the *katun* round (*u kahlay katunob* in Yucatec Maya). It appears in a series of documents called Books of Chilam Balam, which are colonial transcriptions, with variations and emendations, of a possibly single pre-Hispanic hieroglyphic text.[10] Depicted graphically in one of the Chilam Balam texts as a circle or wheel, the *katun* round is an endlessly recurring sequence of thirteen twenty-year periods, called *katuns* (the Maya plural is *katunob*), each with its designation and characteristic events. Thus, the *katun* round serves as both history and prophecy, a guide to the future as well as a record of the past. The same pattern of events discernible for *katun* "10 Ahau" will be repeated each time that the 260-year cycle is completed and 10 Ahau comes around again.[11] The equivalent concept in our own calendrical system would associate, say, droughts with the decade of the 90s in each century and expect that droughts will occur in the 1990s, and the 2090s, as they did in the 1690s, the 1790s, and so on.

The *katun*-round texts are not ahistorical in a strict sense. Unlike the most familiar type of origin myth, in which all specificity has been leached out to leave only the structure, these *katun* rounds often refer to particular people and places, to specific periods of time, and to concrete, nonfantastical actions and events—famines, invasions, political rivalries, exiles, and migrations. But, because of their cyclical structure, the texts are exceedingly frustrating to anyone who seeks to reconstruct from them a linear sequence or even to correlate events with an independently derived chronology (Ball 1974). Take invasions, for example. Various types of material evidence indicate several successive intrusions from central Mexico, or intermediary regions, into pre-Hispanic Yucatan (Miller 1977, 1979). These invasions, called "descents" by the Maya, appear all jumbled together in Chilam Balam texts where logically associated events always occur in the same *katun* regardless of their inner chronology. And references to the most recent descent—the Spanish conquest—are woven into the pre-Hispanic mix to confuse the picture further.

Yet there is also evidence that the Maya were perfectly capable of sorting out chronology, of conceiving of and recording the past in a linear sequence of events, and evidence, moreover, that they could and did operate in both modes at the same time. For the colonial period this evidence comes mainly in the form of chronicles written by members of Maya ruling lineages some decades after the conquest, that is, from the 1560s through the early seventeenth century.[12] They combine genealogies with historical narratives of conquest, migrations, and accessions to power, all in sequential order.

The purpose of the chronicles is clearly to support claims to rulership and territorial jurisdiction, and for the present analysis their accuracy is less important than the historical consciousness that informs them. Whether or not the facts were distorted in the interest of pressing dubious claims, they all based legitimate title on the principles of descent and prior occupation, criteria that rely on a linear construction of the past. Some of the accounts were addressed to the Spanish Crown and emphasize the author's or his ancestors' contributions to Spanish conquest and evangelization as part of his bid for royal recognition. But others were intended solely for Maya consumption. The "Crónica de Calkini," containing the same type of genealogical and historical information and in the same format, refers to the initial resistance by the Canul and related lineages to Spanish rule and only grudging later acceptance (Barrera Vázquez 1957). This and similar accounts eventually entered the Spanish information network as part of land-title records, but their original aim was to convince fellow Maya of the legitimacy of the authors' claims.

Whether addressed to a Spanish or a Maya audience, these documents served as charters for rights and privileges within the colonial context. That does not mean that they should therefore be seen as wholly colonial products. The Maya and other Mesoamerican groups produced numerous local histories shortly after the conquest, containing genealogies and sequential chronologies from pre-Hispanic times. It seems unlikely that they were all concocted to satisfy the new historical criteria introduced by the Spanish. They must have conformed in some way to an indigenous construction of the past, preserved originally either in oral traditions or in pre-Hispanic codices, or in both. And that construction was clearly designed according to a linear code.

We need not rely entirely on colonial records for evidence that the Maya were capable of holding two conceptions of time and the past simultaneously. A large corpus of hieroglyphic inscriptions has survived from the Classic period (ca. A.D. 250–900) that demonstrates two ways of measuring time and dating events. These inscriptions will be discussed in more detail later. Here I need only point out that in one type of calculation, called the Long Count or Initial Series, elapsed time is reckoned in a linear fashion from a fixed starting point, and in the other it is portrayed as recurring in cycles of varying duration (Satterthwaite 1965; Morley and Sharer 1983:555–59). The colonial Maya's dual conception of time may have been a colonial innovation. If so, it was a reinvention of a millennium-old tradition.

Time and the Cosmic Order

The coexistence of linear and cyclical conceptions of time and the past is by no means confined to the Maya and may well be the norm. What needs explaining is *how* they coexist. If we assume a certain minimum congruence among elements in a cognitive system, how do we reconcile within one system the idea of unrepeatable events and repetitive cycles?

In one sense all conceptions of time incorporate both linear and cyclical features, as proposed by Leopold Howe in his study of Balinese time (1981). In a cyclical conception there is linear progression within each cycle. Conversely, cycles (days, months, years, centuries) can be added up to produce a linear sequence. Translated into visual terms it is a question of distance or perspective. The horizon appears as a straight line; the curve of the earth's surface is revealed only from outer space. From an equivalent external perspective the individual circles that accumulate in a sequence will appear as a string of beads, or a spiral, rather like an elongated Slinky toy.

These visual images, and I think logic as well, suggest that while both linear and cyclical concepts can be contained within a single system they cannot coexist on an equal footing. Howe argues that there will be differences in emphasis. I go further and say that one concept must be subordinated to or incorporated into the other.[13]

The western tradition, embodied at the time of the Spanish conquest of America in late medieval Christianity, represents a system—perhaps the only one of its kind—that incorporates cyclical elements into a predominantly linear conception of time. The diurnal and seasonal rhythms of an agrarian society were marked ritually by the daily repetition of the holy offices from matins to vespers and in the annual calendar of festivals. Moreover, the Christian year from Advent through Ascension annually repeats in compressed form the life of Christ. The daily sacrament of the Mass represents the most extreme expression of Christian cyclicity—a radical departure from the linear theodicy of the Old Testament, if not a retrograde step. For Catholics, at least, the Eucharist does not merely commemorate the original sacrifice on Calvary; the crucifixion is reenacted each time through the body and blood of Christ.

Nevertheless, in its major thrust medieval Christianity remained firmly within a historicist cosmology (Berger 1969:122–23; Brandon 1965:148–205). The mystery of transubstantiation lies precisely in its recreation of a unique act, which occurred only once and in historical time. For all the popularity of astrology, with its notion of periodic repetitions, and of millenarian ideologies that promised a return to the past—to the foundation of Christianity—the dominant medieval vision was eschatological. It looked to the future not as a restoration but as a culmination of a relentlessly linear progression, ending in the unrepeatable Last Judgment.[14]

In the Maya system of thought the relationship between the two concepts is inverted: linear time is incorporated into an all-encompassing cyclical pattern. The Maya, in common with the rest of Mesoamerica, saw time as a complex of forces that exert their influence in regular turns marking cycles of different durations. Like a wheel with cogs representing the different units, each cycle will engage with another at the same point only within a larger cycle, and that process will continue in ever-widening dimensions (wheels within wheels). Thus, the 260-day ritual year (called the Sacred Round) compounds two cycles, of 13 days and 20 months, and combines with the 365-day solar year to produce a 52-year cycle called the Calendar Round; a different set produces the characteristically Maya cycle of the *katun* round.

The content of time—what occurs in each unit—is regulated by the conjunction of forces, which gives a particular composite stamp to events. The larger the cycle, the greater the number of forces intermeshing from different subcycles, as in Old World astrology, and thus the greater the similarity of events. If events appear to follow an irreversible sequence, creating the illusion of time as a linear movement, that is only the fault of a limited perspective. No matter how unique a pattern of events may seem, no matter how long the sequence, it will eventually be repeated, when the governing forces of all the cycles of different dimensions coincide in one huge cycle (López Austin 1980:1, 70–71).

We have written records of just such a cycle among the lowland Maya. During the Classic period they developed a calendrical system in which more and more cycles of different durations were incorporated into a single grand computation, expanding in scope until it culminated in a gigantic cycle running into many millions of years (Thompson 1960:317; Satterthwaite 1965:605, 611). This cycle would defy, if not human comprehension, at least the predictive capacity of the calendrical specialists, for it is unlikely that the Maya priests claimed to comprehend all the possible combinations of divine forces governing the subordinate movements of so immense a projection. I suspect that this was precisely the point, a cycle so large that it could in theory encompass any conceivable linear sequence, to account for any contingency and thereby affirm the principle of cyclicity over the appearance of uniqueness and irreversibility. Even if no one could command sufficient knowledge to discern the whole pattern in all its complexity, ultimately the pattern would reveal itself.

In both the western and Mesoamerican systems of thought, concepts of time are intimately bound up with concepts of the sacred to form part of a particular understanding of the way the cosmos works and the way that man relates to it. Both systems attempt to link man's destiny to the sacred, merging human time with cosmic time, to achieve in a sense the timelessness of eternity (Stirrat 1984:203, 209–10). The difference is that in Judaeo-Christian thought cosmic time moves in a straight line, and the point of merger—that is, eternity—comes at the end of it. In Mesoamerican thought cosmic time returns endlessly to the beginning, and human time intersects with cosmic time at each turn of the cycle—or cycles—in "a succession of eternities."[15]

The logic of subordinating human time to cosmic time is clear enough. Less so is why in Mesoamerican thought time should be assigned two different shapes. The key, I think, to the Mesoamericans' conception of time and to their entire cosmology is their preoccupation with order and above all with cosmic order. Paul Ricoeur links this type of preoccupation to a perception of evil as a force external to humans (1982:325, 329–30, 332). Evil is disorder, represented by the chaos that preceded creation and that constantly threatens to reassert itself. The drama of creation is therefore an ongoing one, for the cosmic order must continually be reaffirmed in the face of this ever-looming chaos. Time is part of the cosmic order. Indeed, for the Maya and the rest of Mesoamerica, time is cosmic order, its cyclical patterning the counterforce to the randomness of evil.

Disorder also threatens at the level of human affairs. There, too, evil is characterized as deviation from expected patterns. Accounts of famine in the Books of Chilam Balam, for example, show less concern with starvation than with the fact that people flee into the forest to eat roots and other wild food and the vultures enter the houses. In other words, people cease to live in their customary fashion, at home and subsisting from their milpas, and when they die they are not properly buried. Political conflicts are condemned because they disrupt the established hierarchy, and upstarts and invaders are vilified for breaking the rules. Chaos was not merely socially undesirable but conceptually unsatisfying and psychically disturbing as well.[16]

The idea of a conflict between order and chaos on both the cosmic and human levels would account for the Maya's dual conception of time and the past. They could not help but be aware of the accidental, unpredictable, and therefore threatening nature of human events over the short term. Collective catastrophe in the form of drought comes without warning, as does individual illness, or injury, or death. Rulers can die without succession; upstarts and invaders can challenge the political order. At the same time, a reassuring predictability can be imposed by subsuming these accidental diachronies into the longer-term cyclical pattern of cosmic time. Indeed, events only *appear* accidental from the limited perspective of human, historical time.

It may be that the Maya did not conceive of things as returning to the same temporal point, merely the same logical point, as has been suggested for a similarly dual system of linear and cyclical time in Bali (Howe 1981:231). In either case the rationale is the same, for it is the repetitive pattern of events that counts and not their exact duplication.

The predictable order of cosmic time could not, however, be taken for granted. The idea of a divinely ordained plan is always accompanied by the less comforting caveat that its execution depends largely on human agency. The individual qualms of the Christian who believes in divine providence but cannot be sure where he personally will fit in—whether, through his own actions, he will end on the side of the saved or of the damned—is replaced in

Mesoamerica by a collective anxiety. Evil as disorder might be perceived as being external, coming from outside man rather than from within, but his role in combating it did not therefore become a passive one. On the contrary, the source of energy that kept the whole system going was man's collective rites of renewal. Only through the sacrifices that replenished divine energies were time and the cosmos sustained in orderly motion.[17]

Constructing the Past

Let us now see how the Maya's dual conception of time worked out in practice. I have identified two ways that they represented their past. In one, which corresponds to western concepts of history and what has been called "historical time" (Leach 1954:116), events are presented in linear sequence, while in the other the same or similar events (the texts are too few and the language too allusive to determine the exact degree of overlap) are arranged in the cyclical pattern imposed by cosmic time. The relationship between the two modes can best be understood by looking at the duration and content of historical time as opposed to cosmic time.

A cyclical pattern can, as suggested earlier, incorporate an exceedingly long-term linear progression, as long as the cycle is large enough. Yet Maya historical time had a relatively short depth. The colonial Maya chronicles do not for the most part go back past the time of the League of Mayapan, which collapsed in 1441, barely a century before the arrival of the Spanish. When they do, all precision is lost. An account contained in one of the 1580 "Relaciones de Yucatan" (pt. 2, 160–61) relates the history of a foreign captain, one Ekbalam, who might have been part of the tenth-century Toltec invasion. It is the most circumstantial of the pre-Mayapan accounts, referring to Ekbalam's arrival, his rulership, death, and succession. Yet it is thin in detail and gives no sense of date or duration. The informant confesses that he does not know. This account, the other historical sections of the "Relaciones," and the more extensive chronicles cited previously all appear to have been preserved originally only in oral traditions. Many later colonial documents refer to the recording of genealogies and events in the same way: community "elders" (*ancianos*), as repositories of the collective memory, are gathered together to tell what they know of some aspect of the past "in order to place it on record" (*para que conste*).[18] It might be supposed that the reliance on memory accounts for the short time depth. To some extent this is true; a certain amount of telescoping and loss of detail is inevitable in oral tradition. However, societies are capable of preserving a considerably more detailed and longer-term oral record of their past than the Maya did—if it suits their purposes.

Moreover, the Maya did not have to rely solely on memory. They possessed a system of writing perfectly capable of recording history along with a highly sophisticated calendrical system and a concept of time that could account for long-term sequence while still retaining a cyclical master code. A lack of technological or conceptual capacity will not account for the lack of long-term sequence. If the Maya confined historical or linear time to short duration, they must have done so out of choice. Linear chronology was simply not meaningful or relevant over the short term but was subordinated to the cyclical rhythms of cosmic time represented by the *katun* rounds in the Books of Chilam Balam.

In a recent translation of the Chilam Balam of Tizimin, Munro Edmonson (1982) has reversed this relationship and rearranged the *katun* rounds themselves into a diachronic sequence based on internal evidence. Although his chronology, if correct, will help to sort out Maya history according to our canons, it does not mean that the Maya saw things that way. The unedited long-term record they have left is unequivocally cyclical.

A major clue to the rationale behind this choice lies in the content of Maya historical time as represented in the chronicles. It is not only of relatively short duration but also decidedly this-worldly in subject matter, devoted primarily to genealogical succession within the ruling families and to sequences of migration and conquest. In other words, it encodes information about political power: who wields it, when, where, and, most especially, by what authority. As mentioned earlier, the chronologies of settlement establish territorial rights according to the common principle of "first-comers" (Kopytoff 1986); the genealogies lay claim to rulership according to the equally common principle of descent from previous rulers. These are in the main secular principles based on secular events. Supernatural entities and interventions figure only rarely and more as points of reference than as regulating forces.

The contrast with the *katun* rounds is striking. In the Books of Chilam Balam we have moved from prose to poetry, to a highly charged and allusive language that stresses the quality of time over its factual content. Events, whether political or natural (like drought and pestilence), lose much of their specificity and become types of human experience. They themselves are less important than their effect ("misery" and "affliction" predominate by far over "abundance" and "joy") or their cause, which is invariably the "burden" or nature of the particular *katun*. The cosmic forces dominate not only by imposing their cyclical pattern according to the calendar but also by intervening directly as sacred personages whose identity merges with that of the human actors.[19]

It is commonplace that people construct their past to explain and justify their present. I suggest that the two forms the Maya imposed on their past, historical time and cosmic time, both served as charters for later sociopolitical arrangements, but each in its own way.

The cyclical pattern of cosmic time provided the model for a political system based at least in part on the principle of rotation. This system seems to have functioned during the post-Classic and colonial periods at several levels of political organization. Within each community, municipal authority rotated among the territorial divisions, and presumably their principal lineages, on a four-year cycle. Among ruling families, the twenty-year *katun* period regulated succession; and there is even evidence that "seats of power" shifted around the Yucatan peninsula between the two rival parties of the Itza and the Xiu according to the entire *katun* cycle.[20]

Such a system would help to avoid conflict and disorder in human affairs by providing for a rotational sharing of power: a sort of guarantee to the outs that their time would come. It would also serve to integrate human affairs into the cosmic order. This is seen most clearly in the New Year ceremonies that accompanied the rotational transfer of office in the four-year community cycle. The particular nobleman assuming power undertook to help the god associated with the coming year to bear the burden of the time unit that had been assigned to him (Landa 1941:138–49; Coe 1965).

The weakest point in any political system is the transfer of power. The cyclical pattern of cosmic time could provide a general model for the Maya system along with its sanctioning force—the How and the Why of political power. What it did not provide was the Who—which particular person or persons should wield power at any given moment. This purpose was served by the linear record of historical time with its emphasis on the issue of succession. A rotational system of transferring power would thus explain why careful genealogies and linear chronologies might be important over the short term but not over the long term when linear sequences were subordinated to the repetitions of the *katun* round in practice as well as in theory.

There is, of course, a major exception to the Maya's emphasis on short-term historical time. It appears in the Classic-period stela inscriptions based on the Long Count system of calendrical calculations mentioned earlier. According to this system, linear time is measured over an extremely long span from a theoretical starting point, fixed at 3114 B.C. in the most generally accepted correlation with our calendar (Morley and Sharer 1983:556). Cyclical patterning remained dominant in principle. The Long Count dates would be repeated every 5,200 years,[21] and ever larger cycles were added to the formula as if to see how far the bounds of infinity could be pushed (Thompson 1960:317). But the new calendar represented a radical shift in emphasis, one in which time and the past came to be expressed in linear distance measurements. Moreover, the linear sequence was of such time depth, stretching back well beyond human memory, that it easily dwarfed and indeed incorporated the 260-year cycle of the *katun* round along with the more ancient and more widespread Calendar Round of 52 years.[22] The overall effect was to subordinate cyclical time to a linear scheme.

The noncalendrical content of the stela inscriptions provides a strong clue to the rationale behind this innovation. Epigraphic analysis has increasingly shown that the texts, once thought to be concerned exclusively with time itself and the sacred entities who control its movement, give prominent attention to human affairs. Deities have been identified as earthly rulers (not necessarily wholly separate categories in Maya thought), and the Long Count dates have been found to refer to specific events in their biographies.[23] We are thus dealing with the same kind of historical time represented in later Maya chronicles. The time span is much greater, but the content is virtually identical: conquests, alliances, successions, and the other familiar stuff of political history.

Arthur Miller has suggested that the new emphasis on written genealogies and on long time counts represents a "stela cult" designed to affirm a principle in Maya politics that was also new, that of dynastic rule.[24] Several rulers, for example, Cauac Sky of Quirigua, even had stelae erected with texts claiming genealogical ties far back into the misty past, well before the Long Count starting date. Perhaps the stela cult was also an attempt to wrench cosmic time into historical time, to claim an immensely long ancestry—from the beginning of time or close to it—as a basis for divine kingship. At the very least, by making time irreversible and the past nonrepetitive over a span of several thousand years, a particular lineage could challenge the accepted rotational system and cling to power with the assertion of a permanent claim.

The Classic-period dynasties and their calendrical system came to an end simultaneously. Indeed, the cessation of Long Count dates marks the close of the Classic period in the Maya lowlands. The decline of the stela cult, with its emphasis on long-term linear time, could represent a resurgence of older principles: a reassertion of the primacy of cyclical time and rotating power over linear time and dynastic rule. We have ample evidence from the Maya post-Classic period of deviations from what was considered the established order, of factional struggles and successful bids for power from upstarts and intruders.[25] This does not mean, however, that the rules themselves were scrapped rather than manipulated more or less successfully to make the ideal and the real appear to conform. Though ambitions to establish unbroken dynastic sway over large territories may have been equally strong among post-Classic lineages, perhaps only the great rulers of Classic times could command sufficient power and audacity to rewrite the rules and challenge the cyclical model of time and rulership.

Maya Time under Spanish Rule

Whether best seen as a decline or a resurgence, the change in emphasis in Maya conceptions of time and the past is clear by the time of Spanish contact. While the erection of stone monuments with commemorative dates seems to have persisted on a modest scale, the dates marked intervals in the *katun* round rather than fitting into the extended time reckonings recorded in the earlier inscriptions. Surviving post-Classic codices are devoted primarily to *katun* counts and to lunar and Venus cycles (Thompson 1960:23–26); and the Chilam Balam books, based on other hieroglyphic texts now lost, are firmly rooted in a cyclical conception of the past. In contrast, the Judaeo-Christian system the Spanish brought with them had succeeded in fusing cosmic time into historical time, if not eliminating cyclical patterns altogether, at least subordinating them to a larger model of nonreversible sequence.

Thus, we have two master codes, mirror images of each other, each containing a dual conception of time but with the relationships of superordination and subordination reversed. It remains to be seen what resulted from the confrontation between the two.

I have argued that linear and cyclical conceptions of time can coexist within a single code or system only if one is subordinated to the other. Violation of this hierarchical principle can produce what Frank Salomon has called a "literature of the impossible" (1982). He refers to the postconquest Andean chronicles, compiled by Titu Cusi Yupanqui and other descendants of the Inca nobility, in which a futile attempt is made to combine Andean cyclical notions of the past with European diachronic notions into an integrated, coherent system. The goal is not impossible, only the method. The Andeans sought to reconcile the two conceptions of time on the basis of equality. The Maya were faced with the same conceptual dilemma when confronting the Spanish code, but they seem to have resolved it more conservatively and more successfully.

The postconquest chronicles analyzed earlier may appear innovative. They certainly conform to European rules of constructing history. But they also, as I have argued, conform to a Maya short-term historical patterning that required no redesigning of their cyclical master code. The real conflict would arise in the confrontation between the different conceptions of longer-term cosmic time. Here our best sources are the frequently cited Books of Chilam Balam.

It is worth pointing out here the unusual nature of the Chilam Balam books as expressions of colonial Mesoamerican thought. They share with the highland Maya "Popol Vuh" text the distinction of being the only extant documents outlining Indian postconquest cosmology that were produced exclusively by and for the Indians themselves. The more detailed treatises from central Mexico were either commissioned by the Spanish clergy or interpreted by them in the versions that have survived.[26] They are thus a form of intercultural translation, when not propaganda. The Chilam Balam books served no such purpose; on the contrary, they were deliberately kept secret from the Spanish. They are documents *of* indigenous cosmology, not *about* indigenous cosmology, and whatever European influences they show reflect the Maya's own unedited model of reality.

The Chilam books indicate that the Maya assimilated some of the elements of Hispano-Christian teaching but incorporated them into a wholly Maya framework, a wholly Maya perception of the way the world operates. The Virgin Mary, for example, became associated with the "Rope from Heaven," the sacred umbilical cord that for the Maya symbolized the link between the divinity and the world of man (Roys 1967:82; Farriss 1984:306). Saints, archangels, and Old Testament prophets joined indigenous sacred beings in the repetitive cosmic dramas that continued to follow a wholly Maya script.

The Maya also continued to submit human history to the cyclical rhythms of cosmic time. References to the Spanish conquest, to Christian evangelization, and to subsequent colonial episodes and experiences such as epidemics, idolatry trials, and episcopal visitations recur throughout the Books of Chilam Balam.[27] But instead of following one another in a linear sequence, as in the chronicles, these recent events were interwoven with similar or related events from the pre-Hispanic past, according to the principle of logical association, to form part of the *katun*-round "historical prophecies." Spanish conquest and colonial rule shared the random, threatening quality of all human history, perhaps exceeding previously imagined limits. Yet the Maya could give meaning and predictability to these experiences, also, by merging them into the cyclical pattern of the cosmic order.

To assume that the Maya failed to adopt a European conception of time and the past because they could not understand it would be to underestimate the Maya intelligentsia. The Spanish clergy of the early postconquest period, who knew the Maya elite far better than we do, did not make the same mistake. They might consider the Maya perversely mistaken, perhaps, but not ignorant or confused. All the evidence, both from the Maya's own texts and from what the Spanish tell us, points to a ferment of religious exploration and experimentation after the initial period of evangelization. I have elsewhere explored this experimentation in the sphere of ritual (Farriss 1982). The search for conceptual coherence is less easy to document, but we are told that the Maya elite had a passion for discussions of cosmology and were eager to hear Christian explanations of such issues as the Creation and why eclipses occur.[28] They would meet in nocturnal "juntas" to read and discuss their "holy books." Presumably from these conferences the reformulations evident in the Chilam Balam books emerged. Modern students of these books have found in them evidence that the Maya had studied, and grasped, contemporary European texts on the metaphysics of the Creation (Brotherston 1979) and early understood the Christian calendrical system, as a system (Edmonson 1982:115–30, 168–75, 179–80, 197–98).

Such a scholarly passion should not be surprising, nor should it be ascribed to mere intellectual curiosity. It is almost invariably the case that the dominated understand the dominators better than vice versa. Their need is far greater. The Spanish had redefined reality for the conquered to a degree that was in no way reciprocal. The Spanish might have to readjust some ideas, for example, on the nature of humanity and human differences, to accommodate the newly discovered existence of the Indians (Pagden 1982); or they might wish to learn about Indian religion in order to evangelize them more effectively. They did not have the same intense need to comprehend indigenous modes of thought. Theirs was the dominant ideology with which the conquered would have to contend in theory and in practice.

There was a crucial difference, however, between beliefs and public ritual. In the latter sphere, the Maya had to accommodate themselves to a Christian framework, however much they might infuse the framework with their own meaning. In the sphere of beliefs, they could study Christian cosmology, comprehend it, and ultimately reject its basic tenets without the Spanish being any the wiser.

For the Maya, creation continued to be an ongoing drama; rites of renewal to keep the world going were still required—and performed—now under Christian guises (Farriss 1982; 1984:309–33). Human affairs still needed to be ordered according to this same cyclical rhythm. They might adjust their own calendar slightly to correlate better with the Julian calendar and adjust their ritual cycle so that their own new-year ceremonies could synchronize with the cycle imposed by the Church. But they did not abandon their calendar, nor did they cease to apply its repetitive pattern to human events in the past and the present. Given the premise of ever-threatening chaos, the Judaeo-Christian idea of irreversible progression would be not so much incomprehensible as unthinkable.

How can the ostensibly momentous fact of conquest and all the changes it set in motion have had so little apparent impact on the basic configuration of Maya thought? To say that the Maya were more conservative than the Andeans begs the question. Successful conservatism requires its own explanation. That they could keep their beliefs hidden from the Spanish is only part of the answer. No matter how intellectually and emotionally satisfying the Maya's vision of reality had been in pre-Hispanic times, one might expect that the altered circumstances of colonial life would require some adjustment in perceptions regardless of what the Spanish thought or knew.

But how much had things actually altered? In a larger study of the colonial Maya I have argued that, in many fundamental ways, not very much (Farriss 1984). The Spanish introduced no major changes in the region's economy and human ecology. The Maya remained independent milpa farmers, tuned to the seasonal rhythms of the solar cycle. One might say that the agricultural year and the intimate cycle of the 260-day Sacred Round are one thing, quite another the long-term *katun* round so closely linked to major events and especially to preconquest political organization. Yet modifications in the indigenous political system were also less drastic than one might suppose and in fact tended to reinforce a cyclical conception of time and the past. If, as has been suggested, long-term linear sequence was tied to dynastic succession, the Maya had even less need for it in colonial than in post-Classic times. The Spanish had divided provinces into autonomous communities (Farriss 1984:147–52), carrying the post-Classic process of decentralization even further and permanently dashing any hopes for a dynastic resurgence that the contact-period lineages may have been entertaining. On the community level, the rotational system could continue to function unhindered, and perhaps even on a supraprovincial scale, despite some calendrical manipulations by contending parties (Edmonson 1982:xvi–xvii, 22, 39, 47, 172). Spanish domination did, of course, produce many changes in the lives and perceptions of the conquered Maya but not such that they had to abandon their cyclical model of cosmic order and political action.

Making History from Prophecy

The cyclical structure of cosmic time held more than theoretical significance for the Maya, aside from its expression in ritual and in the rotational principle of the political system. The historical prophecies arranged in the *katun* round provided guidelines for interpreting events in the present as well as in the past and, as such, helped to shape the events as they unfolded.

The tendency to pattern the present on images of the past is a common one, and people with a cyclical view of time would be especially prone to help history to repeat itself, even if in so doing they end by remaking history along new lines (Sahlins 1981, 1985; Peel 1984). The Spanish conquest of Mexico provides one of the more transparent examples of mythical expectation at work, creating a new reality that destroyed the myth as it was reenacted. Faced with the legend of a god-king who was to return to reclaim his kingdom, the Mexica ruler, Moctezuma, fulfilled the prophecy by casting Cortés in the role of Quetzalcoatl, whom he had no choice but to welcome and for whom he must stand aside.[29]

The Maya, despite an equally strong belief in the repetitive nature of human affairs, reacted in a much less fatalistic way to the appearance of the Spaniards. Unlike the empire-building Mexica, who had never themselves experienced foreign conquest and had only the most cataclysmic myth model to guide their interpretations, the Maya could easily assimilate Spanish actions into a familiar pattern of previous invasions. The conflation of past and present conquests, as represented in the Books of Chilam Balam, may have been a later interpretation, but the Spaniards facilitated the association from the beginning by faithfully reproducing—unwittingly, so far as we know—the details of earlier conquests.[30] They arrived as a small band of adventurers, first by sea from the east and then later overland from the west, bringing Mexican auxiliaries with them. Although all this would have been old history to the Maya, it did not incline them to easy capitulation. The conquest of Yucatan, unlike the swift defeat of the Mexica, took almost twenty years. Perhaps resistance was also part of the preordained *katun* pattern.

The influence of the cyclical *katun* prophecies on actual events, as distinct from a possibly ex-post-facto interpretation, is much clearer in the later conquest of the Peten region, at the base of the Yucatan peninsula, where refugees from the north had joined an independent Yucatec Maya group—the Itza—as holdouts against Spanish rule. In 1695 the Itza sent a delegation to notify the Spanish governor that, according to their prophecies, the time was approaching for their conversion to Christianity (an act synonymous with capitulation to

Spanish rule).³¹ They were referring to *katun* 8 Ahau, in which the Itza capital was to fall and be abandoned, duplicating the same fate of Chichen Itza and other former capitals during previous cycles (Roys 1967:136–37; Edmonson 1982:6–11). Earlier in the century, in 1618, they had rejected the overtures of two Franciscan missionaries, on the grounds that the time had not arrived because they were only in *katun* 3 Ahau, and told them to return at a more suitable time (Cogolludo 1688:Lib. 9, caps. 5–10, 12, 13). In 1696 another Franciscan arrived, responding to the veiled invitation that the prophecies had inspired the previous year. Using his thorough knowledge of Maya calendrics and cosmology, he was able to reach an agreement with the Itza ruler Canek that the entire kingdom would accept Christianity in four months: that is, at the start of the appropriate *katun*.³² "Obliging with their usual obtuse alacrity, the Spaniards arrived before Eight Ahau began and forced the Indians into armed opposition" (Edmonson 1982:xix).

To see the conquest of the Itza solely as the fulfillment of the *katun* prophecies would be a serious oversimplification. The kingdom was clearly on the defensive, already encircled by encroaching Spanish settlements and cut off from its source of wealth in the cacao trade (G. Jones 1982, 1983). Pragmatism figured in the Itza responses, much ambivalence, and, among one faction of the nobility, unequivocal opposition. Their spokesman retorted to the Franciscan's arguments on behalf of conversion: "What does it matter that the appointed time has arrived for us to become Christians, if the sharpened point of my lance has not yet become dulled?"³³ In other words, armed warfare and not the *katun* prophecies would decide the issue. Yet the attitude of resignation on the part of other leaders, based largely on the prophecies, was equally genuine. And it has been suggested that much earlier crises in Maya history, during the Classic period and marking its end, were in part brought about by the Maya's own fatalistic expectation of periodic catastrophe in the ill-fated *katun* of 11 Ahau (Puleston 1979).

Once the Maya were brought under Spanish rule, their cyclical conception of time would on the whole have exerted a tranquilizing effect. Spanish rule could be endured under the comforting assurance that all things, however good or bad, have their appointed time to end, according to the pattern already established in the past and to be repeated in the future. At the same time, such a conviction was fundamentally subversive. The widespread perception that Spanish rule had a definite ending point, at which time the Maya would return to a previous era, filled the Spanish with unease, and rightly so. There was always the danger that it could inspire resistance to Spanish rule instead of resignation.

The first recorded link between the Maya calendar and anti-Spanish resistance came at the tail end of the conquest period. The "Great Revolt" of 1546–47 that swept through much of eastern Yucatan was timed to begin on the Maya date 5 Cimi 19 Xul, day names meaning "death" and "end" (Chamberlain 1948:240). Predictions of the imminent end of Spanish rule, based on the type of *katun* prophecies found in the Books of Chilam Balam, began to appear from at least the beginning of the following century.[34] During the latter part of the seventeenth century, these prophecies helped fuel raids and resistance movements that kept Yucatan's southern region in turmoil along the length of the colonial frontier. Fugitives from Spanish rule, who controlled the border between the pacified and unpacified zones, encouraged rebellions and further defections with announcements that the "time had come for all [the Maya] to rise up and return to their ancient rites."[35] The fact that the relevant Chilam Balam prophecies seem to refer to a previous period or periods of alien rule posed no barrier to their colonial application. According to the Maya's cyclical vision of history the same *pattern* of events repeats itself even if the events themselves do not. Just as the recurring pattern of foreign conquest linked the Spaniards to earlier invaders, the chronicle—or prophecy—of downfall would apply to both in the appropriate *katun*.

Although inspired or at least supported by prophecies, these conflicts cannot be considered millenarian movements in the strict sense of the term. They may have the same practical configuration but they respond to different conceptions of the world and time, and of the way that they operate. The one refers to the fulfillment of a promise of a one-time event, the end of the world, the other to the restoration of or a return to a past era. The distinction is important in understanding the practical significance for the Maya of their cyclical conception of time and the past. Otherwise, the Chilam Balam books can too easily take on the quality of intricate intellectual games that the elite used to entertain themselves, according to rules known only to them and having no more connection with the real world than crossword puzzles.

The blending of prophecy and history in practice can be seen in some detail in the Canek revolt of 1761,[36] the only true colonial uprising in Yucatan, as opposed to the border upheavals that more properly belong to the prolonged process of conquest itself. Jacinto Uc, the leader of the revolt that bears his adopted name of Canek, is himself of minor concern here. I have elsewhere outlined certain biographical details that may help to explain his actions (Farriss 1984:100). The real interest lies in his following. Here was a true upstart, with no remotely suitable genealogy—that is, at best a commoner and possibly an orphan—and thus no legitimate political role, who proclaimed himself king of the entire province under the title of King Canek Chan Montezuma.[37] How he was able to enlist anyone in support of this absurd claim poses a puzzle. And he did enlist a far from negligible following—not as numerous as Spanish hysteria first imagined but too numerous to be dismissed as a few deviant malcontents. A large proportion of the towns in the well-populated heartland of the peninsula answered Canek's appeal for troops and sent contingents to the town of Cisteil where the revolt was launched. Equally significant, a number of other towns either sent expressions of support—with manpower possibly to follow—or at least equivocal replies. The point is that, although the Spanish regarded Canek as a charlatan or possibly a lunatic, the Maya did not. Nor did they (with the sole exception of the native functionary of the church in Cisteil) denounce him to the Spanish, despite the risks of not doing so should he fail in his declared intent to "put an end to the Spanish." Why did they risk punishment or, in the case of his active supporters, death? The supporters clearly believed that he would fulfill his promise; the others presumably believed that he had a good chance or they would not even have vacillated. But why?

Canek lay claim to a variety of supernatural powers in which Christian and indigenous elements were mixed. In this respect he was a local version of the "man-gods" who appeared from time to time in central Mexico with equally syncretic visions of divine powers (Gruzinski 1985), and we can find the equivalent in many leaders of nativistic revolts around the globe who have called upon supernatural aid against European rule.[38] The specifically Maya element in his program—and perhaps of most compelling force—was his promise of a return to the past, in fulfillment of the cyclical pattern enshrined in the *katun* prophecies.

One part of his call for support was the almost standard proclamation made by colonial Maya visionaries that the "time had come for the end of the Spaniards."[39] This included the end of tribute, with the less customary twist here that, instead of the Spanish being killed or disappearing, their turn had now come to hand over the wealth to the Indian and pay tribute to *them*. No doubt such an inversion of both the temporal and social orders would have appealed to most of the Maya, even if it had not been so conceptually satisfying—even if it had not conformed to the logic of how things operate. However, as I have said, this general notion was widespread in time and space in colonial Yucatan and was most commonly expressed as a hope—or veiled threat—rather than as a blueprint for immediate action. What Canek managed to do was convince his followers that he was the one designated to fulfill this attractive prophecy. He was to restore Maya rule and do so by summoning up royal authority from the past. He identified himself—not metaphorically but actually—with previous rulers. And his title rested not on the principles of precedence, or priority, or genealogical descent but on the characteristically Maya principle of rotation: by ruling in the past his time had come to rule again.

According to Spanish, and our own, ideas of history, the absurdity of Canek's claim lay precisely in this cyclical principle and the anachronisms that it enveloped. The testimony records that he identified himself with not one but three different rulers or sets of rulers from at least three different periods, all blended together into one source of legitimacy.

1. *Canek,* the Peten Itza ruler. Jacinto Uc assumed this royal name and also asserted that he had come from a place far to the south, that is, the Peten.
2. *Montezuma,* presumably the Mexica ruler (Moctezuma), who had sent an invading force to Yucatan shortly before the Spanish conquest ("Relaciones de Yucatan":Pt. 2, 221–22). It is not clear whether this figure had been incorporated into local historical traditions since that earlier period or was introduced later by the Spaniards who, for example, organized public processions on gala occasions with tableaux featuring Moctezuma (Cogolludo 1688:Lib. 9, cap. 11).
3. *A foreign lord* of undesignated name. His arrival by sea from the east identified him with the "Itza" captains who had entered Yucatan by sea in any of the several pre-Hispanic conquests via the east coast (Miller 1977, 1979).

All three figures share the feature of foreign, specifically central Mexican, origin as the basis of rulership: Moctezuma himself, although the attempt to extend the sway of the Mexica into Yucatan was aborted by their own defeat at the hands of the Spanish; the "Itza" captains, who became local rulers after invading the peninsula; and the Peten Itza King Canek, who himself claimed descent from the earlier "Itza" rulers. (Unlike the most recent, Spanish, intruders, the "Itza" foreigners had acquired a legitimacy hallowed by pre-Hispanic tradition, despite a certain ambivalence on the part of old-guard Maya.)[40] They form part of a pattern that conforms to Maya logic, reinforced rather than vitiated by their separation in time. That this was a collective logic is evident from the statement of Canek's supporters after their defeat and capture. Asked why they had followed him or supposed him to be a legitimate king, they replied that they had done so "because I have heard tell that a king called Montezuma was to appear," and "because it was said that [the king] was to come from the East and [Canek] came from the East."

Canek's claims corresponded with a historical-mythical archetype contained in the Chilam Balam books and recognized, in gross outline at least, throughout Maya society. They refer not to a Second Coming in the millenarian sense but to a third, fourth, or fifth coming—more accurately, to one in a theoretically infinite series of recreations of the same event.

These historical prophecies were transmitted both orally and through the written record of the Books of Chilam Balam. The books themselves would have been inaccessible to all but a tiny literate minority. But the content was widely disseminated, even if greatly compressed and stripped of much of the rich, multilayered symbolism surrounding the main events. The "Chilam Balam prophecies" were specifically mentioned in the trial proceedings following the Canek Revolt, striking terror in the Spaniards then and for many years to come. Their sinister meaning gained added strength from the nineteenth-century Caste War of Yucatan, a much more widespread revolt with the same intention of eliminating the Spanish and restoring the status quo ante the conquest. In an essay on the origins of the Caste War written shortly after its outbreak, a local historian placed partial blame on these "disastrous prophecies," which the Maya throughout the peninsula were so fond of retelling (Sierra O'Reilly 1954:1, 87).

Referring to the Books of Chilam Balam, the same historian lamented that the "vulgar fables and ridiculous counsels" they contained were all that remained of preconquest history in the Indians' memory; the original record had unfortunately suffered "mutilations . . . according to the peculiar criterion of those entrusted with transmitting [the writings] to posterity" (Sierra O'Reilly 1954:1, 32, 35). He failed to recognize in this peculiar criterion the enduring master code by which the Maya continued to remember their future and anticipate their past.

In all societies the past is looked to as a guide to present or future action. Among those with a chain-of-events conception of time, the past will be seen as a signpost pointing in the direction that things are headed. The direction may be irreversible, but it need not be straight. Implicit in the conception is some idea that, if the line is clear, the direction understood, then it can be altered. Even with a subordinate cyclical view to the effect that history repeats itself, we think that by understanding the past we can modify the future. We can avoid another Munich, another Vietnam (or, conversely, multiply them, if that is our goal). In the cyclical conception of the Maya, the pattern cannot be reversed or changed. Or if it could, it should not be, for that would be to disrupt order and thus court disaster. The Maya did not therefore assume the attitude of helpless spectators. They had a crucial role to play in the cyclical drama, through their calendrical computations, their related rituals, and even their wars and revolts. Theirs was the task of helping the gods to carry the burden of the days, the years, and the *katuns* and thereby to keep time and the cosmos in orderly motion.

NOTES

The author wishes to thank Arjun Appadurai, Grant Jones, Alfredo López Austin, and Arthur Miller for their comments and suggestions.

1. See, for example, Arjun Appadurai (1981), J. D. Y. Peel (1984), Renato Rosaldo (1980), and Marshall Sahlins (1981, 1985).

2. But see Edmund Leach (1961) on the "time process" as an "oscillation" or a "zig-zag."

3. This view is advanced most prominently by Jack Goody (1968, 1977) and by Goody and Ian Watt (1968) but builds on earlier studies that stressed the effects of literacy on conceptualization, especially a consciousness of history. See Walter Ong (1982) for a partial summary of the literature on this theme.

4. On cyclical thought in China and India, see Kathleen Gough (1968:75–76), and, for a comparative study drawing on many different cultures, Mircea Eliade (1954:51–92). Samuel Brandon (1965:67–70, 77–79, 81–83) argues that Eliade neglects evidence of linear, historical thought among the Egyptians and Mesopotamians.

5. Many elements in western thought, along with the alphabet, can be traced to Ancient Greece, where a cyclical conception of time was dominant (Brandon 1965:85–96); both the alphabet and linear thought coincide in medieval Europe but long preceded the shift from a largely oral to a written culture (Clanchy 1979).

6. Ong (1982:92–93) gives the example of Korea, but the shift there from Chinese characters to an alphabetic script is still not complete.

7. I refer here to Yucatec Maya, in which a large corpus of colonial documents has been preserved, some of them transcribed from post-Classic glyphic texts (now lost) and probably closely related to the language or languages represented in the hieroglyphic inscriptions of the Classic period (Thompson 1960:16; Morley and Sharer 1983:504).

8. See Nancy Farriss (1984:401–2) for a discussion of colonial Yucatec sources.

9. Basic works on the complex subject of Maya calendrics are Sylvanus Morley (1915), J. E. S. Thompson (1960), and Linton Satterthwaite (1965). For a recent and

admirably lucid summary, see Sylvanus Morley and Robert Sharer (1983:548–63); and on the "afanes cronológicos" of the Maya, see Miguel León-Portilla (1968:17–28).

10. There are three known versions of the Chilam Balam that contain *katun* counts. Ralph Roys (1967) and Munro Edmonson (1982) have published the Chumayel and Tizimin texts (respectively), with English translations and extensive commentaries; Eugene Craine and Reginald Reindorp (1979) provide an English version of the Mani, translated from an earlier Spanish translation, and without the Maya text; Alfredo Barrera Vázquez and Silvia Rendón (1948) have made a composite text, in Spanish translation, of the three books plus a fourth that lacks a *katun* count. The Chumayel text has been published in facsimile (*Book of Chilam Balam* 1913).

11. The *katun* count is based on the 360-day *tun* rather than the 365-day solar year (*haab*) so that the cycle of 260 *tun* years is actually ca. 256 solar years.

12. Barrera Vázquez (1957), Daniel Brinton (1969), and Juan Martínez Hernández (1926) publish original Maya texts with translations. France Scholes and Ralph Roys (1968:367–82) publish a similar chronicle, with translation, in Chontal Maya. For the same kind of chronicle, but written in Spanish, see "Relaciones de Yucatan": Pt. II, 42–45; Archivo General de Indias (hereinafter referred to as AGI), Mexico 104, Probanza de Gaspar Antonio Chi, 1580 (Chi was also the author of the published *relación*); AGI, Guatemala 111, Probanza de don Francisco, cacique de Xicalanco, 1552; and AGI, Mexico 140, Probanza de don Juan Chan, 1601. See "Títulos de Tabi," in the Latin American Library, Tulane University, for a family chronicle that spans the late post-Classic and early colonial periods.

13. Howe's Balinese evidence supports this proposition; even what appear to be long-term linear progressions are subsumed under one-hundred-year cycles (1981:232, 233).

14. Marc Bloch (1961:84–85). On millenarian ideologies, see Norman Cohn (1970:29–36 et passim); and, on nonrevolutionary millennial ideas among Spanish missionaries in Mexico, see John Phelan (1970). On the persisting strength of astrological beliefs in seventeenth-century England, see Keith Thomas (1971:283–385).

15. Eliade (1959:88), quoting from M. Mauss and H. Hubert, "La représentation du temps dans la religion et la magie," in *Mélanges d'histoire des religions* (1960).

16. On famines in the Chumayel version of the Chilam Balam, see Roys (1967:122, 133, 134). On upstarts and invaders, see ibid.: 83, 93, 102, 106, 112, 149, 153, 156–60.

17. For the Maya sense of a "collective purchase of survival," see Farriss (1984:320–43); and, for a similar Aztec view, see Alfredo López Austin (1980:1, 72–74).

18. Compare, for example, the early colonial "Crónica de Calkini" (Barrera Vázquez 1957) with, among many similar procedures for substantiating land titles and other community rights, those recorded in AGI, Mexico 3066, Informaciones instruídas sobre cofradías, 1872, cuadernos 5–10.

19. Edmonson (1982) often identifies these figures as priests and rulers rather than deities, stressing the historical content of the texts, as epigraphers have also done with the inscriptions of the Classic period (see esp. pp. 4, 16, 22, 25–26, 40). But the identities in the Maya text are often ambiguous, and, I suspect, deliberately so.

20. Edmonson (1982) sees rotation of seats of power as the dominant theme in both the post-Classic and the colonial material of the Tizimin Chilam Balam (see, e.g., pp. 6, 20–21, 38–39, 42–43). The system, as originally outlined by Michael Coe (1965), referred to intracommunity divisional rotation. For evidence of its persistence into the colonial period and indications of rulership (*batab*) rotation, see Philip Thompson (1978:267–82).

21. Satterthwaite (1965:615). The cycle, composed of thirteen *baktuns* (one *baktun* equals 400 *tun* years of 360 days each), is closer to 5,120 calendar years of 365 days each.

22. The antiquity of the *katun* round (also called Short Count) is not known, but it may have preceded and even served as a model for the Long Count sequence, with its multiple of thirteen *baktuns* (Satterthwaite 1965:626–77). And the *katun* round itself may be a stretched-out version of the Sacred Round, with the same numerical combination of thirteen times twenty, translated from days to *tuns*.

23. The list of recent studies, starting with Tatiana Proskouriakoff's pioneering work (1960, 1964), is extensive. For a sample, see Christopher Jones (1977), David Kelley (1976), Joyce Marcus (1976), and Linda Schele (1976).

24. Miller (1983a, 1983b). Miller (1986) expands and elaborates this thesis with particular reference to Tikal. For further evidence of the rise of dynastic rule, see the works cited in note 2 and, for Tikal in particular, Clemency Coggins (1975), and Jones and Satterthwaite (1982).

25. These struggles form the major theme of the chronicles cited in note 12 and of the historical sections of the "Relaciones de Yucatan" and also figure prominently in the Chilam Balam books.

26. John Glass (1975:15–16). A useful brief discussion of these texts is David Carrasco (1982:13–19, 41–50).

27. Some of these references have been misinterpreted in edited versions. The "ropes," "nooses," and "hanging" in the Chumayel (Roys 1967:138, 152, 153) and Tizimin (Edmonson 1982:60, 67, 162) refer not to punishment (in the 1562 idolatry trials directed by Fray Diego de Landa) but to the system of judicial torture used by the friars (Farriss 1984:503). The "giving of heads to the archbishop" (Roys 1967:123; Edmonson 1982:189) refers not to submission or execution but to the sacrament of confirmation; and the *bula* (Edmonson 1982:188) refers to the Bull of the Holy Crusade, a type of tax (Farriss 1984:40).

28. Pedro Sánchez de Aguilar (1937:122–23, 181). For colonial Maya incorporations of European teachings on eclipses and rotation of the earth, see Roys (1967:87–88), and Edmonson (1982:73).

29. The encounter is outlined in Carrasco (1982:191–204). In the myth as model, Carrasco emphasizes Quetzalcoatl's downfall and sees Moctezuma as fearing the same fate (pp. 174–78), while López Austin (1973:135–39, 172–85) and Serge Gruzinski (1985:20–23) suggest that what Moctezuma feared was Quetzalcoatl's *return*, regarding himself as a usurper. Sahlins (1981:17–23; 1985:104–35) recounts a similar response by an eighteenth-century Hawaiian chief to the arrival of Captain James Cook, but with a crucial twist: Cook, seen as the lost god-chief who returned to reclaim the land, was sacrificed at the end of the welcoming ritual instead of conquering the kingdom.

30. See Robert Chamberlain (1948) for a history of the conquest, and Farriss (1984:20–25) on Maya interpretations of the conquest.

31. AGI, Patronato 237-1-3, Governor of Yucatan to King Canek, 8 December 1695.

32. Fray Diego de Avendaño, Relación de las dos entradas que hice en las conversión de los gentiles Itzaes, 6 April 1696 (manuscript in Newberry Library, Chicago), fols. 34–36ᵛ.

33. Ibid. fol. 36.

34. See Diego López de Cogolludo (1688:Lib. 12, cap. 14); and AGI, Mexico 3048, Diligencias que se hicieron sobre la junta y pláticas de algunos indios, 1607.

35. AGI, Mexico 362, Governor of Yucatan to Fray Payo de Rivera, 10 June 1678. Grant Jones provided this citation and alerted me in general to the prophetic element in the border conflicts. For other references, see especially AGI, Mexico 307, Sobre las diligencias . . . para la reducción de los indios de Sahcabchen y otros pueblos, 1670; and AGI, Escribanía de Cámara 317B, Cuaderno 8, Autos fechos por Pedro García Ricalde, 1668. On the southern frontier and refugees, see Scholes and Roys (1968:305–15), G. Jones (1983), and Farriss (1984:72–76).

36. See Farriss (1984:68–72) for a discussion of the Canek revolt. Documents on the revolt and the ensuing trial, including the testimony of Canek and his followers quoted below, are in AGI, Mexico 3050, Testimonio de autos sobre la sublevación que hicieron varios pueblos . . . 1761, and Autos criminales seguidos de oficio de la Real Justicia . . . 1761.

37. Montezuma, rather than Moctezuma, is the form used in the transcript of the testimony of Canek and his followers.

38. Among numerous examples that could be cited, see James Mooney (1965) on the North American Ghost Dance rebellion.

39. See, for example, AGI, Mexico 3048, Diligencias que se hicieron sobre la junta y pláticas de algunos indios, 1607.

40. See Farriss (1984:22–23, 244–47) on these distinctions. Edmonson (1982) sees the Itza as old guard compared to the Xiu, and they may have been, but there are also references to the Itza themselves as newcomers and foreigners (Roys 1967:83–84, 93, 106, 153, 169; Craine and Reindorp 1979:98–116).

REFERENCES

Appadurai, Arjun. 1981. "The Past as a Scarce Resource." *Man,* 16:2, 201–19.
Ball, Joseph W. 1974. "A Coordinate Approach to Northern Maya Prehistory: A.D. 700–1200." *American Antiquity,* 39:1, 85–93.
Barrera Vázquez, Alfredo. 1975. *Códice de Calkini.* Campeche: Talleres Gráficos del Estado.
Barrera Vázquez, Alfredo, and Rendón, Silvia. 1948. *El libro de los libros de Chilam Balam.* Mexico City: Fondo de Cultura Económica.
Berger, Peter L. 1969. *The Sacred Canopy: Elements of a Sociological Theory of Religion.* Garden City, N.Y.: Doubleday, Anchor Press.
Bloch, Marc. 1961. *Feudal Society.* Chicago: University of Chicago Press.
Book of Chilam Balam of Chumayel. 1913. Facsimile ed. Anthropological Publications, vol. 5. Philadelphia: University Museum, University of Pennsylvania.
Brandon, Samuel G. F. 1965. *History, Time, and Deity.* Manchester: Manchester University Press.
Brinton, Daniel G. 1969. "The Chronicle of Chac-Xulub-Chen." In *The Maya Chronicles,* Daniel G. Brinton, ed. (reprint of 1882 ed.), 189–259. New York: AMS Press.
Brotherston, Gordon. 1979. "Continuity in Maya Writing: New Readings of Two Passages in the Book of Chilam Balam of Chumayel." In *Maya Archaeology and Ethnohistory,* Norman Hammond and Gordon R. Willey, eds., 241–58. Austin: University of Texas Press.
Carrasco, David. 1982. *Quetzalcoatl and the Irony of Empire: Myths and Prophecies in the Aztec Tradition.* Chicago: University of Chicago Press.
Chamberlain, Robert S. 1948. *The Conquest and Colonization of Yucatan, 1515–1550.* Washington, D.C.: Carnegie Institution.
Clanchy, M. T. 1979. *From Memory to Written Record: England, 1066–1307.* London: Edward Arnold.
Coe, Michael D. 1965. "A Model of Ancient Community Structure in the Maya Lowlands." *Southwestern Journal of Anthropology,* 21:2, 97–114.
Coggins, Clemency C. 1975. "Painting and Drawing Styles at Tikal: An Historical and Iconographic Reconstruction." Ph.D. diss., Harvard University.
Cogolludo, Diego López de. 1688. *Historia de Yucatan.* Madrid: J. García Infanzón.
Cohn, Norman. 1970. *The Pursuit of the Millennium,* 2d ed. New York: Oxford University Press.
Craine, Eugene R., and Reindorp, Reginald C. 1979. *The Codex Perez and the Book of Chilam Balam of Mani.* Norman: University of Oklahoma Press.
Edmonson, Munro, ed. 1982. *The Ancient Future of the Itza: The Book of Chilam Balam of Tizimin.* Austin: University of Texas Press.
Eliade, Mircea. 1954. *The Myth of the Eternal Return, or Cosmos and History.* Princeton: Princeton University Press.

---. 1959. *The Sacred and the Profane: The Nature of Religion.* New York: Harcourt, Brace and World.
Farriss, Nancy M. 1982. "Sacrifice and Communion in Early Colonial Yucatan." Paper presented at the 44th International Congress of Americanists, Manchester, England, 5-11 September.
---. 1984. *Maya Society under Colonial Rule: The Collective Enterprise of Survival.* Princeton: Princeton University Press.
Glass, John B. 1975. "A Survey of Native Middle American Pictorial Manuscripts." In *Handbook of Middle American Indians,* Robert Wauchope, gen. ed., vol. 14, pt. 3, 3-80. Austin: University of Texas Press.
Goody, Jack. 1968. Introduction to *Literacy in Traditional Societies,* J. Goody, ed., 1-26. Cambridge: Cambridge University Press.
---. 1977. *Domestication of the Savage Mind.* Cambridge: Cambridge University Press.
Goody, Jack, and Watt, Ian. 1968. "The Consequences of Literacy." In *Literacy in Traditional Societies,* J. Goody, ed., 27-68. Cambridge: Cambridge University Press.
Gough, Kathleen. 1968. "The Implications of Literacy in Traditional China and India." In *Literacy in Traditional Societies,* J. Goody, ed., 69-84. Cambridge: Cambridge University Press.
Gruzinski, Serge. 1985. *Les hommes-dieux du Mexique: Pouvoir indien et société coloniale, XVIe–XVIIIe siècles.* Paris: Editions des Archives Contemporaines.
Howe, Leopold E. A. 1981. "The Social Determination of Knowledge: Maurice Bloch and Balinese Time." *Man,* 16:2, 220-34.
Jones, Christopher. 1977. "Inauguration Dates of Three Late Classic Rulers of Tikal, Guatemala." *American Antiquity,* 42:1, 28-60.
Jones, Christopher, and Satterthwaite, Linton. 1982. *The Monuments and Inscriptions: Tikal Reports, No. 33.* Philadelphia: University Museum, University of Pennsylvania.
Jones, Grant D. 1982. "Agriculture and Trade in the Colonial Period Southern Maya Lowlands." In *Maya Subsistence: Studies in Memory of Dennis E. Puleston,* Kent V. Flannery, ed., 275-93. New York: Academic Press.
---. 1983. "The Last Maya Frontiers of Colonial Yucatan." In *Spaniards and Indians in Southeastern Mesoamerica,* Murdo J. MacLeod and Robert Wasserstrom, eds., 64-91. Lincoln: University of Nebraska Press.
Kelley, David H. 1976. *Deciphering the Maya Script.* Austin: University of Texas Press.
Kopytoff, Igor. 1986. "The Internal African Frontier: The Making of African Political Culture." In *The African Frontier: The Reproduction of Traditional African Societies,* Igor Kopytoff, ed. Bloomington: Indiana University Press.
Landa, Diego de. 1941. *Landa's Relación de las cosas de Yucatan,* Alfred M. Tozzer,

ed. and trans. Cambridge: Peabody Museum of American Archaeology and Ethnology, Harvard University.

Leach, Edmund R. 1954. "Primitive Time Reckoning." In *A History of Technology,* Charles Singer, E. J. Holmyard, and A. R. Hall, eds. 1:110–27. Oxford: Clarendon Press.

———. 1961. "Two Essays Concerning the Symbolic Representation of Time." In *Rethinking Anthropology,* 124–36. London: Athlone Press.

León-Portilla, Miguel. 1968. *Tiempo y realidad en el pensamiento maya.* Mexico City: Universidad Nacional Autónoma de Mexico.

López Austin, Alfredo. 1973. *Hombre-dios: Religión y política en el mundo náhuatl.* Mexico City: Universidad Nacional Autónoma de Mexico.

———. 1980. *Cuerpo humano e ideología: Las concepciones de los antiguos nahuas.* 2 vols. Mexico City: Universidad Nacional Autónoma de Mexico.

Marcus, Joyce. 1976. *Emblem and State in the Classic Maya Lowlands: An Epigraphic Approach to Territorial Organization.* Washington, D.C.: Dumbarton Oaks.

Martínez Hernández, Juan, ed. 1926. *Crónica de Yaxkukul.* Merida: Talleres de la Compañía Tipográfica Yucateca.

Miller, Arthur G. 1977. "Captains of the Itzá: Unpublished Mural Evidence from Chichén Itzá." In *Social Process in Maya Prehistory,* Norman Hammond, ed., 197–225. London: Academic Press.

———. 1979. "The Little Descent: Manifest Destiny from the East." In *Actes du XLIIe Congrès des Américanistes. Paris, 1976,* 8:221–36. Paris: Société des Américanistes.

———. 1983a. "The Communication of Power among the Lowland Maya: Images and Texts." Paper presented at the annual meeting of the American Anthropological Association, Chicago, 16–20 November.

———. 1983b. "Image and Text in Pre-Hispanic Art: Apples and Oranges." In *Text and Image in Pre-Columbian Art: Essays on the Interpretation of the Visual and Verbal Arts,* Janet C. Berlo, ed., 4–54. Oxford: British Archaeological Reports.

———. 1986. *Maya Rulers of Time: A Study of Architectural Sculpture at Tikal, Guatemala.* Philadelphia: University Museum, University of Pennsylvania.

Mooney, James. 1965. *The Ghost Dance Religion,* 2d ed., Anthony F. C. Wallace, ed. Chicago: University of Chicago Press.

Morley, Sylvanus G. 1915. *An Introduction to the Study of Maya Hieroglyphics.* Washington, D.C.: Bureau of American Ethnography, Smithsonian Institution.

Morley, Sylvanus, and Sharer, Robert W. 1983. *The Ancient Maya.* Stanford: Stanford University Press.

Ong, Walter. 1982. *Orality and Literacy: The Technologizing of the Word.* London: Methuen.

Pagden, Anthony. 1982. *The Fall of Natural Man: The American Indian and the Origins of Comparative Ethnology.* Cambridge: Cambridge University Press.

Peel, J. D. Y. 1984. "Making History: The Past in the Ijesha Present." *Man*, 19:1, 111–33.
Phelan, John L. 1970. *The Millennial Kingdom of the Franciscans in the New World*, 2d ed. Berkeley: University of California Press.
Proskouriakoff, Tatiana. 1960. "Historical Implications of a Pattern of Dates at Piedras Negras, Guatemala." *American Antiquity*, 25:4, 454–75.
———. 1963–64. "Historical Data in the Inscriptions of Yaxchilan," pts. 1 and 2. *Estudios de Cultura Maya*, 3:149–67; 4:177–201.
Puleston, Dennis E. 1979. "An Epistemological Pathology and the Collapse, or Why the Maya Kept the Short Count." In *Maya Archaeology and Ethnohistory*, Norman Hammond and Gordon R. Willey, eds., 63–71. Austin: University of Texas Press.
"Relaciones de Yucatan" (pts. 1 and 2). 1898–1900. Vols. 11 and 13 (n.s.) of *Colección de documentos inéditos relativos al descubrimiento, conquista y organización de las antiguas posesiones de Ultramar*. Madrid: Real Academia de la Historia.
Ricoeur, Paul. 1982. *Finitud y culpabilidad* (trans. of *Philosophie de la volonté: Finitude et culpabilité*). Madrid: Taurus.
Rosaldo, Renato. 1980. *Ilongot Headhunting, 1883–1974: A Study in Society and History*. Stanford: Stanford University Press.
Roys, Ralph. 1960. "The Maya Katun Prophecies of the Books of Chilam Balam, Series I." *Contributions to American Anthropology and History*, 12:1–60. Washington, D.C.: Carnegie Institution.
Roys, Ralph, ed. *The Book of Chilam Balam of Chumayel*. 2d ed. Norman: University of Oklahoma Press.
Sahlins, Marshall. 1981. *Historical Metaphors and Mythical Realities: Structure in the Early History of the Sandwich Islands Kingdom*. Ann Arbor: University of Michigan Press and Association for Social Anthropology in Oceania.
———. 1985. *Islands of History*. Chicago: University of Chicago Press.
Salomon, Frank. 1982. "Chronicles of the Impossible: Notes on Three Peruvian Indigenous Historians." In *From Oral to Written Expression: Native Andean Chronicles of the Colonial Period*, Rolena Adorno, ed., 9–39. Syracuse: Maxwell School of Citizenship and Public Affairs, Syracuse University.
Sánchez de Aguilar, Pedro. 1937 [ca. 1613]. *Informe contra idolorum cultores del obispado de Yucatan*. 3d ed. Merida: Imprenta Triay e Hijos.
Satterthwaite, Linton. 1965. "Calendrics of the Maya Lowlands." In *Handbook of Middle American Indians*, Robert Wauchope, gen. ed., vol. 3, pt. 2, 603–31. Austin: University of Texas Press.
Schele, Linda. 1976. "Accession Iconography of Chan-Bahlum in the Group of the Cross at Palenque." In *The Art, Iconography, and Dynastic History of Palenque*, Merle G. Robertson, ed., 9–34. Pebble Beach, Calif.: Robert Louis Stevenson School.

Scholes, France V., and Roys, Ralph L. 1968. *The Maya Chontal Indians of Acalan-Tixchel: A Contribution to the History and Ethnography of the Yucatan Peninsula.* 2d ed. Norman: University of Oklahoma Press.

Sierra O'Reilly, Justo. 1954–57 [1848–51]. *Los indios de Yucatan,* Carlos R. Menéndez, ed. 2 vols. Merida: Compañía Tipográfica Yucateca.

Stirrat, R. L. 1984. "Sacred Models." *Man,* 19:2, 199–215.

Thomas, Keith. 1971. *Religion and the Decline of Magic.* New York: Scribners.

Thompson, J. Eric S. 1960. *Maya Hieroglyphic Writing: An Introduction.* 2d ed. Norman: University of Oklahoma Press.

Thompson, Philip C. 1978. "Tekanto in the Eighteenth Century." Ph.D. diss., Tulane University.

Chronology and Its Discontents in Renaissance Europe: The Vicissitudes of a Tradition

Anthony T. Grafton

A Machine That Works Like a Clock?

The night sky of strange disciplines at which the scholars of the late sixteenth century gazed was dominated by a bright new star: chronology, the study of calendars and dates. So historians of scholarship and science have said for the last two centuries and more. For the most part, they have insisted on the densely technical and sharply modern quality of this undoubtedly fashionable discipline. And their point of view seems entirely reasonable. Consciousness of time has been a central element in Europe's experience of modernity. The French Revolution and the Napoleonic Wars revealed to intellectuals time's power to change the textures as well as the colors of the social fabric. The new forms of transportation created by the Industrial Revolution required the regulation of clocks by a uniform standard, still remembered as "railway time." The speed of journeys on trains, steamships, and balloons—and their snapping of the old connection between spatial and temporal distance—fascinated nineteenth-century novelists. Time lived, time remembered, time lost obsessed Europe's modernists in the years around 1900, as time itself seemed to accelerate.[1] No wonder, then, that modern historians have seen Renaissance intellectuals' efforts to master time as proof that they were, in Jacob Burckhardt's phrase, the firstborn sons of modern Europe.[2] The chronologer, they have held, inhabited an intellectual landscape radically different from the timeless one of the medieval thinker (or peasant).[3]

No single piece of evidence evokes this view more strikingly than the magnificent structure that the Strasbourg scholar and scientist Conrad Dasypodius reared inside the local cathedral in the early 1570s. This gorgeous mechanism still, in updated form, delights the tourists who line up every day to hear (and watch) it strike noon. Gorgeous and complex as a Mannerist grotto, it was far more than a clock. It showed not only the time of day but "eternity, the century, the periods of the planets, the yearly and monthly revolutions of the sun and moon." Its devices included models of the planets and mechanical automata. And its iconography depicted, Dasypodius claimed, "everything from history and poetry, sacred texts and profane ones, in which there is or can be a description of time."[4]

Dasypodius's account was not exaggerated. The face of what Thomas Coryat called "the famous Clock of Strasbourg" was as crowded and busy as a sixteenth-century street market.[5] An astronomical mechanism showed the motions of the planets around the earth. Another display predicted eclipses, while a perpetual calendar laid out the moveable feasts, leap years, and the twenty-eight-year solar cycle of the ecclesiastical calendar for the next century. Automata represented the pagan gods for whom the days were named, Time, and Death. They both served as an animated almanac and embodied the all-destroying force of time and change. The clock as a whole, finally, presented itself as a piece of history—modern history. A portrait of the great innovator Copernicus on the smaller tower forcefully suggested that time brought to light new truths and new inventions. Any educated visitor would take the point. For the clock, like gunpowder and the compass, was one of the first distinctively modern technologies, and late medieval and Renaissance intellectuals loved to cite it when arguing that the ancients had not exhausted all fields of knowledge and invention.

The Strasbourg clock, in turn, was only a particularly glamorous case in point of a phenomenon that cropped up from Italy to Scandinavia. From the later Middle Ages on, city squares and cathedrals across Europe sprouted impressive escapement clocks, many of them equipped with elegant mechanical models representing "the approaches and retreats of the planets, the sun and the moon."[6] Thousands of smaller clocks swarming with similar details decorated mantelpieces across Europe.[7] And the fascination with time that they reveal permeated European culture; for the guilty, secret obsession of early modern society was neither sex nor money but the desperate desire to use time well and the pervasive fear that wasted time would waste those who abused it.

This was only natural. Mercantile society both assigned time a new value and gave intellectuals a new language for measuring and assessing it. The successful merchant must count his minutes as obsessively as his money. Otherwise, as Uncle Giannozzo emphasized in Leon Battista Alberti's *Della famiglia,* easy tasks would become difficult and disaster inevitable. But the moral man must weigh his waking hours with the same obsessive care as *homo oeconomicus,* according to another Giannozzo (Manetti). Otherwise he would never be able to render a satisfactory accounting to the celestial shopkeeper who scrutinizes all men's books.[8] The mercantile memorandum book of fourteenth- and fifteenth-century Florence, with its neatly dated births, deaths, partnerships, and bankruptcies, traces the contours of a secular life lived, for the first time, by the clock.

Yet Renaissance chronology—even in the crystalline, lucid form that Dasypodius gave it—was not wholly of a piece or wholly modern. The Strasbourg clock consisted not only of modern, mechanical devices but also of traditional emblems and paintings. One set of these divided time into the great scenes of the Christian drama: Creation, Original Sin, Redemption, Resurrection, Last Judgment. Another cut secular time into the four empires of traditional apocalyptic prophecy: Assyria, Persia, Greece, and Rome. On the top of the smaller tower, a statue of a pelican symbolized Christ the Redeemer and eternity. Dasypodius's work, in short, was not simply a display of the powers of modern science and technology; it was also a miniature cathedral of its own, in which the pious visitor could read the most traditional lessons about the past, the present, and the (rather abbreviated) future of the human race. Nothing mattered more about time, even here, than its end. And no evidence suggests that these images formed a less organic part of Dasypodius's design than the clockwork they concealed and interpreted. After all, the clock—as Gerhard Dohrn-Van Rossum has shown in a masterly book—governed the silences and liturgies of the monastic day as well as the conversations and transactions that made up the merchant's daily round.[9]

Images of time proliferated throughout the period, often contradicting one another. The ease with which time could be lost inspired lyric and tragic poets. But the regular action, complexity, and reciprocal interplay of parts characteristic of clocks provided tragedians and philosophers with a new set of mechanical metaphors for describing both the larger body politic and the individual human body.[10] The irregular and inescapable damage that time inevitably inflicts inspired artists and poets; but so did the truths that time inevitably reveals. Illustrated manuscripts and printed editions of Petrarch's *Trionfi* and commented editions of Ovid's *Metamorphoses,* emblem books and lyric poems, collections of classical inscriptions, and suites of modern prints mourned the effects of *tempus edax rerum* and celebrated those of *veritas filia temporis.* Texts and images show not only the inescapable fascination of time but also the range and richness of the palette with which early modern individuals tried to depict it.[11]

Time posed technical as well as artistic problems, which a wide range of intellectuals confronted and tried to resolve. But these, too, were complex and stratified. The ecclesiastical calendar was in disarray, since its year of 365¼ days was slightly too long, and in its more than a millennium of use the cumulative error had become serious. Throughout the later Middle Ages and the sixteenth century, calendar reformers quarreled and agonized while, as they complained, Jews mocked the Christians for their inability to celebrate Easter on the correct day.[12] After the Reformation, inhabitants of Protestant countries had to find meaning in a year whose sacred landmarks had been torn away en bloc. Old rhythms of work and leisure, formal church and civic rituals, and traditional agrarian celebrations were transformed or challenged. New holidays, drawn from recent history, gradually marked the passage of time with new high points.[13] Even in Catholic countries traditional festivities like the Ambrosian carnival in Milan were transformed to meet new standards of rigor and austerity. From the 1570s on, the preparations for the Gregorian Calendar Reform, the debates it provoked, and the changed practices it brought about divided both scholars and nations.[14] Chronology, in short, both murdered the time and preserved it, both demolished old rituals and created new ones, both challenged and restored the ordinary Christian's sense of rootedness in time and place.

No discipline was more central to the study of time, and none was more prominently displayed on Dasypodius's clock, than the meticulous study of time past. Technical chronology concerns itself with reconstructing the calendars and dating the main events of ancient and medieval history. From the Renaissance to the Enlightenment, it won the interest of many of the most innovative European thinkers. Europeans plastered their walls and filled their purses with calendars and charts of world history. They produced and consumed majestic books, tiny monographs, and even *"machines chronologiques"* that laid out the structure and the details of historic time.[15] They argued heatedly over the nature of lost calendars, the dates of great events, and the reliability of the available sources. By the early seventeenth century they had identified what remain the central sources and data for the study of technical chronology. They had constructed the armature of precise Julian dates to which the events and movements of ancient and medieval history are still affixed. And they had studied, sometimes fragmentarily but sometimes presciently, the rich body of religious beliefs and practices that ancient calendars had rested on and regulated. No Renaissance science, perhaps, achieved more dramatic or more lasting results.

By the middle of the sixteenth century, a stately series of great quartos and folios synthesized traditional methods and argued for new results in chronology. Intellectuals of very different kinds regarded it as a particularly fascinating and important subject—indeed, as part of the core of civilization. When Ogier Busbecq wished to describe the Turks as barbarous, he pointed out that they "have no chronology and conflate all histories in a bizarre way. If it occurs to them, they will not hesitate to claim that Job was King Solomon's *magister curiae,* that Alexander the Great was his general, or things even sillier than these."[16] When Montaigne read Francisco López de Gomara's account of another nonwestern culture, that of the Aztecs, he came to the opposite conclusion. Their complex system of chronology, which partly matched the western theory of the great conjunctions, indicated that "The people of the kingdom of Mexico were somewhat more civilized and skilled in the arts than the other nations over there."[17] Competence in chronology revealed civility.

This conviction was widely shared by men who in other respects had little in common with Montaigne—or Scaliger. When the Jesuit Juan de Tovar needed to reassure his confrère Joseph de Acosta about the reliability of Aztec records, for example, he pointed out that they had regulated their calendar by a fifty-two-year cycle as accurate as the Julian ecclesiastical calendar. He had read Diego Durán's careful account of the ancient calendar, which showed that

> the symbols representing each day of the month functioned as letters. In general, these painted characters were used as picture writing, describing native history and lore, memorable events in war, victories, famines and plagues, prosperous and adverse times. All was written down, painted in books and on long papers, indicating the year, month and date on which each event had occurred. Also recorded in these painted documents were the laws and ordinances, the census, and so forth. All this was set down painstakingly and carefully by the most competent historians, who by means of these paintings recorded extensive chronicles regarding the men of the past.[18]

The Aztecs, Tovar told Acosta, had recorded great events by depicting them at the appropriate places on the *ruedas* that represented each fifty-two-year period. The arrival of the Spaniards, for example, they had denoted by a properly situated figure of "un español con un sombrero y sayo colorado." Their histories, accordingly, were not the fantasies of barbarians but the well-kept records of an organized society.[19]

Though Acosta included Tovar's information in his *Natural and Moral History of the Indies* (6.2), he continued to doubt that one could extract a long-term chronology in the European style from Aztec records; and, given the vast distance that separated Native American from European preoccupations about the ordering of time, he may well have been right.[20] But Durán and many other observers found in the precision and elegance of Mesoamerican chronicles one unequivocal area of high technical achievement, which deserved only admiration and should never have been suppressed: "These writings would have enlightened us considerably, had not ignorant zeal destroyed them. Ignorant men ordered them burned, believing them idols, while actually they were history books worthy of being preserved instead of being buried in oblivion, as was to occur."[21] Aztec religion might be the work of the devil, but Aztec history deserved the attentive study of Christians. Its firm chronological backbone proved its value.

The prestige that chronology enjoyed in Renaissance Europe may surprise the inhabitants of a world in which the whole field has become a lost continent of erudition. But in Renaissance Europe respectful interest was the attitude naturally inspired not only by the frightening figure of Old Father Time but by the dizzying columns of numbers and staggering assemblies of rebarbative texts that one encountered on passing the theatrical title-page and eloquent front matter of any normal treatise on chronology. Educated people knew the night sky of the human past, as they did the night sky of nature, far more intimately than we do. Every event had a determined place in a history that began with Creation and proceeded in an orderly way, through the Flood and the Exile, the fall of Troy and the rise of the European nations, to modern times. When Diogo Câo raised the Cape Cross monument (in what is now Namibia) in 1485, he had no trouble locating his voyage in universal time, "in the year 6685 of the creation of the earth and 1485 after the birth of Christ."[22] A century later, after Portuguese voyages had laid open the East and Spanish ones the West, after Reformation and religious war, Jean Bodin still believed that time had had a clear beginning. He began the education of his sons—aged three and four—in 1582 by teaching them to talk in Latin about angels, darkness, light, "the world and how old it was, to wit 5534 years."[23] Bodin used a world era drawn from the Hebrew Bible, which set Creation 3,948 years before the birth of Christ, whereas Câo had used one drawn from the Greek Bible, or the *Chronicle* of Eusebius, which made the interval an even 5,200 years. But both men clearly saw themselves as inhabiting the same comfortable, clearly defined, long-term history. Both knew where they lived, at points clearly marked on the crisply calibrated scale of biblical and historical time.

Even more than the eras of world history, the technical components of chronology formed part of the fabric of common life. The study of calendars, their relations to one another, and their always inexact coordination with the movements of the sun and moon belonged to the art of the computus, which the vast majority of intellectuals and many others knew intimately until the eighteenth century.[24] After the spring equinox many Europeans turned naturally to reckoning by the moon until Easter had come and gone, and then turned as naturally back to their normal, solar calendar for the rest of the year.[25] Men whose rituals and even travel arrangements were still governed by the moon readily took an interest in her ancient complaint, recorded by Aristophanes, that the Athenians had failed to observe her phases and starved the gods by celebrating their festivals on the wrong days.[26]

Historical chronology, moreover, was more than an austere and technical discipline. Like an anatomist, the chronologer provided as the foundation of all further study a set of skeletons—rigid, fleshless, easily memorized dates for events, the intervals between them, and calendrical tables. But, just as the articulated skeletons in up-to-date Renaissance anatomy theaters took poses and held banners that taught traditional religious and aesthetic lessons, so the chronologer's skeleton of dates was embedded in a rich interpretative background—one so old, solidly established, and dizzyingly varied in its sources as to be almost identical with the literary and historical traditions of the West.

In the first place, chronology provided the keys to the kingdom of providential history. Persians, Jews, Christians, and Muslims believed that the hand of the Creator had shaped history to fit an elegant, symmetrical last. But the techniques by which chronologers tried to discover the neat rhythms that underlay messy events varied from culture to culture and period to period. From the Hellenistic period, many Jewish writers followed Greek precedent and cut biblical and later time up into a limited number of periods of distinct character, normally in the hope of predicting when history would end. Jewish calculators wove a nightmarish variety of clashing chronological schemes, often retailoring biblical and postbiblical history to make it fit numerological symmetries that were really too neat to match the facts. These, they hoped, would reveal the divine order that lay behind the terrible chaos of the destruction of the Temple, the crushing of Bar Kochba, the pogroms of the Crusaders, and the expulsion of the Jews from Iberia.[27] The fathers of the Church offered elaborate (and often conflicting) numerologies of their own, and through the twelfth century medieval chroniclers en masse computed the intervals and the end of history. Further dozens of scholars and prophets, many of them inspired by Joachim of Fiore, carried this tradition forward through the later Middle Ages and into the Renaissance. In the years just before 1500, in particular, computations of the end of time were shouted in the streets, multiplied by the presses, carried westward on Columbus's ships, and into battle by rebellious peasants.[28] Arabic astrologers made earthly history dance to the music of the spheres, as revealed by the conjunctions of Jupiter and Saturn that took place every twenty years.[29] In the Renaissance, Jewish and Christian readers tried to fuse all of these methods and more, as cultural contact made any given chronologer the master of as many schemes as a modern hacker has programs.[30]

Chronology, in the second place, offered rich information as well as pointed instruction. Since the third and fourth centuries A.D., when Christian chronography took shape, world chronicles had spiced the rigorous tedium of their names and dates with general information, much of it not quantitative or technical, about the human past. Werner Rolewinck's *Fasciculus temporum* and Hartmann Schedel's *Nuremberg Chronicle,* like Eusebius's *Chronicle* and many others before them, gave rapid access to much of what was known about the barbarian sages of the ancient Near East and the pagan kingdoms among which the ancient Jews had wandered and suffered. They listed the inventors of arts, a subject that fascinated Renaissance humanists as it had fascinated Hellenistic grammarians long before. They explicated the Greek myths. And they laid out the lives and achievements of the great men and women of Nineveh and Babylon, Greece and Rome, in dense, readily assimilable detail, stuffed with bizarre anecdotes and illustrated with charmingly inappropriate cuts of ancient heroes and cities. Chronologies could link secular to sacred time, showing that the rise and fall of the great kingdoms of the world fitted the patterns of biblical prophecy. But often they found readers less for any powerful message than for their overwhelming mass of facts and pseudofacts. Such glittering jewels and pyrites attracted Petrarch, among many other early Renaissance readers. Jerome's Latin adaptation of the *Chronicle* of Eusebius became one of his favorite books. He filled its margins with fascinated marginalia; and legions of humanist readers followed his example, making old and new chronologies their favorite sources for curious stories and recondite histories.[31] As late as 1606, when Laurentius Fabricius wanted to know whether the Magi who visited Jesus were "private individuals, or were officially sent by their king to greet the new king of the Jews and to present him with gifts, in the manner of Orientals and, in general, of kings," he addressed his question to what seemed its natural recipient: the pigeonhole of the chronologer Seth Calvisius whose expertise on obscure matters of Near Eastern history evidently guaranteed his mastery of the institutions of the Persian court.[32]

At once a hermeneutics of the past and a handbook of names and dates, a stimulus to thought and devotion and a source of fascinating trivia, chronology offered rich rewards to both its practitioners and its readers. No wonder, then, that the preeminent chronologer Joseph Scaliger worried as his chosen field became fashionable. As always in academic life, popularity bred inferior work. Scaliger complained to Calvisius, one of the few chronologers he respected, that every Frankfurt book fair brought a new crop of chronologies, each more incompetent than the last.[33] The shelves of libraries formed in the sixteenth and seventeenth centuries, from Oxford to Wolfenbüttel, groan under the weight of the synthetic folios and argumentative pamphlets about which Scaliger waxed querulous. The structures that he and his rivals reared on paper rivaled Dasypodius's in complexity and profusion of detail—if not in aesthetic appeal. To that extent—as in its rich fusion of the traditional and the up-to-date, the iconographical and the mechanical—the Strasbourg clock nicely exemplifies the larger enterprise of chronology.

Argument without End

In other respects, however, the image may mislead. Dasypodius worried that not everyone would understand his recondite devices and images. But he never raised the question of whether everyone agreed about the simple harmonies and rigid periods that he imposed on world history, far less whether any agreement was possible. True, the general value of chronology was not in doubt. When someone challenged Philipp Melanchthon, one dinnertime in Wittenberg, insisting that chronology was useless, he dismissed his opponent with contempt.

> I said in reply: "That answer is clearly unworthy of a doctor." What a fine doctor he is, that uneducated fool; one should shit a turd into his doctor's hat and put it back on his head. What madness! It is one of God's great gifts that everyone can have the weekday letters on his wall.[34]

Almost everyone would have agreed with Melanchthon. Chronology was essential to civilized life. One of the eyes of history (geography being the other), it gave man's past order and coherence. It offered essential help to the theologian reading his Bible, the doctor reading his Galen, and the naturalist reading his Pliny. "Once you have grasped what needs to be known about the year, the months, and the days," Pierre Haguelon promised the readers of his *Trilingual Calendar* of 1557,

> you will find it far easier to understand the points on which lawyers consult Hippocrates. You will find it far easier to understand Aristotle's view on when the salpa, the sar, the ray and the angel-fish give birth. You will find it far easier to understand the passage—not to mention others—where Aristotle says that quail migrate in [the month of] Boedromion, but cranes in [the month of] Maemacterion.[35]

Who could ask for anything more? And anyone who did could be silenced by the even more authoritative testimony of the early Church. "Those whose chronology is confused," so ran Tatian's memorable phrase, "cannot give a true account of history."[36] The greatest of the fathers, from Eusebius to Jerome and Augustine, evidently agreed, since they had eagerly occupied themselves in trying to work out tiny chronological details as well as sorting out the meaning of history as a whole.[37]

But the widespread agreement about the ends and merits of chronology was not accompanied by a similar agreement about its methods and results. Every aspect of the field excited controversy. Should the chronologer frame elaborate, philosophical definitions of time and eternity or stick to quantitative measurements and precise, soluble problems? Should he cast his results in eloquent prose or austere tables? Should he draw on the Bible alone or on the classics as well? Every imaginable position on these questions had adherents and opponents. And even sharper arguments attended those problems of historical detail that the chronologer was expected to solve—for example, the determination of the number, names, and order of the kings of ancient Israel. Scaliger said that no one of sound mind could hope to solve this in a fully satisfactory way—an admission that dismayed his friend and ally Calvisius who hated to admit that so central a problem had no rigorous solution.[38]

Chronology, in other words, was not only familiar but contested. Controversies bloomed in every corner of the field. Westerners, as we have seen, agreed that the complexity and sophistication of the Aztec calendar were astonishing. But they disagreed sharply about its actual nature. Toribio de Motolinía criticized the Indians' knowledge of the tropical year: "They were at fault only as to the leap year." Their ignorance he excused by a leap of the historical imagination, arguing that they had shared it with that of the most learned of all pagans: "But this was the case also as to the great philosopher Aristotle and his teacher Plato, and many other wise men who did not attain [knowledge of] it." Motolinía took the analogy with classical calendrics very far. He even claimed that the Indians, like the ancients, had had "olympiads," "lustra," and "indictions." And he insisted that "in this calendar there is nothing idolatrous," though knowledge of "the names of the days and the weeks" had been the preserve of a few.

Bernardino de Sahagún, by contrast, argued that the Indians had known the tropical year perfectly well: "because in the count which may be called a true calendar they count three hundred and sixty-five days, and once every four years they counted three hundred and sixty-six days." The Aztec calendar was thus as technically solid as the Julian ecclesiastical calendar—and, presumably, more so than the ancient Greek lunar calendar. However, some Aztec ways of computing time were theologically problematic, and even diabolical. Sahagún denounced their 260-day divinatory cycle: "For they do not carry out this count according to any natural order; for it was an invention of the devil and an art of soothsaying." The Indians' technical proficiency did not prove that they had been "natural philosophers," as Motolinía thought.[39]

Two strategies appear. Motolinía, like many Renaissance chronologers, evidently found himself baffled by the precise details of solar and lunar movements and the small differences that distinguished one calendar from another. He offered fuzzy, complex descriptions in place of concrete, precise analyses. The style of his discussion of the Indians' knowledge of the calendar is typical.

> The Indians who well understood the secrets of these wheels and calendar showed and explained them to but few, because through their knowledge they gained their livelihood and were held in esteem and considered wise and intelligent. Nevertheless, almost all adult Indians knew and were informed about the year, as to both the numbers and house in which they were. But of the names of the days and weeks, and many other secrets and counts which they possessed, only those masters who worked with them attained knowledge.[40]

Sahagún, by contrast, set out fearlessly to make distinctions and justify exclusions. He ruthlessly denied Motolinía's analogy between classical and Indian calendrics: "What he saith of olympiads, five-year periods, indictions, and the like, is false and pure invention."[41] He carefully laid out the differences between divination by days, which had been the job of a few specialists, and the general knowledge of the real calendar that ordinary people had shared with the priests. And he described in detail the methods and ceremonies with which—thanks, perhaps, to diabolic inspiration—the Indian diviners had predicted the future.[42]

Durán took a third tack. He noted the analogy between their efforts to identify propitious and impropitious days for sowing, tilling, and reaping and "our almanacs," which "record the signs of the zodiac, where we are told that some influences are good, others bad, and others indifferent regarding the sowing of crops and even the health of our bodies." Then he used his matchless experience of life in the field to trace the continued existence of divinatory practices in Christian New Spain and to prove that where western almanacs rested on "experience," Aztec ones relied on "superstition."

> I see that they are still kept strictly, and I base my opinion upon the fact that one day I asked an old man why he was sowing a certain type of small bean so late in the year, considering that they are usually frostbitten at that time. He answered that everything has a count, a reason, and a special day. I shall give another example. Let us suppose that the maize of a certain field is already dry, ripe, and ready to be harvested. It will deteriorate if it is not reaped. Here and in many other places they will not harvest it, even though it is lost entirely, until the elders decide that it is time to reap. I dare to swear to these things because in church I myself have heard the public announcement, all the people being present, that the time of the harvest has come. They all rush off to the fields with such haste that neither young nor old remain behind. They could have gathered the crop earlier, at their leisure; but since the old sorcerer found in his book or almanac that the day had come, he proclaimed it to the people, and they went off in great speed.[43]

It would be hard to determine which of the three made the best observer. My own preference probably goes to Motolinía, whose evident lack of calendrical expertise went along with a singular lack of prejudice. He showed far more tolerance than his rivals when comparing European with Indian calendrics.

> And this [Aztec] calendar is exceedingly ancient, and if the names of the days, weeks, and years, and their representations, are of animals and beasts, and other creatures, it is nothing to wonder at; for if we look at

ours, they also are of planets and of gods which the pagans had. Even if many rites, falsehoods, and ancient sacrifices are written of here [in connection with] a thing so good, of such excellence and truth as these natives possessed, there is no reason to condemn it. For we know that all good and truth, no matter who expresseth it, is of the Holy Ghost.[44]

For Motolinía, then, the high quality of the Aztec calendar proved that it was divine, not devilish: the result of an independent, primeval revelation to the Indians, like those that Renaissance philosophers had long ascribed to the ancient sages of Egypt and Greece.[45] Motolinía's sense that a calendar often has multiple layers of diverse origin imposes respect; so does the remarkable correspondence between his description of the status of calendar diviners, however vague its phrasing, and those produced, in the twentieth century, by anthropologists who have studied their professional descendants.[46] His historical insight is as striking as his theological openness. It was usually only when Renaissance scholars investigated the practices of Christians—as Olaus Magnus did when he watched, admired, and reported in detail on the Scandinavian peasants who could use their rune sticks, originally a pagan invention, to predict the dates of Christian moveable feasts—that they showed so much sympathetic interest for the unfamiliar.[47]

In the end, however, awarding retrospective marks is an exercise in pointmissing. No sixteenth-century observer could know the immense variety of Mesoamerican calendrical practice unearthed by modern archaeologists or grasp the very alien cultural assumptions about time and space that underlay them.[48] Naturally they fitted what they found to such western models as the sphere, the circle, and the almanac. What matters, for present purposes, is simply the radically disparate assumptions, points of view, and bits of information that these experts brought to bear on a single calendar, which all of them had actually seen in operation. These made it inevitable that they would assess its nature, value, and historical import rather differently.

It was all the more natural, then, that calendars and eras attested only by historical sources, rather than live experts, generated even sharper disagreements.[49] In these cases, after all, not only present-day polemics but traditions of dispute that stretched back for millennia sharpened nibs and warmed tempers. By the beginning of the seventeenth century, everyone knew that chronologers, like clocks, never agreed—a proverb that makes modern historians' assertions about the cultural meaning of the clock in early modern Europe sound just a little ironic.[50] In the course of the next century and a half, the chronologers' proverbial inability to reach accord would bring their discipline as a whole into discredit. Chronologers incurred the mockery of enlightened thinkers as diverse as Vico, whose *Scienza Nuova* set out to show that their systems rested on faulty premises, and Voltaire, who mocked "the sterile science of facts and dates, which confines itself to establishing when some man of no importance died."[51] Yet at the same time their debates and data would delight the same *philosophes* by proving—or so many intellectuals thought—that the Bible did not tell the whole story of the human past. Two further case studies may illuminate this anfractuous history. In particular, they may suggest how dense, complex, and durable the tradition of chronological argument proved—and how little permeability it showed to new evidence and new methods of analysis.

Time Changes Shape

In the last third of the sixteenth century—so one story goes—time past underwent a radical change. It lost its traditional shape. Several scholars had a hand in the surgery that removed such traditional joints and tendons as the Four Monarchies and the six days of a thousand years. But the chief attending scholar was the French Renaissance jurist Jean Bodin, whom students of the English Renaissance remember from Philip Sidney's advice to his brother that "For the method of writing history Boden hath written at large; you may read him and gather out of many words some matter."[52] As Sidney indicates, in 1566 Bodin published a massive book about the uses of the past. He called it, not without some false advertising, a *Method for the Easy Comprehension of History*. In fact, this sprawling, difficult book instructed the young jurist on how to read all historians systematically, taking notes and assessing the events described by a system of marginal notes (CH, for example, stood for *consilium honestum*, CTV for *consilium turpe utile*). Bodin taught source criticism and comparative sociology as well as moral philosophy and politics. Before choosing which historian to believe, the student of Bodin will test all of them for credibility. He will follow, for preference, those who wrote on the basis of political experience but were not connected with (and partisan about) the events. After collecting all their information about kings and constitutions, the scholar will be able to see why each nation has the institutions appropriate to the factor that chiefly determines its character—the climate of its original home (Asians are contemplative and fit for monarchy, Europeans good constitutionalists, and so on).

Bodin presented readers terrified by the European religious wars of the late sixteenth century with a massive program of social reconstruction through historical scholarship. His work combined humanist source criticism with the comparative political analysis of Machiavelli and his followers. Naturally, his book was often reprinted, widely assigned in universities, and heatedly discussed by readers for at least a century—despite its occasional lapses into quaintness, as when he proved the original French habitat of the Walloons from their Latin name, Ouallones, which derived from their habit of wandering about asking "Où allons-nous?" Intellectuals and statesmen from Montaigne to Gabriel Harvey followed Bodin's instructions as they explored the margins of their ancient and modern historians—even when these led to conclusions that contradicted Bodin's.[53] In the 1960s and after, when historiography attracted widespread interest, the *Methodus* came to be seen as a work of revolutionary modernity.[54]

No chapter in Bodin's method attracted more abuse in his own day—or has won more praise in recent times—than the short one in which he attacked the historical theory of the four empires, so popular in Protestant universities, and the associated belief that history had begun with a golden age of poverty and purity and degenerated ever since. Bodin argued that the earliest times had in fact been an age of lead, not gold, a period haunted by the superstition and cruelty that had given rise to the practice of human sacrifice. He insisted that the moderns had produced inventions—like the compass and the printing press—incomparably superior to the technology of the ancients. He used the vast power and extent of one modern empire, that of the Turks, to show that the four empires of Daniel's prophecy could not possibly encompass all of history. And he concluded that man and nature retained the same power of creation that they had always had. From Bodin to Bacon was only a step; this partial but powerful attack on the authority of the past has long been presented as a vital skirmish in the sixteenth century's historical revolution. Bodin, after all, describes the ancient past not only as radically different from the present, but also as distinctly primitive—an insight soon to be given brilliantly vivid form by the English artist John White, who pictured the Picts who had once inhabited Britain with the same forms he used for the Indians who lived in Virginia in his own time. Bodin's arguments seem a splendid case in point of the impact of new experience on the brittle structures of inherited theory. And they have played that heroic role in many accounts of early modern chronology.[55]

In fact, however, Bodin's refutation of the Golden Age was no attack on classical authority. Consider, in the first place, Bodin's evidence that the Golden Age was a time of fear and superstition. He said a good bit about the Golden Age. It lasted 250 years. It was dominated by people of dubious character—Ham, who rendered his father impotent, and Nimrod, the mighty hunter, whose name alone (meaning 'rebel' in Hebrew) revealed his bad character. Its denizens lived a life of unremitting savagery, a war of all against all. And yet this age of barbarism and bad conduct was the Golden Age—for Nimrod, "as everyone agrees," was the ruler called Saturn by the Greeks and Romans, the very one whose Saturnian kingdom was the Golden Age of the ancient poets.[56]

Bodin's sources for these facts (or factoids) are not far to seek. He found them, as he said, in the "Defloratio" of the Babylonian priest Berosus and the "Origines" of the Roman scholar Cato.[57] And that already suggests a problem, for these were not the genuine fragments that Scaliger studied, and about which modern classicists so enjoy framing wild hypotheses. They were forged texts published in 1498 by the Dominican theologian Annius of Viterbo. These works and their companions, embedded in Annius's rich commentary, fooled Erasmus and Luther as well as Bodin. They far outsold such prosaic rivals as Herodotus and Thucydides throughout the sixteenth century—even though, as Walter Stephens has shown in a massive, absorbing book, they in fact manipulated well-known ancient and medieval sources and topoi.[58] Annius forged his texts for a highly specific reason. He detested the humanists of his time and the Greek and Latin classics that they saw as central to a new culture. He wished to convict the Greeks and Romans as liars and to replace them, as authorities on the past, with his own creations. One obvious way to do so was to include in his texts ample denunciations of the supposed Golden Age of early man that featured so prominently in classical literature. Thus, Bodin, the eminent modern theorist of historical criticism, took basic facts about the ancient world and their interpretation from a forger of his own day.[59]

Annius, of course, was a good Dominican—so good that he received both a miraculous cure from the Virgin and a death by poisoning at the hands of Cesare Borgia. He nowhere adumbrated Bodin's idea of progress in the arts and sciences (in fact, he argued that God had revealed these fully to Jewish Patriarchs, Etruscan priests, and Druids, among others). Neither, of course, did Martin Luther—from whose *Address to the Christian Nobility of the German Nation* Bodin seems to have lifted the notion that the existence of the Turks provided a stumbling block to those who wished to fit all modern history inside Daniel's statue. But this, too, came from a source—and a more respectable one. In the course of his attack on the nostalgia of classicism, Bodin described the prevalence of piracy in the ancient world. He cited Thucydides' story that the ancient Greeks were so accustomed to crime that they routinely asked all passersby if they were robbers or pirates. And he then remarked that the early Greek cities had had no fortifications or walls and were accordingly very vulnerable. Barbarians—those who lived farther from true *humanitas* than the Greeks—had retained these conditions even longer.[60]

This nod is at least as good as a wink. Though Bodin cited Thucydides only for one specific story, he used him heavily. In the so-called archaeology with which book 1 of Thucydides' history begins, Bodin found the intellectual armature for his own argument. Thucydides brilliantly demonstrates there, drawing on the sophists of his time, that the Peloponnesian War was far greater than any previous one. Population, fortification, and all the appurtenances of civilized life were far larger and better developed in the fifth century than they had been in the age of Homer. Hence the Peloponnesian War far outranked the mere skirmish fought by Greek chieftains at Troy. In the course of this elegant—if sophistical—argument Thucydides even adumbrated Bodin's comparative approach, since he argued that the barbarians of his own time lived a life similar to that of the Greeks of Homeric times.[61] Like Bodin, he thus treated geographical and historical distance as convertible—an idea that José de Acosta, unlike Bodin, would soon apply to the primitive societies of the New World.[62] Thucydides thus provided the positive model for Bodin's argument about modernity, as Annius's forgeries provided his negative information about the Golden Age. Gibbon said of Augustine that his learning was too often borrowed and his arguments too often his own. At least the latter criticism does not apply to Bodin. He was a highly traditional intellectual. He found the implements with which he attacked old ways of dividing history within, not outside, the tradition of chronological scholarship. And he therefore was entirely consistent when he not only attacked the stereotyped periodizations of others but offered numerological ones of his own, derived from Plato, to replace them (like the 496 years, $7 \times 70 +$ the perfect number 6, which he took to be the normal duration of a national history).[63] Bodin's case illustrates the adaptability, not the fragility, of normal chronology (so, of course, do the refutations of his work, which far outnumbered the texts written in his support).

Early Times

Between 1605 and 1607, Joseph Scaliger was mugged, painfully and unexpectedly. The mugger, David Pareus, should have been a friend. Like Scaliger, he belonged to a liberal Protestant university, that of Heidelberg. Like Scaliger, he devoted himself to the study of both the Bible and the classics.

But, unlike Scaliger, he came to the conclusion that only the Bible deserved credence as a source for ancient history. Pareus insisted that the ancient Greeks had had no certain knowledge of any event before the fifth century B.C. For earlier periods they had boldly invented the facts they lacked, as the absurd myths with which they had filled the millennium before their own, their stories of monsters, gods, and superhuman heroes, clearly showed. Their own time, to be sure, they had known, but they had systematically lied about it. Herodotus, for example, wrote at length about one Xerxes, king of Persia; the Bible did not mention this obviously nonexistent figure. Scaliger had built his vast chronological works, the *De emendatione temporum* of 1583 and the *Thesaurus temporum* of 1606, with their two-thousand folio pages of polyglot erudition, on the assumption that he could reconcile biblical with classical histories. He was as misguided as he was learned. Pareus enjoined him to abandon his faith in mendacious Greeks and repose all his trust in the Hebrew Bible—especially, of course, the book of Daniel.[64]

Scaliger fought back, hard and tenaciously. He was, of course, the greatest classical scholar of the age—an age in which recondite scholarship, based on learning in many obscure languages and now-forgotten skills, dominated the intellectual firmament like a vast spiral nebula of erudition and error. His interests originally centered on the editing of Latin texts but gradually expanded to embrace chronology. Scaliger's successes at this second daunting enterprise took him from poverty and isolation in southern France to success and fame at Leiden, the most innovative university in the world. There he spent his last years, from 1593 to 1609, in a full-time research position, engaged in a vast inquiry into what he himself called "the history of 8000 years, according to the pagans."[65]

Scaliger had no trouble dealing with his overmatched adversary from the south. In a brilliant *Elenchus* he excoriated what he referred to as Pareus's *stercoreae orationes* ('shitty speeches') against him. Pareus failed to realize that the Greeks had had very precise knowledge of their past, based on datable events in the heavens (like the "facts" that the moon reached its third quarter on the night of Troy's fall and that a solar eclipse accompanied Xerxes' crossing into Greece). By using these they had constructed a rigorous and coherent chronology as early as the Hellenistic age—something no Jew of the time, even Judah Maccabee himself, could have done, much less a Pareus or some other "prophet" of the present day. A full history of the ancient world must use the Greeks' data to complete the begettings and king lists of the Bible.[66]

Scaliger's position seems prescient now. His insistence on studying both

Jews and Greeks, his trust in alien, alarming writers like Herodotus, his understanding that the Biblical account of the ancient past was not sufficient or complete—all these positions adumbrate those of modern scholarship. They seem remarkable in a faithful Calvinist who chose the rigorous predestinarian Gomarus as his literary executor (true, Scaliger respected the liberal Arminius more, and he remarked that Gomarus understood chronology as well as he himself understood counterfeiting).[67] How can we account for his prodigiously original historical criticism?

Standard accounts of the history of history offer a standard answer. Medieval scholars—so they taught us—spun myths about the ancient world because they lacked a sense of history. Powered by a horror of the vacuum, like the similar force they attributed to nature, they filled every gap in the historical record with boldly invented legends and connections. The imaginary genealogies they spun for modern royal and noble families stretched back to the dawn of historic time. They thus showed—as Johannes Trithemius bitterly complained around 1500—that everyone in Europe had Trojan ancestors, "as if there hadn't been older peoples than the Trojans, and as if there had been no Trojan rascals."[68] Hampered by theological prejudices, they could not accept the bold and sensual classical myths. So they removed the sting from these by elaborate allegorical exegesis, encrusting classical stories of rape and transformation with Gothic ornaments that distorted them and made them harmless. This double effort to modernize the ancient world by rationalist retrofitting naturally resulted in a massive falsification of its records.

To be sure, medieval students of the ancient world, like the historian Petrus Comestor, did follow classical precedents to some extent. Their favorite way of dealing with ancient myth, for example, was invented not by a Christian but by the Hellenistic scholar Euhemerus, and it was fully elaborated by another pagan, Varro, the most learned of the Romans. These men treated the gods of classical myth as great men and women who had been accorded divine status after their deaths because of their achievements—rather like Alexander the Great. They interpreted Atlas, the giant who held up the heavens, as a great astronomer; Vulcan, the god of the forge, as the inventor of the blacksmith's craft; Venus, the goddess of love, as the greatest trainer of courtesans; and so on. This method was classical, not Christian—however crude the versions of it that most medieval scholars knew, especially that found in Isidore's *Etymologies*. But in Christian hands it necessarily increased the distance that separated modern thinkers and readers from the classical past.[69]

Where medieval humanists accumulated, excused, and invented, Renaissance ones discriminated, condemned, and analyzed. They spent their time expunging from the record errors and misunderstandings—including the allegorizations that had peopled earlier visions of Greece and Rome with medieval people acting in Christian ways. By their insistence on the need to recreate the ancient world by an exclusively historical approach to ancient texts, the humanists arrived for the first time at a fully modern vision of the past and a fully operational historical method. And, by their ability to frame and apply the core methods of modern scholarship—like the use of astronomical evidence to provide absolute dates for past events—Renaissance scholars replaced the mythical past of the Middle Ages with the dull but reliable pegs on which we still hang our histories of the ancient and medieval worlds. The sense that the past is another country had to arise before intellectuals could develop techniques for mapping that country's borders and features in accurate detail. Only by facing up to the ancient world as it really was could a Scaliger hone his historical criticism to a cutting edge.[70]

These comforting bromides unfortunately shatter on contact with the rebarbative structure of Scaliger's chronological work. The fourth-century world chronicle of Eusebius—which Scaliger spent six years editing and explicating—forced him to confront matters far more complex and problematic than mere solar eclipses. In a Byzantine world chronicle that contained many fragments of Eusebius's original Greek text, Scaliger also found the fragments of the Babylonian historian Berosus and the Egyptian historian Manetho. Berosus told a Babylonian Creation story that began with the appearance of a monstrous figure, one Oannes, with the head of a fish and the body of a man, who taught humanity the arts and sciences. Manetho offered a Greek version of the dynastic history of Egypt that stretched back not only before the Flood but before Creation itself.[71] Scaliger had every reason to reject both texts as Hellenistic forgeries; their contradictions of the Bible proved their falsity. Instead, he insisted that they must somehow be genuine records of their two Near Eastern cultures—records so alien that he could neither interpret them nor explain them away. Of Berosus's wild cosmogony he said virtually nothing. Manetho's wild chronology—which was precise, coherent, and in its later parts matched other records—required discussion. At first he relegated the early dynasties to the age of myth. Eventually, however, he became dissatisfied with that simple, traditional formulation and set them in what he cheerfully but unhelpfully called "proleptic time"—a period that began long before the Creation.[72] His solutions pleased none of his early readers, from his loving Calvinist friend Casaubon to his bitter Jesuit opponent Petavius. They reveal an openness to the otherness of the past, a willingness to expand the boundaries of the thinkable, that seems almost shocking in an inhabitant of a strict Calvinist milieu. Certainly, Scaliger's response to these texts was both in some sense modern and in every way his own.

At other points, however, both the problems and Scaliger's solutions were almost shockingly traditional. Eusebius retold, briefly but cogently, the traditional stories of Theseus, Heracles, and the Argonauts. Christian though he was, he treated these lurid tales as clues to datable events in human history. And Scaliger reacted to these passages in a clear-cut, if slightly paradoxical, way. He agreed that Eusebius was quite right to treat myths as histories. But he added that to analyze myths properly one must apply the proper reagents to them, in order to stain and identify their hard core of fact. The basic ingredient, happily, was not a recondite method but simple common sense. This taught, as Scaliger explained, that "it is equally certain that Heracles really existed, and that the Hydra, eternally reborn with its innumerable heads, did not." By shearing off the obviously fabulous, one could transform Greek mythology as a whole into a *"certum Chronicon,"* generation by generation, of the events that had preceded the Trojan War. Myths were not lies but disguised histories.[73]

When it came to specific myths, Scaliger did not mind defining common sense a little more fully and precisely. Consider, for example, the story of Daedalus who not only equipped his son and himself with wings but made statues that moved. Eusebius explained this story euhemeristically: Daedalus had really been the first to make statues with their feet and legs separate. Their appearance, rather than a magical ability to move, formed the factual core of the story. Scaliger disagreed with Eusebius—but only on the details, not the substance, of his allegoresis. In fact, he argued, Daedalus had made statues filled with quicksilver, which moved of their own accord. They were really, not just apparently, *autokineta*.[74] In rewriting myth into history, in other words, Scaliger took his method directly from one of the same late antique sources that Petrus Comestor and every other medieval allegorist had used.

In some respects Scaliger did go beyond his predecessors. Eusebius drew many of his euhemerist allegories from one earlier Greek writer, Palaephatus. His little book *On Incredible Things* explained not only Daedalus's moving statues but Pandora's box—she was really a lady overly fond of cosmetics, and the box was her vanity case—and Pegasus the winged horse—this was really a fast ship that had a horse as its figurehead. Medieval scholars and early humanists like Petrarch depended on Eusebius's brief extracts, as translated by Jerome, for their knowledge of Palaephatus. Scaliger, too, used Palaephatus, but he read the whole text in the original Greek rather than in Eusebius's Latin summaries. Yet in doing so he added only highlights and shadows, not new outlines, to the sketchy history of Greece's mythical times that a Petrus Comestor could have set out without the learned help of Renaissance philology.[75]

Scaliger's resort to classical euhemerist allegoresis was not idiosyncratic. Almost all seventeenth-century historians and students of myth made the same evasive maneuvers when confronted by the same lurid monsters and heroes. From Scaliger's younger Protestant friend G. J. Vossius to his Catholic critic P. D. Huet, they agreed that Greek myth was not a tradition to be studied for its own sake but an arbitrarily altered factual tale that could be reconstituted by a simple application of Palaephatus's razor. Most of them, indeed, argued one thesis from which even Scaliger had shrunk: that the Greek myths were mostly allegories of one specific set of ancient events, those recounted more accurately and more literally in the Hebrew bible. Vulcan was Tubal Cain, Deucalion was Moses, Pandora was Eve. The whole Homeric *Iliad* and *Odyssey*, more than one polymath argued, retold the biblical history of Israel in dactylic hexameter. The most advanced historical scholars of the seventeenth century mistreated Greek myth exactly as the classicizing friars they despised had done three hundred years before.[76] Like the rest, Scaliger reposed exactly as much faith in these—traditional—arguments as he did in his novel ones about eclipses. Indeed, he devoted the bulk of his reply to Pareus to showing that Greek mythology deserved as much credence as the Greek historians themselves. And he worked hard to prove that some "mythical events"—like Hercules' celebration of the Olympic games—could actually be assigned rigorous dates of their own based on astronomical evidence.[77] Scaliger's defense of Greek myth and use of it to recreate early human history could be more plausibly represented as a relapse into medieval mythography than as a triumph of humanist historical criticism. At the late Renaissance zenith of chronology, then, historical critics still went to work carrying a toolbox that included both up-to-date cutting tools, powered by the most modern of attitudes, and museum pieces that they found in the very sources they sought to understand and explicate. It would be a mistake to consider their results trivial or short-lived. More than a century later, Vico still wove the dated Olympic games of Hercules and other benchmarks established by Scaliger into the supposedly modern fabric of his *De constantia philologia* and *Scienza nuova*.[78]

The Pasts of a Village Immobile

A real history of early modern chronology would evidently not make a pretty sight. It would replace the traditional Mantegnesque frieze of the triumph of modernist virtue and precision with a messy little cartoon strip, rather like *Mad Magazine*'s old feature "Spy vs. Spy," in which history and allegory, insight and prejudice, new and old methods wrestle, deceive one another, and blow one another up. The scene may seem less edifying than the one it replaces. But it better represents the Renaissance's real combination of vision and illusion, penetration and error, historical criticism and creative fantasy. And it does enable us to follow the example of a very great Renaissance scholar—though one whose contributions to scholarship, because she was a woman, were made not in writing but in conversation. John Selden remarks that

> T'was an excellent question of my Lady Cotton, when Sr. Robert Cotton was magnifying of a shoe, wch was Moses's or Noahs, & Wondering att the strange shape & fashion of it: But, Mr. Cotton, sayes shee, are you sure 'tis a shooe.[79]

This question may represent Renaissance historical criticism at its sharpest. It certainly affords an example of healthy skepticism, which we—like our Renaissance predecessors—would do well to imitate. And it suggests that students of historiography would do well to examine the texture and flavor of conversation in the Indian and Italian villages whose pasts still lie open to their inhabitants, as well as to scrutinize the heavy, moldering folios in the libraries they frequent.[80] Comparisons may do much to reveal the contradictions that lie concealed at so many points in the apparently solid, modern structures of paper and iron that Scaliger and Dasypodius fabricated.

NOTES

1. See, e.g., R. Koselleck, *Vergangene Zukunft*, 3d ed. (Frankfurt, 1984), esp. chap. 1; W. Schivelbusch, *The Railway Journey* (Berkeley, 1986); and S. Kern, *The Culture of Time and Space, 1880–1918* (Cambridge, Mass., 1983).

2. See the classic studies of E. Panofsky and F. Saxl, *La mythologie classique dans l'art médiéval*, trans. S. Girard (Saint-Pierre-de-Salerne, 1990); and T. E. Mommsen, "Petrarch's Conception of the 'Dark Ages,'" in *Medieval and Renaissance Studies*, ed. E. F. Rice, Jr. (Ithaca, N.Y., 1959), 106–29.

3. All accounts of chronology since Jacob Bernay's classic life of the supposed founder of the discipline, *Joseph Justus Scaliger* (Berlin, 1855), have emphasized its modernity—even such well-informed ones as the Nachbericht by F. Hammer, in J. Kepler, *Gesammelte Werke*, 5: *Chronologische Schriften* (Munich, 1953), 397–495. For a very different treatment, parts of which are reused or summarized here in

heavily revised form, see A. Grafton, *Joseph Scaliger: A Study in the History of Classical Scholarship*, vol. 2: *Technical Chronology* (Oxford, 1993).

4. C. Dasypodius, *Heron mechanicus: Seu de mechanicis artibus atque disciplinis. Eiusdem horologii astronomici Argentorati in summo Templo erecti descriptio* (Strasbourg, 1580), sig. [F. iiii recto-verso].

5. See O. Mayr, *Authority, Liberty and Automatic Machinery in Early Modern Europe* (Baltimore and London, 1986), 10–13.

6. O. Magnus, *Historia de gentibus septentrionalibus* (Rome, 1555), I.32, p. 52.

7. See C. Cipolla, *Clocks and Culture, 1300–1700* (New York, 1967); Mayr, *Authority, Liberty and Automatic Machinery;* and G. Dohrn–Van Rossum, *Die Geschichte der Stunde* (Munich and Vienna, 1992).

8. H. Baron, *In Search of Florentine Civic Humanism* (Princeton, 1988), 2:52–53. See also R. Glasser, *Time in French Life and Thought*, trans. C. G. Pearson (Manchester, 1972), chaps. 4–5.

9. Dohrn-Van Rossum, *Geschichte der Stunde*.

10. See Mayr, *Authority, Liberty and Automatic Machinery*.

11. See, e.g., the classic studies of E. Panofsky, *Studies in Iconology* (New York, 1962), 69–93; F. Saxl, "Veritas filia temporis," in *Philosophy and History: Essays Presented to Ernst Cassirer* (Oxford, 1936), 197–222; and R. Wittkower, *Allegory and the Migration of Symbols* (London, 1977), 97–106. More recent studies include R. Quinones, *The Renaissance Discovery of Time* (Cambridge, Mass., 1972); S. K. Heninger, Jr., *Touches of Sweet Harmony* (San Marino, 1974); Heninger, *The Cosmographical Glass* (San Marino, 1977); S. Cohen, *The Image of Time in Renaissance Depictions of Petrarch's "Trionfo del Tempo,"* (Tel Aviv, 1982); R. Doggett et al., *Time: The Greatest Innovator* (Washington, D.C., 1986); S. L. Macey, *Patriarchs of Time* (Athens and London, 1987); and G. J. Whitrow, *Time in History* (Oxford, 1988). Two distinguished studies of a closely related subject are J. A. Burrow, *The Ages of Man* (Oxford, 1986), and E. Sears, *The Ages of Man* (Princeton, 1986).

12. See the studies collected in G. V. Coyne et al., eds. *Gregorian Reform of the Calendar* (Vatican City, 1983), especially O. Pederson, "The Ecclesiastical Calendar and the Life of the Church," 17–74; and J. D. North, "The Western Calendar—'Intolerabilis, horribilis, et derisibilis': Four Centuries of Discontent," 75–113. The Jews' mockery appears as a prime motive for reform in Paul of Middelburg, *Exhortatio pro calendarii emendatione*, Vat. lat. 3684, fols. 2 verso–3 recto; and in Cardinal Bessarion's letter to Pope Paul, published in L. Mohler, *Kardinal Bessarion als Theologe, Humanist und Staatsmann*, vol. 3: *Aus Bessarions Gelehrtenkreis* (Paderborn, 1942), 546–48.

13. C. Phythian-Adams, "Ceremony and the Citizen: The Communal Year at Coventry, 1450–1550," in *Crisis and Order in English Towns, 1500–1700: Essays in English History*, ed. P. Clark and P. Slack (London, 1972), 57–85; D. Cressy, *Bonfires and Bells* (Berkeley and Los Angeles, 1989).

14. See, most recently, Coyne et al., *Gregorian Reform of the Calendar*.

15. For a detailed description of an eighteenth-century "chronology machine" see S. Ferguson, "The 1753 *Carte chronographique* of Jacques Barbeu-Dubourg," *Princeton University Library Chronicle* 52 (1991): 190–230.

16. A. *Gislenii Busbequii Omnia quae extant* (Amsterdam, 1660), ep. 1, p. 94.

17. M. de Montaigne, *Essais* III.6 ("Des Coches"), in *Oeuvres complètes*, ed. M. Rat (Paris, 1962), 892, in the translation by D. M. Frame, *The Complete Works of Montaigne* (Stanford, 1957), 697–98.

18. D. Durán, *The Ancient Calendar*, chap. 2, in his *Book of the Gods and Rites and the Ancient Calendar*, trans. and ed. F. Horcasitas and D. Heyden (Norman, Okla., 1971), 395–96.

19. J. G. Icazbalceta, *Don Fray Juan de Zumarrága*, ed. R. A. Spencer and A. C. Leal (Mexico City, 1947), 4:89–95.

20. S. Gruzinski, *La colonisation de l'imaginaire* (Paris, 1988), chap. 2. On the interaction between Maya and European ways of organizing time, see the classic study by N. Farriss, "Remembering the Future, Anticipating the Past: History, Time, and Cosmology among the Maya of Yucatan," *Comparative Studies in Society and History* 29 (1987): 566–93 (also in this volume).

21. Durán, *The Ancient Calendar*, 396.

22. F. M. Rogers, "Exploring the Atlantic," in *Portugal-Brazil: The Age of Atlantic Discoveries* (New York, 1990), 67.

23. A. Blair, "Restaging Jean Bodin: The *Universae naturae theatrum* (1596) in Its Cultural Context," Ph.D. diss., Princeton University, 1990, 105–6.

24. See F. H. Maiello, "Il tempo dei calendari in Francia (1484–1805)," *Studi Storici* (1990), 413–36.

25. See E. Le Roy Ladurie, *Carnival in Romans*, trans. M. Feeney (New York, 1979), 189.

26. *Clouds*, 615–26.

27. A. D. Momigliano, *Settimo contributo alla storia degli studi classici e del mondo antico* (Rome, 1984), 77–103, 297–304; A. Hillel Silver, *A History of Messianic Speculation in Israel from the First through the Seventeenth Centuries* (1927; repr. Boston, 1959).

28. B. Guenée, *Histoire et culture historique dans l'Occident médiéval* (Paris, 1980), 148–54; D. Cantimore, *Eretici italiani del Cinquecento* (repr. Florence, 1970), 10–17; M. Reeves, *The Influence of Prophecy in the Later Middle Ages* (Oxford, 1969); Reeves, *Joachim of Fiore and the Prophetic Future* (New York, Hagerstown, San Francisco, and London, 1977); M. Haeusler, *Das Ende der Geschichte in der Mittelalterlichen Weltchronistik* (Cologne and Vienna, 1980); R. Lerner, *The Powers of Prophecy* (Berkeley, Los Angeles, and London, 1983); P. Moffitt Watts, "Prophecy and Discovery: On the Spiritual Origins of Christopher Columbus's 'Enterprise of the Indies,'" *American Historical Review* 90 (1985): 73–102; O. Niccoli, *Prophecy and People in Renaissance Italy*, trans. L. G. Cochrane (Princeton, 1990); M. Reeves, ed., *Prophetic Rome in the High Renaissance Period* (Oxford, 1992).

29. M. Sondheim, *Thomas Murner als Astrolog* (Strasbourg, 1938); J. North, "Astrology and the Fortunes of Churches," *Centaurus* 24 (1980): 181–211; K. M. Woody, "Dante and the Doctrine of the Great Conjunctions," *Dante Studies* 95 (1977): 119–34.

30. See G. D. Cohen, *A Critical Edition with a Translation and Notes of the Book of Tradition (Sefer Ha-Qabbalah)* (Philadelphia, 1967), for a particularly learned and perceptive study of the desire to find numerical patterns in past history.

And for Ibn Daud the surest sign of Providence, or as he calls it, of divine consolation, was to be found in the rhythmic workings of history: construction, destruction, and reconstruction of the two Temples were decreed from Heaven to occur in periods that were equal in length and, therefore, symmetrical. It is in the symmetrical periodization of history, then, that we may discern the transcendent plan of history. It matters not that Scripture did not supply all the necessary data for such a symmetry, or even that Ibn Daud's symmetry contradicted the manifest data of Scripture. The function of the historian was to find the plan and rewrite the chronological facts where necessary [p. 191]. . . . Schematology always betrays a very superficial interest in the events themselves, but a deep desire to unravel their meaning and their place in the plan of history as a whole. If the gates of prophecy had been closed, the avenues of exegesis and wisdom were still wide open. The perspicacious student of the past could gain a reliable glimpse into the future. (p. 213).

31. See Grafton, *Scaliger*, vol. 2, sect. 1.5.

32. M. Laurentius Fabricius to S. Calvisius, 17 January 1606, Göttingen, Niedersächsische Staats- und Universitätsbibliothek, MS Philos. 103, II, p. 9.

33. Scaliger to Calvisius, 3 October 1605, *Epistolae,* 611: "Nullae Francofurtenses nundinae sine Chronologorum proventu."

34. K. Hartfelder, *Melanchthoniana paedogogica* (Leipzig, 1892), 182.

35. P. Haguelon, *Calendarium trilinque* (Paris, 1557), fol. 5 verso.

36. Tatian *Oratio ad Graecos* 31, quoted on the title page of Scaliger's *De emendatione temporum* (Paris, 1583), and by Edo Hildericus in his translation of Geminus (Altdorf, 1590), sig. a iiii.

37. See, e.g., Mommsen, "Petrarch's Conception of the 'Dark Ages'"; and A. C. Momigliano, "Pagan and Christian Historiography in the Fourth Century A.D.," in *Essays in Ancient and Modern Historiography* (Oxford, 1977), 107–26.

38. Scaliger to Calvisius, 21 May 1607, Göttingen, Niedersächsische Staats- und Universitätsbibliothek, MS Philos. 103, II, 26: "De regibus Israelis, quis sanus promiserit se aliquid dicere, quod certum esse jurare possit?"

39. B. de Sahagún, *Florentine Codex: General History of the Things of New Spain,* vols. 5–6: *Book 4—The Soothsayers; Book 5—The Omens,* trans. C. E. Dibble and A. J. O. Anderson (Santa Fe, 1957), 137–46. Sahagún quotes the passages by Motolinía that he attacks.

40. Ibid., 141.

41. Ibid., 140.
42. Ibid., 142–46.
43. Durán, *The Ancient Calendar*, 396–97.
44. Sahagún, *Florentine Codex*, 140.
45. See, e.g., D. P. Walker, *The Ancient Theology* (London, 1972).
46. See, above all, the remarkable study by B. Tedlock, *Time and the Highland Maya* (Albuquerque, 1982).
47. Magnus, *Historia de gentibus septentrionalibus*, 1.33–34, pp. 54–55.
48. On the former point, see the very up-to-date synthesis by J. Marcus, *Meso-American Writing Systems* (Princeton, 1992); on the latter, the forthcoming study by W. Mingolo, to whose brilliant commentary on an earlier draft of this essay I am much indebted.
49. Cf. Tedlock, *Time and the Highland Maya*.
50. H. Treutler to S. Calvisius, Göttingen, Niedersächsische Staats- und Universitätsbibliothek, MS Philos. 103, II, 58.
51. For Vico's chronology, see P. Rossi, *Le sterminate antichità* (Pisa, 1969); and P. Rossi, *The Dark Abyss of Time,* trans. L. G. Cochrane (Chicago and London, 1984); on Voltaire, see K. Pomian, *L'ordre du temps* (Paris, 1984).
52. Philip Sidney to Robert Sidney, 10 October 1580, *The Correspondence of Sir Philip Sidney and Hubert Lanquet,* ed. S. A. Pears (London, 1845), 199.
53. L. Jardine and A. Grafton, "'Studied for Action': How Gabriel Harvey Read his Livy," *Past & Present* 129 (1990): 30–78.
54. An argument most eloquently put in the classic work of J. Franklin, *Jean Bodin and the Sixteenth-Century Revolution in the Methodology of Law and History* (New York and London, 1963).
55. See, e.g., J. L. Brown, *The Methodus ad facilem historiarum cognitionem of Jean Bodin: A Critical Study* (Washington, D.C., 1939), chap. 4; and C.-G. Dubois, *La conception de l'histoire en France au XVIe siècle (1560–1610)* (Paris, 1977), 486–95.
56. J. Bodin, *Methodus ad facilem historiarum cognitionem* (Amsterdam, 1650), chap. 7, 310–23.
57. See, e.g., ibid., 317.
58. W. Stephens, *Giants in Those Days* (Lincoln, Neb., and London, 1989).
59. On Annius's program, see also A. Grafton, *Defenders of the Text* (Cambridge, Mass., and London, 1991), chap. 3, with references to earlier studies.
60. Bodin, *Methodus,* 318.
61. Cf. E. R. Dodds, *The Ancient Concept of Progress and Other Essays on Greek Literature and Belief* (Oxford, 1971).
62. A. Pagden, *The Fall of Natural Man* (Cambridge, 1982).
63. For the prevalence of such schemes over empirical research, see Grafton, *Defenders of the Text,* chap. 4.
64. Bernays, *Scaliger,* 127–28, 295.

65. *Secunda Scaligerana*, s. v. "Joseph Scaliger," in *Scaligerana, Thuana, Perroniana, Pithoeana et Colomesiana*, ed. P. Desmaizeaux (Amsterdam, 1740), 2,554.

66. J. J. Scaliger, *Elenchus utriusque orationis chronologicae D. Davidis Parei* (Leiden, 1607), 41–42.

67. *Secunda Scaligerana*, s.vv. "Arminius," 207; "Gomarus," 352.

68. J. Trithemius, "Chronologia mystica," *Opera historica*, ed. M. Freher (Frankfurt, 1601), 1, sig. ** 5 verso.

69. The best survey of this tradition remains J. Seznec, *The Survival of the Pagan Gods*, trans. B. F. Sessions (New York, 1953).

70. Cf. Grafton, *Defenders of the Text*, chap. 1.

71. The chronicle in question was the *Ecloga chronographica* of George Syncellus (ca. 800 A.D.), for which see the recent edition by A. A. Mosshammer (Leipzig, 1984), and the study by W. Adler, *Time Immemorial* (Washington, D.C., 1989), which cites earlier literature. For the texts attributed to Berosus, see *Die Fragmente der griechischen Historiker* 680; for those attributed to Manetho, see ibid., 609; and *Manetho*, ed. and trans. W. G. Waddell (London and Cambridge, Mass., 1940; repr. 1980).

72. J. J. Scaliger, "Isagogici chronologiae canones," *Thesaurus temporum*, 2d ed. (Leiden, 1658), bk. 3, esp. pp. 278–79, 316–17. The change from "mythical" to "proleptic" time occurs in the surviving partial holograph of the *Thesaurus*, Leiden University Library, MS BPL 1909.

73. Scaliger, *Elenchus Parei*, 80–81.

74. Scaliger, "Animadversiones in chronologica Eusebii," *Thesaurus temporum*, 45.

75. The most recent account of the classical sources for the myth of Daedalus is S. P. Morris, *Daidalos and the Origins of Greek Art* (Princeton, 1992), pt. 1 (for Palaephatus, see 242); and, more generally, P. Veyne, *Did the Greeks Believe in their Myths?*, trans. P. Wissing (Chicago and London, 1988), 67–69.

76. See the rich survey of D. C. Allen, *Mysteriously Meant* (Baltimore and London, 1970). See also E. Iversen, *The Myth of Egypt and its Hieroglyphs in European Tradition* (Copenhagen, 1961); and R. J. W. Evans, *The Making of the Habsburg Monarchy, 1550–1700* (Oxford, 1979), chap. 12.

77. Scaliger, "Animadversiones," *Thesaurus temporum*, 51.

78. G. B. Vico, *De constantia iurisprudentis*, p. 2: *De constantia philologiae*, in *Opere giuridiche*, ed. P. Cristofolini (Florence, 1974), 599; G. B. Vico, *Scienza nuova* (1744), para. 87.

79. J. Selden, *Table Talk*, ed. F. Pollock (London, 1927), 116.

80. See especially the seminal paper by B. S. Cohn, "The Pasts of an Indian Village," *Comparative Studies in Society and History* 3 (1961): 241–49 (also in this volume).

Indian Time, European Time

Thomas R. Trautmann

In the master narrative of modernity, India finds a place alongside other non-European civilizations as the embodiment of the premodern. The construction of the master narrative appropriated the results of the new Orientalism that followed the British conquest of Bengal in the mid-eighteenth century, and ever after ancient Indian views of time have played a prominent role in the definition of archaic or nonprogressive cultures. Although nineteenth-century European discussion of Indian time is prefigured in significant ways by Christian time and its apologetics, it is with Hegel, Marx, and James Mill that India's place is finally fixed in a narrative of progress. India and Indian time continue to play the same role in the twentieth century, in spite of the fact that a robust progressivism proves more difficult to sustain, even in writers such as the prominent historian of religion Mircea Eliade, whose point of view could not properly be described as progressivist. Thus, Eliade's discussion of Indian time puts India in the category of traditional societies, as opposed to modern ones, and constructs of their sense of time the unity of the Australian Aborigine, the Chinese, the Hindu, and the European peasant.[1]

One of the more salient aspects of the situation we now find ourselves in is that the master narrative of modernity has been thrown into question and subjected to critical scrutiny with an intensity that seems to surpass any previous attacks. I think especially of such recent works as Johannes Fabian's fine critique of anthropological concepts of primitiveness, *Time and the Other* (1983), and Robert Young's rather wild *White Mythologies* (1990), the Derridean title of which deflates the universal pretensions of the narrative of modernity by giving it a racializing name. Whatever the outcome of that critique may be in the long run, it offers us a few degrees of freedom from the master narrative that we can use opportunistically to reconsider Indian time—what it is, what its place has been in European discussion, and how we should understand its place in Indian culture.

I

The lawbook of Manu is a convenient text from which to illustrate the ancient Indian conception of time and an apt one for the study of western reactions to the Indian conception. It is convenient because its exposition of the divisions of time is comprehensive and succinct. It is apt because this text has had great authority in Hindu culture ("Everything Manu has said is medicine," is a maxim among the pandits) and because it has been among the most cited of proof texts in western discussions of Hindu culture, since 1794 when its first translation by Sir William Jones was published.[2]

"Time and the divisions of time" are among the phenomena of this world that proceed from the body of the Creator in the Hindu cosmogeny with which the book opens (Manu 1.12), and the units of time, from the very small to the very large, are specified and explained in the course of the creation story. The basic unit of the creation narrative is the day-and-night (*ahoratra*), and its fundamental trope is the metonymical relation it establishes between the human day-and-night and the creative and inactive phases of the Creator, Brahma. The two halves of the day-and-night pair are associated with light and darkness, activity and sleep, subject-object consciousness, and the drawing in of consciousness. These opposed pairs are identified with the cosmic alternation of creation and destruction, such that the day-and-night of Brahma consists of the bodying-forth of the creation, followed by its dissolution and reabsorption, in continuing cycles.

There are, however, intermediate grades of beings between humans and Brahma, and Manu's account of the divisions of time follows the scale of being. Different time units (Manu 1.64–80) appertain to five kinds of beings, each of which has its own scale of time: humans, ancestors, gods, the Creator, and the Manus, a class of being to which the putative author of the text belongs.

Beginning, then, with human time, our text opens the account of time divisions with its version of the infinitesimal and builds to the basic unit. Eighteen twinklings of the eye (*nimeshas*) make one "second" (*kastha*); thirty seconds make one "minute" (*kala*); thirty minutes one "hour" (*muhurta*); and thirty hours a day-and-night. The sun divides days and nights, human and divine, we are told, the night for repose and the day for exertion. The days and nights of different grades of divinities are then explained. A month is the day-and-night of the ancestors, consisting of a dark fortnight for their active exertion and a bright fortnight for their sleep (it is interesting that for the dead the associations of bright and dark are interchanged). A year is the day-and-night of the gods, the half-year of the north-going sun being their day and the half-year of the south-going sun their night.

We then come to the day-and-night of Brahma, which is more complicated, involving the explication of the four world ages, or *yugas*. The first age, the Krita, consists of four thousand years, with a "twilight" preceding it and another following it, of as many hundreds, being homologized to the daytime and its two twilights, dawn and dusk. The three succeeding ages and their twilights are diminished by one in each.[3]

Krita	4,000	+	400	+	400	=	4,800
Treta	3,000	+	300	+	300	=	3,600
Dvapara	2,000	+	200	+	200	=	2,400
Kali	1,000	+	100	+	100	=	1,200
Total							12,000

These years, however, are years of the gods; and since a human year is a day of the gods, and the days in the year are taken to be 360, we can work out the equivalence in human time as follows.

Krita	4,800	×	360	=	1,728,000
Treta	3,600	×	360	=	1,296,000
Dvapara	2,400	×	360	=	864,000
Kali	1,200	×	360	=	432,000
Total	12,000	×	360	=	4,320,000

The twelve thousand divine years, which are the total of four human ages, make one age of the gods, a *mahayuga,* or "great age." A thousand of these ages of the gods makes a day of Brahma, called a *kalpa,* his night being of equal length. This gives us a kalpa of 12 million years of the gods, or 4.32 thousand million human years. Waking at the end of his day-and-night, Brahma creates mind, which performs the work of creation by modifying itself, impelled by Brahma's creative desire.

Finally, we come to the time of another class of beings, the Manus. Their age, the *manvantara,* is seventy-one *mahayugas,* fourteen of them occupying each day of Brahma or *kalpa.*[4] "The manvantaras, creations and destructions of the world are numberless; as if playing, Brahma makes it again and again" (Manu 1.80).

Thus, in Manu's account of time each grade of being has its periodicity, based on the cycle of the day-and-night: for humans the solar day-and-night, for ancestors the dark and bright fortnight of the lunar month, for the gods the north-going and south-going paths of the sun in the solar year, for Brahma a pair of *kalpas;* the day and night of each grade of being standing in fixed relation to that of the other grades. (The text does not give the Manus a determinate day-and-night.) Or, again, each grade of being (ancestors excepted) has its "age."

humans	*yuga*
gods	*mahayuga*
Brahma	*kalpa*
Manus	*manvantara*

What matters to us on the human scale is the character of the *yugas*. The text does not need to tell its readers what the names of the *yugas* say of themselves, that each succeeding *yuga* is inferior to the last. The names of the four ages are taken from the four possible throws of the dice. The ancient Indian dice were oblong and had four faces, not six, the winning throw being a four (Krita, "made" or perfect), followed in decreasing order of value by the trey (Treta), deuce (Dvapara), and ace (Kali, "black"), the losing throw.

Manu's text goes on to elaborate upon the significance (1.81–86) of the *yugas*. In the Krita age, Dharma and Truth are four-footed and whole, and no one profits from unrighteousness; but in each of the succeeding three ages, because of unjust gains, Dharma is deprived of one foot, and because of theft, falsehood, and fraud the merit gained by pious deeds is diminished by a fourth. Here Dharma is figured, as a cow perhaps, standing on four legs, then three, two, and finally one. Again, in the Krita age humans are free of disease and live four hundred years, but the span of life diminishes by a hundred years in each succeeding age. Because of human decline, moral and physical, different duties are prescribed in the different *yugas;* more especially, in the Krita age the highest duty is the performance of austerities, in the Treta religious knowledge, in the Dvapara sacrifice, and in the Kali the giving of religious gifts.

Though Manu's text does not say so, it clearly presumes what the testimony of Sanskrit literature uniformly holds, that the age in which we now live is the most degraded, the Kali age. Ancient Indian texts are generally agreed that the war of the great epic (the *Mahabharata*) is the juncture at which the Kali age begins. Traditions as to its date vary; one of the more widely held puts it at the equivalent of 3102 B.C. Since the whole duration of the Kali age will be 432,000 years, by far the greater part lies before us, and we may expect a continual worsening of the human condition. A pious monarch can, it is true, arrest the decay, or even bring about for his time the return of the Krita; the king, as they say, makes the age, not the age the king. However, the general shape of the historical process is decidedly one of long and inexorable decline from a golden age until, at the end of the Kali, there is a general dissolution followed by the restoration of the golden age, and the recommencement of the four ages. It is a cyclical view of time in the long, long run, but in the merely long run (the last four million years of human history, give or take), it is degenerationist. What it is not on any scale is a developmental or law-of-progress view of history.

To this brief account from Manu a great deal more could be added from the *Mahabharata* and the Puranas, but, while that would add much detail and some variation, the structure would remain essentially the same throughout. Jaina and Buddhist texts, on the other hand, have schemes of world ages with distinctly divergent nomenclatures and numbers, and some have the look of deliberate distortions or redefinitions of Hindu schemes that read like parodies, but even here gross structural similarities with the Hindu system are immediately apparent. We give only one passage, from Buddhist literature, for its compelling image: suppose a mountain of iron is brushed by a piece of muslin once every hundred years. The mountain will be worn away before the end of the "incalculable" (*asankheyya*), one of the four divisions of a "great period" (*mahakappa,* Skt. *mahakalpa*).[5]

Names and numbers differ among the different religions, but it is plain that we are dealing with variants of a single pattern, a unitary Indian intellectual culture of time. Its tendency is to multiply cycles of world ages without limit, to make of time an eternity.

II

Europeans acquired accurate knowledge of Indian time conceptions following the conquest of Bengal by the East India Company in the middle of the eighteenth century, which created the conditions under which access to Sanskrit literature became possible. The older Orientalism of Europe had been based on Hebrew, Arabic, and Persian, and the study of an East that was Near. The work of the British Orientalists, stimulated by the formation in 1784 of the Asiatic Society of Bengal at Calcutta under Sir William Jones, made possible a new Orientalism in which the study of Sanskrit and of ancient India was central. In the first flush of European enthusiasm for the newly revealed Indian past, the idea gained currency that knowledge of Sanskrit and its literature would bring about a quantum expansion of European consciousness, a second Renaissance, as knowledge of Greek had brought about the first. The new India-centered Orientalism supplied the materials with which enthusiasts for the Oriental Renaissance constructed their version of ancient wisdom, as Raymond Schwab has so thickly described.[6] Their progressivist critics, holding a "future wisdom" view, used the same materials to answer the exam question that they felt history itself had posed: "Explain the causes of the superiority of European civilization to the civilization of India." What is striking is the unanimity of opinion that comes to prevail among the most divergent participants in the discussion of the West respecting Indian time.

Beginning with the Calcutta Orientalists, European representations of Indian civilization invariably feature discussion of the *yugas* as a key giving privileged access to its inner character; and, in comparative discussions of time cultures, Indian time is almost always represented and often figures as the defining type of cyclical time conceptions. Very early on, a set of ideas about Indian time is formed and becomes general. These include the idea that ancient Indians wrote no, or little, history, that history was weakly developed or absent as a department of intellectual activity, that there was no sense of history as we, of Europe, know it. They consist also of ideas about linkages between the lack of history and Indian conceptions of time: that ancient Indian scales of time were so fabulously large that they drowned the significance of human history, that the cyclical character of Indian time was inimical to a true (linear, developmental) historical sense, and that the endless cycle of rebirth (*samsara*) tended in the same direction—any combination of these in the work of many different authors constitutes a core of ideas about India from the eighteenth century onward.

In all such discussions of Indian time the sense of amazed and sometimes appalled fascination that attends draws its energy from the feeling that in this respect India gives privileged access to what it is, and what Europe is not, at some deep level, and by the contemplation of this deep not-Europe, of what it is that is deeply and distinctively European. One gets a strong dose of this overplus of feeling in Hegel and in James Mill, both of them well read in the information supplied by the new Orientalism, both of them critics of the enthusiasm for India.

Hegel's long chapter on India in the *Philosophy of History* dwells on the absence of history as a department of ancient Indian literature. He finds a number of causes of which the most general is the lack of cumulative development, without which (he argues) history writing does not arise. This implicates the cyclical/linear contrast, which he theorizes in the following way.[7]

> The mutations which history presents have been long characterized in general, as an advance to something better, more perfect. The changes that take place in Nature—how infinitely manifold soever they may be—exhibit only a perpetually self-repeating cycle; in Nature there happens 'nothing new under the sun,' and the multiform play of its phenomena so far induces a feeling of *ennui;* only in those changes which take place in the region of Spirit does anything new arise.

In the language of the structuralists, nature is to history as repetition is to development.

Marx's inversion of Hegel leaves India exactly where Hegel had put it.[8]

> Indian society has no history at all, at least no known history. What we call its history is but the history of its successive intruders who founded their empires on the passive basis of that unresisting and unchanging society.... England has to fulfil a double mission in India: one destructive, the other regenerating—the annihilation of old Asiatic society, and the laying of the material foundations of Western society in Asia.

If anything, the excess of affect in Hegel's treatment of India, hostile to the idea of Indian civilization and of an Oriental Renaissance based on the study of Sanskrit, is even more evident in James Mill's *History of British India* (first published in 1818); indeed, it lies on the surface and makes no attempt to conceal itself. The long essay "On the Hindus" therein represents itself as a scientific inquiry into a question of great public moment for Britain, namely, the place of India in the "scale of civilization." Mill finds it to be very low, whereby British rule, though despotic, is shown to serve India's good. The essay opens with the issue of Indian time and history, the first count, as it were, in the bill of indictment. I resist the temptation to quote him on these matters; it is so easy to hold Mill up to the ridicule of his own words that there is no pleasure in it. But I need to give the opening passage, which frames the discussion of Indian time and history in a larger context.[9]

> Rude nations seem to derive a peculiar gratification from pretensions to a remote antiquity. As a boastful and turgid vanity distinguishes remarkably the oriental nations, they have in most instances carried their claims extravagantly high. We are informed, in a fragment of Chaldaic history, that there were written accounts, preserved at Babylon, with the greatest care, comprehending a term of fifteen myriads of years. The pretended duration

of the Chinese monarchy, is still more extraordinary. A single king of Egypt was believed to have reigned three myriads of years.

The criticism of Chaldean, Chinese, and Egyptian time in the same breath, so to say, with the Indian, has its own history, which is well worth tracing, for it articulates a certain structure of thought about time in which different non-European nations are interchangeable in the sense that they occupy an identical relation to the European sense of time. More especially, European criticism of the time of the Chaldeans, Chinese, and Egyptians predates the criticism of Indian time, so that the criticism of Indian time may be seen as the bringing to bear on a new object (India) of an existing critique of non-European time or—what comes to the same thing—defense of European time against the time of other nations. Let us see what that defense is, by situating it in history and identifying the subject and the object of it.

III

We may readily see what and where it is not. The defense of European time against that of the Indians et al. could not have come about for the first time in the present century, nor did it come about in early Greek encounters with India. I hope I may be permitted to violate the ancient taboos of history writing, and draw significance from the double negative of something the Greeks did not say, and an event that did not take place in the twentieth century.

The best of Greek knowledge of India—though one should not exaggerate its quality—came about as a result of Alexander's expedition in the fourth century B.C. and, after his death, the sending of ambassadors to the Mauryan imperial court by the Hellenistic successor-states such as the embassy of Megasthenes, sent by Seleucus to Chandragupta Maurya. In this literature the only item that concerns the Indian time-sense is a statement from Megasthenes, appearing in Arrian's *Indica,* that from Dionysus to Sandracottus (Chandragupta) the Indians counted 153 kings over 6,042 years, and that Dionysus came to India fifteen generations before Heracles.[10] So far from offending the Greek sense of the possible by exaggeration, the Indian deep past is assimilated to the Greek by the identification of Dionysus and Heracles with characters of Indian mythology.

The rest is silence, and silence is notoriously difficult to evaluate. Let me at least indicate how the alternatives appear to me. In the first place, since our Sanskrit literary sources for Indian time in its developed state are not datable prior to the second century B.C., it is possible that the doctrines of world ages were not yet formulated by the time of the Hellenistic ambassadors. However, the edicts of Ashoka, grandson of Chandragupta Maurya, refer to the idea of the *kalpa* as a very long period of time and use a term of art that occurs in later Buddhist texts to speak of "the age of cosmic destruction."[11] This is within the period of close interchanges between India and the Hellenistic states. The evidence is thin, but it is enough to encourage us to think that large time periods, and the dissolution of the world that marks their boundaries, were sufficiently well developed in Ashoka's day to have become part of the fund of common knowledge upon which the emperor could draw in his public pronouncements.

Since, then, the time doctrines seem to have been established by the time of the Greek ambassadors, a second alternative is to suppose that the Greeks did not inquire about such matters. The Alexander historians wrote about military matters, and the ambassadors to the Mauryan court were practical men who also were especially interested in matters that had a military bearing such as the methods of capturing and training elephants for warfare. They did not have Indian languages, and their knowledge of India and of Indian learning was accordingly quite shallow, and sometimes downright mistaken, as when Megasthenes says that the Indians have no slaves and no written laws. And they were susceptible to reports of monstrous races that had been a staple of Greek writing on India for centuries previous. However, it seems to me unlikely that the Greek ambassadors should not have had some encounter with Indian ideas of large time periods if they were matters of public discourse, as the Ashokan edicts imply, but it is entirely possible that their orientation toward the practical and the exotic was not engaged by it.

If, then—and this is the third alternative, which seems to me the likely one—Indian accounts of world ages were known to the Greek travelers but did not excite their interest in the practical and the exotic, the reason may well be that such ideas were after all not so terribly different from Greek ideas of the eternality of the world and of cyclic time.

In the twentieth century, too, the ancient Indian conceptions seem more familiar than exotic. Four million years—the length of the four yugas—may be too much for the span of human history but, over the two million years or so through which hominid development may now be traced, not vastly so. And for the age of the earth it is, if anything, much too short; at 4 billion years, it is the *kalpa* that is more nearly its measure according to modern notions. It has often seemed to me that the Indian time units, whose magnitude has long been the object of criticism in the West, should, if anything, have been a more congenial matrix for the development of a modern scientific cosmology than the cramped confines of the time sense traditional in Europe. Indeed, it grows more congenial as science develops. Since the discovery of the shift of stellar light to the red end of the spectrum, it has been known that the universe is expanding, but whether the rate of expansion is steady or slowing remains uncertain. If it is slowing, the expansion may reach a limit after which the universe will begin to contract. One might then imagine the world process as an oscillation between phases of expansion and contraction—a cosmic heartbeat, as it were. It should be obvious that ancient Indian time conceptions are favorable to such ways of thinking about the universe.

I am not suggesting that the ancient Indian conceptions of time influenced the actual course of modern science, or even that they anticipated it in any real sense, since the empirical element in them is small, and the element of learned invention is obvious. What I am suggesting is that the western response to ancient Indian time conceptions would not have been so uniformly negative had the close encounter with Indian thought begun in the twentieth century with its expanded time horizons for the history of humans, of life, of the earth, and of the cosmos. Of course, it did not; it arose before the vast expansions of traditional European conceptions of time, more especially before the expansion, in the mid-nineteenth century, of human time from the narrow limits of the biblical chronology.

Which leads us to the object of our quest; for, if the European critique that encompasses Indian time has its source neither in modern science nor in ancient Greece, it must surely come from Christianity. And, indeed, the two features of Indian time on which the critique fastens, the vast length of the ages and their cyclicity, with which Greek and modern science fail to engage, are reflexes of the specific characters of Christian, or rather biblical, time. The chronological materials in the Bible do not form a finished scheme and pose many thorny problems of interpretation, which results in an irresolvable plurality of solutions, but the solutions fall into a band of variation that puts the Creation between something less than four thousand years to somewhat more than five thousand years before Christ. World chronologies that derive from the Bible show a strong resemblance among themselves and collectively are very short compared with those of the Chaldeans and Egyptians, to say nothing of Greek notions of an eternal world. Moreover, the content of that short time of creation is strongly directional and hardly cyclical, being filled with a sequence of unique events; at best we could say that the shape of biblical history forms a single cycle, of fall and redemption, Paradise lost and Paradise regained.

The immensely influential Christian philosophy of history that St. Augustine constructs in his *City of God* is the most important prior text to the passage we have quoted from James Mill, for in it Augustine confronts and attacks pagan views of time in very similar terms.[12] In his view, time is created by God and is characterized by movement and change, situated between the two eternities of the past and of the future state at the end of history. He adopts the idea of the Great Week. As "One day is with the Lord as a thousand years, and a thousand years as one day" (2 Pet. 3.8; *City of God* 20.16, p. 719), the six days of Creation stand for a duration of six thousand years before the eternal sabbath. Thus, Augustine says more than once that "not 6,000 years have yet passed" since the Creation (12.10, 12; 18.40, pp. 390, 392, 648). This short chronology he defends against histories, characterized as "false" and "mendacious," which allow many thousands of years to the world's past, of which he singles out the Egyptian as the type (12.10, 12). Within that short chronology he defends the superior antiquity, and therefore the authority, of the Bible against exaggerated claims—at 100,000 years!—for the wisdom of the Egyptians. Hebrew was preserved and transmitted to Abraham by Heber, from whose name the word Hebrew is derived, as a written language, he argues, whereas such wisdom as the ancient Egyptians had could not antedate Isis, who gave them letters, since he lived after Abraham, only about two thousand years ago (18.37–40, pp. 646–48). He defends as well the createdness of the world and of time against Greek views as to the eternality of the world and time, or numberless worlds, or the cyclical destruction and renewal of the world (12.11, 13).

Christian chronographers devised several methods of dealing with the challenge posed by the great antiquity claimed by the Egyptian and the Chaldean national traditions in addition to dismissing them as mere lies. Augustine, for example, considers the possible resolution offered by the reports that the Egyptian year was formerly only four months long, three such years being equivalent to one of ours (12.10, p. 391). But critiques and resolutions went largely unanswered, and their ingenuity was lost upon their object, which no longer existed. For the priesthoods and religions of ancient Egypt and Chaldea were long extinct, having been replaced by Christianity and, later, Islam; and with them was lost the ability to read the hieroglyphs and cuneiform tablets until the great decipherments of the nineteenth century. For a very long time the Christian defense of biblical time was a dialogue among the converted, unchallenged from without.

That condition prevailed until the voyages of discovery, which brought to Europeans new knowledge of other national traditions, in Asia and the New

World, with long chronologies. Of these the most frequently mentioned was that of China, in the reports of the Jesuits, the third player in the dramatis personae of James Mill's attack on non-European time. Chronology, in the Age of Enlightenment, became a battleground in which skeptics invoked the newly recovered national histories and the ancient records of Egyptian and Chaldean historical traditions to attack Christianity, and believers sought rational arguments to defend biblical time against skeptics and to attack other national histories or try to reconcile them with the Bible. Nineteenth-century treatments of the issue, including those of Hegel and James Mill, it seems to me, should be thought of as secularized renderings of the Christian culture of time, in which the subject and object of the Christian discourse are replaced by Europe and not-Europe. A biblical culture of human time, then, time that is short and linear, directional, possibly progressive, possibly unicyclic but in any case not recursive, is perpetuated in a secularized transformation, surviving at least until the Brixham Cave excavations of 1859, which made the short chronology for human history difficult to sustain.[13]

For completeness' sake, and for its relevance to the problem of Indian time, I must add a word to what has already been said about the subject of biblical time. The subject, of course, is not only Christian but Jewish and Muslim as well. Lacking a ready-made term more comprehensive than Judeo-Christian, we might adopt a Muslim expression and say that among the Peoples of the Book there is a common culture of time, and that the opposing object of its apologists is defined reflexively as the Gentiles, "the Nations" of the Bible. The geography of biblical time and its limits are made very visible in the works of the eleventh-century scholar al-Biruni who was one of the finest of the Muslim chronologers, and the greatest of Muslim Indologists, who learned Sanskrit and studied the Indian sciences at their source.[14] In his *Book of India* he gives us an extensive account of the Indian time-doctrine. His work comparing the chronologies of the peoples known to him offers unusually clear evidence of the siting of a major cultural fault line. His discussion of Jewish and Christian chronologies criticizes them for the many discrepancies flowing, as he sees it, from their corruption of the biblical text. But the differences among them are small compared to what he encounters in India. For not only are Indian stretches of time unimaginably long, but by virtue of cyclical conceptions the Hindus do not even believe in creation properly so-called; for them God reworks the same clay to make the world from one age to the next. We would be justified in saying that in confronting Indian conceptions of time the biblical peoples sense that they have touched a deep core of difference.

IV

Having established its context in the broadest terms, we come to that other intimate contact with India by Peoples of the Book, in the eighteenth century, the one that is decisive for the formulation of a unitary European view of Indian time. Like that of al-Biruni, the new Orientalism of the West was made possible by conquest, and it required for its success the learning of Sanskrit.

Before they learned Sanskrit, Europeans approached India through the existing, older Orientalism of Arabic and, above all, Persian. Thus, Anquetil Duperron's translations into French of the *Avesta* (1771) and the Upanishads (1801) were mediated by Persian; Alexander Dow's contribution, one of the earliest British contributions to follow the conquest of Bengal, was a translation of Firishtah's Persian history of India (1768); Nathaniel Halhed's translation of the so-called *Code of Gentoo Law* (1776) was from a Persian translation of the Sanskrit original compiled by pandits and commissioned by the East India Company for its use; and Sir William Jones, who consolidated the new Orientalism by founding the Asiatic Society at Calcutta (1784), came to India, before he had learned Sanskrit, with a well-established reputation in Arabic and Persian scholarship.

The particular angle of entry into Hindu antiquities provided by Perso-Muslim literature tended to reinforce the biblical frame through which Europeans would incline to view the long past of India, and in this frame the vastness of the *yugas* could not be fitted and accepted as true. On the other hand, Enlightenment values of reason and skepticism of religious authority gave another angle of entry, which inclined some to take seriously Hindu claims to great antiquity, to the detriment of the biblical frame, and use Indian time to gain critical perspective upon received ideas. These two points of view and the problem of their conflict are made explicit in the works of Dow, Halhed, and Jones. We examine the three of them, for in them we see the beginnings of European discussion of Indian time at a moment when the assessment of it was as yet fluid and in process of being fixed.

Dow's translation of Firishtah opens with an account of the four *yugas* and Brahma's day-and-night, followed by a specimen of the early history of the Hindus, which amounts to a summary of the epic, the *Mahabharata*. But Firishtah does not credit this supposed history, and to recover the deep past of India he abandons Hindu sources for the biblical frame.[15]

> As the best and most authentic historians agree that Adam was the father of mankind, whose creation they place about five thousand years before the

Higerah, the sensible part of mankind, who love the plainness of truth better than the extravagance of fable, have rejected the marvellous traditions of the Hindoos, concerning the transactions of a hundred thousand years, and are of opinion that they, like other nations, are the descendants of the sons of Noo, who peopled the world.

The Hindus "pretend to know nothing of the flood" of Noah, but the testimony of other nations proves its truth, whence their origins may be traced from that era according to the best authorities. Firishtah then gives a listing of the sons of Noah's sons, Shem, Ham, and Japhet, which amounts to a brief ethnology of the world: the sons of Shem are the progenitors of the Arabs and Persians, among others; of Japhet, Turc (father of the Turks), Chin (the Chinese), and Rus (the Russians); of Ham, Hind (the Hindus) and others. Thus, India is drawn into the "Mosaic ethnology" of the biblical tradition at the expense of Hindu claims to a fathomless past.

But, though Firishtah rejects Indian time for biblical time, his translator Dow is not so certain, and he adopts a stance of enlightened skepticism.[16] "The Mohammedans know nothing of the Hindoo learning," Dow claims, and he speaks of Muslim ignorance of Sanskrit. Circumstances prevented Dow from carrying out his plans to learn Sanskrit, but he consulted pandits in Persian and Hindustani, acquiring as best he could a skeletal knowledge of the contents of Sanskrit literature. His estimate of what was yet to be found therein was highly favorable to Indian time. He thought there were many hundreds of volumes of Sanskrit prose (aside, that is, from the *Mahabharata*) treating of ancient India. Moreover, "the Hindoos carry their authentic history farther back into antiquity, than any other nation now existing." Sanskrit records give accounts of the history of western Asia very different from those in Arabic (which is tantamount, is it not, to saying that they differ from the Bible?), and it is likely that examination will show them to be older and more authentic than the latter. "But whether the Hindoos possess any true history of greater antiquity than other nations" is for the moment an open question, which "must altogether rest upon the authority of the Brahmins, till we shall become better acquainted with their records." Their pretensions are very high, and they regard the Jewish and Mohammedan religions as heresies from the religion of the Vedas, the Jewish religion being the creation, in the beginning of the Kali *yuga,* of the apostate son of a Hindu raja banished to the western region. Elsewhere Dow gives credence to a tradition that the Vedas were collected together by Vyasa at the commencement of the Kali *yuga,* "of which æra the present year 1768, is the 4886th year."[17] The danger Indian time poses for the biblical is made very plain without stating it, and the view that the Vedas predate the Bible became a matter of considerable controversy in Europe.

Halhed, the second of our trio, worked much like Dow, through Persian and Bengali, to get information about Sanskrit literature, but he acquired a distinctly better knowledge of the structure of Sanskrit, including particulars of its similarity to Greek. Like Dow he tended to look with favor upon Indian claims of a high antiquity, going even further in using Indian time to try out the possibilities of a rational skepticism. The conflict of times is made explicit at the outset of the discussion: "The Hindoos as well as the Chinese have ever laid claim to an Antiquity infinitely more remote than is authorized by the Belief of the rest of Mankind."[18] But he is careful to put his argument tentatively and to punctuate it with expressions of solidarity with biblical truths. Even so, he got into a certain amount of hot water for it.

Although there is nothing of special interest in Halhed's rendering of the doctrine of the *yugas,* a couple of arguments he makes in the course of his treatment of Indian time are worth noting. First, Halhed invokes an emerging conflict between geological doctrines and the Mosaic chronology in the West as a point in favor of Indian time—which tends to sustain the notion of the congeniality of Indian time to science.[19] Second, he finds the very long life spans of the antediluvial patriarchs in the Book of Genesis as difficult to accept as the Brahmanical ones, a curiously backhanded way of delivering a favorable verdict. The two chronologies are then partially reconciled, such that the beginning of the Kali *yuga* (about three thousand years before Christ), when the Hindu canonical life span reaches its present dimension of a hundred years, coincides nearly with the Flood of the Bible, at which point also the biblical human life span loses the vast dimensions that are so troubling to the faith in an age of reason. There is, however, the problem for this partial reconciliation of the chronologies of the Brahmans and of Moses, namely, that "some excellent authors"—doubtless including Dow—believe that most of the Hindu scriptures were composed at about the beginning of the Kali,[20]

> for then we at once come to the immediate Æra of the Flood, which Calamity is never once mentioned in those Shasters, and which we must think infinitely too remarkable to have been even but slightly spoken of, much less to have been totally omitted, had it even been known in that Part of the World.

The Brahmins remove this objection by two assertions: one, that all their scriptures were written before the Flood; the other, that the Flood did not reach India. Reaction to these ideas in England was strong, and negative, shaped by George Costard, clergyman chronologer and historian of astronomy, who wrote a pamphlet against them.[21]

The third and most important of our early British Orientalists is Sir William Jones whose mastery of Sanskrit and whose reputation for Orientalist scholarship, the latter achieved well before he went to India, gave his deliverances on Indian time an authority that was unrivaled.[22] He turned the matter entirely around. The drift of Dow's and Halhed's treatments of Indian time had been to align it with rational inquiry and, at least by implication, against the Bible story. Jones's project, to the contrary, was to find a completely rationalistic proof of the truth of the Mosaic narrative through study of the historical traditions of Asiatic nations, and in doing so he shows (as he believes) that ancient Sanskrit literature is an ally, not an enemy, of the Bible. The drift of his work is to provide a rationalist answer to Voltaire and others who were using the new knowledge of the non-European world to attack Christianity. But in doing so he was obliged to reject Indian time. Like Dow and Halhed, he was inclined to treat the learning of the Brahmins very favorably and to credit their historical traditions. But his guiding conception was that all national traditions were partially corrupt remembrances of a shared past, which was preserved uncorrupted only in the account of Moses. The authenticity of the Mosaic account was proved by points of agreement between it and the independent historical traditions of Indians and other Asiatic nations. He accepted the authenticity of Sanskrit accounts of the past when they could be interpreted as reflections of the same events recounted in Genesis or when they did not conflict with it; but when they positively conflicted with the Mosaic account he rejected them. He took the biblical chronology to be integral to the account and fundamental for a rational world history, and Indian time had to make way for it.

Though the execution of it was new, and made more effective by direct access to Sanskrit literature and various improved tools, Jones was in effect carrying on a project that has a long and venerable history among biblical peoples, the project of universal history or, as I like to call it, Mosaic ethnology (we see, indeed, a compressed version of it in Firishtah). For the Peoples of the Book the most distant nations and their histories find their ideal unity in the Mosaic narrative of the descendants of Noah and their dispersal. The account in Genesis of Noah's sons, Shem, Ham, and Japhet, and their various sons and sons' sons bearing ethnonyms like Eber (Hebrew), Misraim (Misr, Egypt) and Javan (Ionian, Greek) gives us a branching, treelike structure in which we recognize what anthropologists call a "segmentary" way of thinking about the relations among peoples, that is, in terms of genealogical distance among the patrilineal descendants of a patriarch. This narrative, and the segmentary principle that it embodies, unites history and ethnology at the global level and provides the overall frame for universal histories. But it is only a frame, in which we see the mere outline of a Great Story that unites all particular histories, but not the substance of it, since sacred history limits itself to the history of the line that descends through Noah and Shem through Eber and Abraham and does not recount the histories of "the Nations" as such. Thus, the Mosaic narrative gives us an incomplete structure, a family tree of mankind in which only one of the branches is carried forward in time and only the beginnings of the other branches are sketched in, the very incompleteness of which cries out to be completed. The project of universal history, or of Mosaic ethnology, then, consists of fitting the national histories of the Gentiles onto the segmentary structure of the Genesis account of the descent of Noah. The method is to find names and events in Gentile history that can be identified with names and events in sacred history. This requires, especially, synchronisms to make the match-ups and to integrate the separate national traditions into an overall history. Eusebius, in the fourth century, first constructed synchronic tables of national traditions as a grid for universal history. We can think of Jones as trying to find the place of the Indian tradition in the Eusebian grid for world chronology.[23]

Jones finds the match-up he needs in the Hindu doctrine of the ten avatars, or "descents," of Vishnu, in which he finds the Indian flood narrative that Halhed, and Dow's Firishtah, had unaccountably missed. The first three of the ten avatars of Vishnu are especially concerned with stories of worldwide flood: the Fish, the Tortoise, and the Boar. The Fish incarnation in particular seems ready-made for identification with the biblical Flood, carrying Manu (the first human), his family, and the seven sages (*rishis*) in a ship (the Ark of Noah!) fastened to a horn on his head. Jones identifies Manu with Noah (distinguishing this Manu from an earlier one, the progenitor of the human race, whom he identifies with Adam) and he and the seven sages with the eight humans aboard the Ark in the Bible story (Noah, his three sons, and the four wives) from whom the entire human race has since been propagated. Jones further identifies the fourth avatar of Vishnu, Narasimha, with the biblical Nimrod, descendant of Ham; and Bali, the demon who was overcome by the fifth avatar, the Dwarf, with the biblical Bel. The seventh incarnation, King Rama of Ayodhya, Jones identifies with the biblical Raamah, also in the line of Ham. With him begins "civil government" in India, or, as we would say, civilization. The Buddha ushers in the Kali age, and the king lists of historic figures such as Chandragupta Maurya follow thereafter. The four *yugas* are squeezed into the Ussherite chronology, putting the Creation at 4004 B.C., rejecting the traditional figures for them, or the traditional dating of the beginning of the Kali (3102 B.C.). In outline, Jones's Mosaic reading of Indian chronology may be represented as follows.[24]

Adam	Manu I	Krita *yuga*
Noah	Manu II	
The Flood	Fish, Tortoise, and Boar avatars	
Nimrod	Narasimha	Treta *yuga*
Bel	Bali	
Raamah	King Rama	Dvapara *yuga*
	The Buddha	Kali *yuga*

Today nothing remains of this construct, and the cross-cultural summing of flood narratives is carried on only by fundamentalists and those who attribute the rise of civilizations to extraterrestrial visitations. But the project of Mosaic ethnology provided Jones with the framework within which he made his most enduring achievements. The thing for which he is best known to us, of course, is his recognition of the historical relationship among Sanskrit, Latin, Greek, Gothic, Celtic, and Persian and his proposal of their common descent from a lost ancestor language, the conception of what we call the Indo-European language family. The Mosaic frame gave him the segmentary way of conceptualizing these relations; the nations speaking Indo-European languages are Hamians, codescendants of Ham.[25] Moreover, "On the Chronology of the Hindus" was not his last word on the subject, and problems of Indian chronology (as well as astronomical periods that might account for the vast periods of Indian time) continued to occupy him. In these efforts he comes up with a brilliant hit, a proof that the Sandracottus of the Greek accounts of Alexander is the Mauryan King Chandragupta, who displaced the Nandas in about 324 B.C. This synchronism provided a fixed point from which the rest of ancient Indian chronology could be revised and determined within the Eusebian space of world chronology.

The response in the European metropole to the examination of Indian time by the Calcutta Orientalists was surprisingly uniform. The tentative efforts of Dow and Halhed to align Indian time with a rationalist and skeptical perspective was greeted with some overt hostility, and in the aftermath of the French Revolution it came to nothing. Jones's rejection of Indian time prevailed, uniting the most divergent constituencies. Jones was the architect of the view that became the foundation for the India enthusiasm in Europe and among the

Transcendentalists of New England, the view that put India at the center of the conception of primitive monotheism. He posited the unity of the Indians, Greeks, and Romans (and also the Egyptians) as codescendants of Ham, as the inventors of the arts and sciences of civilization, and as adherents of a pure, natural religion from which, however, they were the first to decline into idolatry. The Veda was the oldest and purest draught of that original natural theology, and Hinduism was the still-living relative of the long-extinct Greek paganism, within which one could still find vestiges of the most ancient wisdom. Proponents of the Oriental Renaissance, therefore, looked upon Sanskrit literature as a source of European religious renewal but within a biblical time horizon that put the Veda close to the time of Noah. British Evangelicals and Calvinist missionaries, on the other hand, generally took the view that Hinduism was without merit, and they found in Jones's treatment of diluvial history and his rejection of Indian time the materials for a very un-Jonesean attack on Hinduism.[26] Progressivists such as Hegel, Marx, and the Utilitarian Mill found in it material for a parallel attack on Indian civilization. Given the various hostilities among the parties to this agreement (James Mill, for example, singled out Jones as the very type of the wooly-minded Orientalist who was soft on Hinduism) and the range of political positions they occupy, the unanimity of the European response to Indian time is impressive. Biblical views of heathen time conform themselves to the shape of the new mold of the master narrative of modernity and its march-of-civilization logic, readily and without residue.

The irony of the early nineteenth-century European chorus of opinion rejecting Indian time is that at mid-century the discovery of human artifacts in association with the remains of long-extinct animals in Brixham Cave and elsewhere forced a major reappraisal of the antiquity of human existence that burst out of the constraints of the Mosaic chronology. The "revolution in ethnological time," as I have called it, undermined the short biblical span for human history and replaced it with an archaeological time span receding backward to an extent that was, at the outset, immeasurable; and, as its measure was taken, it resembled that of the *yugas* the more as it resembled that of biblical time the less.[27]

In the end, European reactions to Indian time reveal more about Europe than about India. What they reveal is a striking continuity among Christian and secular views of Indian time, views that by their own momentum persist into the twentieth century where they are not quite at home and where they could not have arisen ab origine. They reveal as well, in the intensity of feeling that accompanies them, that there is more at issue than time alone. It is a question, we sense, of time being an integument that joins the individual to the object of a quest, a source of knowledge and meaning, that lies beyond time: the pilgrim's progress toward salvation, or the nation's progress from rudeness to civilization. It is this sense that is baffled and affronted before the immensity of Indian time.

V

If this is correct, we may hope to press home the gains of the finding in order to come to a better understanding of Indian time and its place in Indian culture, and in the world. For the analysis thus far shows that in respect of Indian time the master narrative of modernity is cracked beyond repair, and a fresh look is needed. And it suggests, as a way to move forward, that we examine Indian time in relation to ideas of person and authority.[28]

Analysis must take as its starting point the recognition that the European discussion of Indian time, though motivated by extrinsic concerns in its valuations, has not fabricated the sense of difference from which it sets out or the finding that history is relatively underdeveloped as a branch of ancient Indian literature.[29] It is, of course, perfectly valid to take a different approach, according to which "history" is replaced by a broader notion such as "representations of the past." We could then say that ancient Indians (and perhaps others) have a well-developed sense of their past and a branch of literature (*itihasa, purana*) to match, but that that sense was very different from the European. Romila Thapar has used this approach with good effect.[30] But, since to do so only shifts the sense of difference from the subject to the predicate, it tends to confirm that it is an inescapable datum for any discussion of the matter. While there are enough counterexamples to show conclusively that ancient India was perfectly capable of producing what can truly be called histories, the deficiency relative, let us say, to China and Greece is noticeable and must have a meaning for us to discover.

With this as our starting point, we begin by observing the historical-social location of the Indian doctrines of world ages. They are socially locatable as the work of religious specialists, the brahmins and the Buddhist and Jain monastics. Historically we locate them in a somewhat imprecise interval between the end of the Vedic period and the times of the emperor Ashoka (third century B.C.) when, as we have seen, they appear to be established in public culture. The textual location confirms the post-Vedic character of these doctrines: they are found in the literature of the anti-Vedic religions of Buddhism and Jainism and in the post-Vedic brahmin productions of Hinduism comprised of the lawbooks of Manu and others, the great epic (the *Mahabharata*), and the compendia of mythology and theology called the Puranas. The doctrines of world ages therefore appear to have a definite beginning in time—however difficult it would be in practice to say exactly when that was—and not to go back to a fathomless past. And they are part of what is common to the post-Vedic religions in spite of their antagonisms inter se. That common ground is constituted principally by the doctrine of transmigration. The doctrine has two aspects: *samsara*, the wheel of rebirth, according to which one acquires another body after death, and *karma*, the law of retribution by which the moral quality of one's deeds brings about an eventual reward or punishment, that is, in another life. Together they constitute an impersonal moral mechanism governing all of life, even the gods. All the Indian high religions take *samsara* to be the fundamental ill and give to their conception of the highest good the character of escape from it. It is from this agreed problematic that the great Indian religions set out, but they differentiate themselves from one another by the escape routes they offer.

The transmigration doctrine introduces a conception of the person extended before birth and after death into a series of states that must necessarily be very long if it is to have the negative valence it has in the post-Vedic religions. By contrast, the earlier Vedic belief in a heaven or world of the ancestors has no intrinsic time requirement. *Samsara*, figured as a wheel, contains in itself the salient features of length and circularity that characterize the doctrines of world ages, and it is certainly the starting point and motive force for their elaboration.

Indeed, the doctrine of world ages shows an unmistakable tendency in the texts to expand without limit, creating, as I have said, an eternity within time through which the transmigrating soul must be endlessly reborn. This expansive tendency expresses not only the endlessness and pointlessness of *samsara* but the location of truth and liberation from *samsara* in an eternity outside time. The transmigration doctrine establishes a principle superior to the gods, a moral law that is impersonal and to which all persons, whether divine or human, are subject. This contributes to the formation of an idea of Truth (or the religious law, or the teaching of the Buddha, and so forth) that is uncreated and eternal, having ultimately no person as its source and lying beyond time. Authority, and more particularly the authority of the religious specialists of Hinduism, Buddhism, and Jainism, is founded upon claims of access to this ultimate level of Truth. Insofar as the world ages tend to multiply without limit, there must be a corresponding multiplication of authoritative transmitters—one for each age perhaps—whence Hinduism has its many

Manus, Buddhism its Buddhas, and Jainism its Tirthankaras. The players in the world drama, then, are the transmigrating subjects in time, seeking an authorless Truth, the source of which lies beyond time, which is mediated in time by a series of transmitters. At the ends of the transmitter chains are the brahmins and monks of time present.

We return to Manu's lawbook, where we began, to illustrate the relation of time to concepts of person and authority. We begin with the person of Manu. Who is he? His is a story in the postmodern mode, of the disappearing author—or, we might say, in the ancient Indian mode, of the multiplying author. Manu is, to start with, the son of the Self-existent Brahma, the creator—or, again, the son of Vivasvat. He is the lord of created beings, Prajapati, who fathers mankind, the Hindu Adam, as the Orientalists say. Already there is ambiguity about whether Manu is one. Sir William Jones, reconciling Hindu chronology to the biblical, felt free to identify two Manus with Adam and Noah. But Manu himself tells us that there are many Manus, seven in all, each generating and protecting all beings during the age (*manvantara*) belonging to him, the multiplication of Manus, then, being a concomitant of the theory of world ages.

And who, then, is the author of the *Laws of Manu*, the text that bears his name? Apparently he is one of the seven, the son of the Self-existent, Svayambhuva Manu. But the frame story with which the text opens leaves the authorship question hopelessly ambiguous. In it the great sages approach Manu and ask him to declare the sacred laws to them, whereupon he recites the story of the creation of the world, including the time doctrine we have been discussing. The text of the lawbook, Manu says, was composed by the Creator himself, Brahma, who taught it to him in the beginning; Manu taught it to the ancient sage Bhrigu (among others). He then passes on to Bhrigu the task of reciting the lawbook to the sages who had approached him, which he proceeds to do. In one sense, then, the author of the Laws of Manu is the Creator; in another it is Manu, who taught it to Bhrigu; in yet another it is Bhrigu, since the body of the *Laws of Manu* is represented as being voiced by him. But, although we are invited to think that the bulk of the *Laws of Manu* are the words of Bhrigu voicing Manu's teaching, there is yet another narrative level, that of the anonymous voice that describes the framing scene of petitioning sages answered by Manu and Bhrigu. To account for this faceless speaker one commentator says that the text is compiled by someone in the unbroken pupilary succession from Bhrigu.[31] But, wherever on this line of transmitters the ambiguous voice of the text is located, its authority is unimpaired because in the last analysis it does not derive from a person.

There are other complications, outside the problem of voice, having to do with its supposed location in an immemorial past, the time of the sages. It is notable that the text refers to no datable historical event, as, indeed, what purports to be an originary text must not. But looking closely we find it often wanders from its proper time horizon, that of Manu's and Bhrigu's colloquy with the sages, by referring to the views of other authorities whose time horizon is long after, and to the legal literature as a class, of which the *Laws of Manu* is supposed to be the first item. Moreover, many verses of the text are also found in the *Mahabharata* (including those on the world ages we have been discussing) and there are many verses quoted in other lawbooks attributed to Manu (including an "Old Manu" and a "Great Manu" whose texts have not survived) which are not found in the *Laws of Manu*. "All that Manu has said is medicine," is the jurist's adage, but who is Manu and what has he said? Manu functions not as an author, a creative intelligence in the here and now, but as an authoritative transmitter, located at the beginning of time, of an uncreated, eternal truth of which fragments are preserved by a teacher-pupil chain of reciters. The transmitters in the period of the compilation of the *Laws of Manu* are the brahmin schools of the law.

In the end, Indian time is best understood as a part of the intellectual formation of which it was a part. In doing so it is important to attempt to see that intellectual formation as a whole, in its strengths as well as its weaknesses. Within a configuration of ideas joining transmigrating subjects with authorless, remote truth and multiplied truth-transmitters, the intellectual product of the brahmins and Buddhist and Jain monks who wrote the texts on which we rely for our knowledge of ancient India has a characteristic bent, a signature note that cannot be missed. We may say that it values and fastens upon the typical and general at the expense of the local and particular. The ancient Indian sciences, viewed globally, show as their negative side a relative inattention to history, biography, geography, mapping, and the like, and as their positive side a strong preference for a kind of timeless, placeless conceptualizing with a search for durable types that are taken to be normative. The strengths of this intellectual tradition lay especially in structural linguistics, in which the ancient Indians were centuries ahead of the rest of the world, though not, paradoxically, in historical linguistics, which the encounter with Sanskrit so powerfully stimulated in the West. It lay also in the astronomy-astrology-mathematics area, with which Indians were major participants in an extended dialogue involving Chaldeans, Greeks, and Arabs. In the brahmin science of the law the overall orientation was toward the eternal, impersonal dharma. While it recognized that both king-made law and local custom have legal force, both are limited geographically and in duration, and the jurists show no interest in recording them. In poetry, history (i.e., *prashasti* 'eulogy'), and royal biography we find a preference for the type over the individual and a highly developed theory of literature or poetics.[32] In this tradition the passage of time bears a negative sign, and the particular is seen as an imperfect embodiment of the normative type, immune to the ravages of time.

It is something of this that Eliade catches in his idea of an antihistoric ontology that informs the Vedic ritual order: the ritual reenacts the creation, thus, as he so memorably says, abolishing time.[33] But stressing Vedic antecedents for Indian time tends to paint it as timelessly Indian, and Indianly timeless. In doing so it obscures the essentially post-Vedic timing and the suddenness of onset of the time doctrines. The post-Vedic brahmin mentality, while accentuating the bias against the particular, is anything but archaic or primitive; it is the product of a thoroughgoing and highly self-conscious reinterpretation of the Vedic hymns and ritual texts by a scholarly class in the light of the emergent doctrine of reincarnation.[34] The resultant devaluation of the particular and the assignment of value to abstract formulations not limited in place and time was decisive. Brahmin intellectual activity, revolutionized by the transmigration doctrine and its effects, sets off on a new course. The intellectual production that follows is not the perpetuation of an archaic mentality as Eliade would have it, or as it is represented in the master narrative of modernity. It was, in its own way, one of the antecedents of the modern, in ways we don't yet quite know how to conceptualize. As an approximation, we may say that India did not teach the modern world to think historically, but it did teach it algebra and the number system.

NOTES

Raymond Grew's comments on a draft of this essay were most helpful, and I am glad to be able to record my thanks to him. I am also grateful for help received from Matt Herbst and Juan Cole.

1. Mircea Eliade, "Time and Eternity in Indian Thought" (1957), 174. To the same effect is the unity of "archaic" and "primitive" societies, as species of "traditional society," in his very influential *The Myth of the Eternal Return, or, Cosmos and History* (first published in 1949, in French). Though it is a general study, Eliade's early training as an Indianist has shaped the traditional society concept. The point of view is Christian and critical of post-Hegelian philosophical developments that do not, he believes, resolve the "terror of history," so that his is not a story of progressive

modernity. But insofar as he makes biblical historicism the necessary precursor of modern secular historicism it is recognizable as a variant of the master narrative of modernity without its valorizations.

2. Jones's 1794 translation is cited in the bibliography. In this essay I rely on the Gurumandalagranthamala edition of the Sanskrit text (2 vols., Calcutta 1967–71). Translations follow, with some changes, those of Georg Bühler, *The Laws of Manu* (1886). Bühler's long prefatory discussion of the text is still indispensible. For general treatments of the facts of Indian time, there are a number of publications. It is always worthwhile consulting P. V. Kane (chaps. on Kalivarjya and on Kalpa, Manvantara, Mahayuga, and Yuga), *History of Dharmasastra*. Beyond that the best summary treatment I have found is in Spanish: Luis González Reimann's fine book, *Tiempo cíclico y eras del mundo en la India*. Articles on "Ages of the World," in Hastings's *Encyclopaedia of Religion and Ethics* are showing their age but still have merit, especially that of Louis de La Vallée Poussin on Buddhism. On Jainism, see P. S. Jaini, *The Jaina Path of Purification*, 29–34. Philosophical treatments abound. I omit diacritics in the rendering of Sanskrit words.

3. The calculations and tables of this paragraph follow those in Reimann, *Tiempo cíclico*, chap. 6, esp. pp. 96–97, 101.

4. The *kalpa* does not divide equally into fourteen *manvantaras;* see Reimann, 116–21.

5. *Samyutta Nikaya* 2.181–82, cited by Louis de La Vallée Poussin, "Ages of the World (Buddhist)."

6. Raymond Schwab, *The Oriental Renaissance* (1978; first published 1950). It is impossible to overestimate the importance of this book. Schwab insists on the India-centered character of the new Orientalism of the eighteenth and nineteenth centuries.

7. G. W. F. Hegel, *Philosophy of History*, 54.

8. Karl Marx, "The Future Results of the British Rule in India" (1853), cited in Young, *White Mythologies*, 176, n. 4.

9. James Mill, *The History of British India*, 1:107. This passage opens Mill's long condemnatory essay "On the Hindus." In it he attacks Halhed for his more favorable reception of Indian time (which I discuss further on in this paper) as an example of European "love of the marvelous" and uncritical acceptance of Hindu chronology (p. 113).

10. Schwanbeck, fragment 50=Arrian, *Indica* (i.e., the eighth book of his *Anabasis of Alexander*), 8.9, pp. 332–33, of the Loeb edition. In the same passage he adds that besides Dionysus and Heracles "no one else ever invaded India." A variant of the same statement appears in Pliny, 6.23: 153 kings from Father Bacchus to Alexander, over 6,451 years and three months.

11. *Samvattakappa*, Skt. *samvartakalpa*, Rock Edict 4, Girnar, 1.9; E. Hultzsch, ed., *Inscriptions of Asoka*, 9.

12. Augustine's critique of pagan (especially Egyptian) histories is mainly to be found in books 12 and 18 of *The City of God*.

13. On the Brixham Cave excavations, see Jacob W. Gruber, "Brixham Cave and the Antiquity of Man"; and T. Trautmann, *Lewis Henry Morgan and the Invention of Kinship*, chap. 9, and "The Revolution in Ethnological Time."

14. *Alberuni's India*, 321–22; and his *Chronology of Ancient Nations*, chap. 3, 16ff. I have discussed the biblical chronology at greater length in *Lewis Henry Morgan*, 205–13.

15. Alexander Dow's translation of Firishtah, *The History of Hindostan*, 7–8. The account of the *yugas* occurs at pages 2–3.

16. See Dow's preface to the Firishtah, vii.

17. Dow's prefatory "Dissertation Concerning the Customs, Manners, Language, Religion and Philosophy of the Hindoos," in his Firishtah, xxvii. This important text was for a time the most authoritative English account of the Sanskrit language and its literature.

18. Nathaniel Brassey Halhed, translator's preface to *A Code of Gentoo Laws*, xxxvii. Rosane Rocher's excellent biography, *Orientalism, Poetry, and the Millennium*, is indispensible.

19. Halhed cites Brydone's *Tour Through Sicily and Malta*, reporting the geological arguments of Giuseppe Recupero, which Halhed characterizes as "conscientious Scruples [concerning the Mosaic chronology] of the Historiographer of Mount Ætna." See Rocher, *Orientalism*, 58.

20. Ibid., xxxix.

21. Rocher traces the mostly hostile reception of Halhed's deliverances on Hindu chronology at pages 57–60. The pamphlet of George Costard (who was vicar of Twickenham and a family friend) is titled, *A Letter to Nathaniel Brassey Halhed, Containing Some Remarks on His Preface to the Code of Gentoo Laws Lately Published* (1778). Halhed's reply is published in Rocher, *Orientalism*, appendix B.

22. Jones's Indological writings appeared initially in the first four volumes of *Asiatick Researches*, published at Calcutta by the Asiatic Society (and pirated several times), and have been reprinted in his *Works*, edited by Ann Maria Jones and published in 1799 (6 vols.), 1807 (13 vols.), and elsewhere. The ethnological project is the subject of the "Anniversary Discourses" to the society, in the third of which he proposes the conception of what we now call the Indo-European family of languages. What I have to say about that project and its "Mosaic" character in this essay derives from research in progress. Here I discuss his chronography, largely from "On the Chronology of the Hindus" *Asiatick Researches* 2 (1788): 111–47, which is the other wing of his project of defending the Mosaic narrative. The most up-to-date and comprehensive biography of Jones is that of Garland Cannon, *Oriental Jones* (1990), but S. N. Mukherjee's *Sir William Jones* (1968) continues to be the best study of his Indological work, now joined by O. P. Kejariwal's *The Asiatic Society of Bengal and the Discovery of India's Past* (1988), chap. 2, an excellent general study of the Calcutta Orientalists.

23. Stephen Toulmin and June Goodfield draw attention to the importance of

Eusebius in *The Discovery of Time*, 60. I discuss the "Eusebian grid" in *Lewis Henry Morgan*, 208–9. On the revival of chronography in the Renaissance, Anthony T. Grafton's "Joseph Scaliger and Historical Chronology" and his contribution to this volume are essential reading, together with the recently published second volume, on "Historical Chronology," of his monumental intellectual biography, *Joseph Scaliger: A Study in the History of Classical Scholarship* (1983–93). On "segmentary" thinking, I have found Paul Dresch, "Segmentation: Its Roots in Arabia and Its Flowering Elsewhere" (1988), most helpful.

24. This table is abstracted from a somewhat longer one at the end of Jones's essay "On the Chronology of the Hindus," 147.

25. Egyptians (the Mizraim) are, of course, Hamians in Genesis, and Firishtah's view that Indians were Hamians was the common Muslim view. Greeks, however, are generally identified with Javan, a descendant of Japhet not of Ham. Jones followed Jacob Bryant (*Analysis of Antient Mythology*, 1774–76) in the unusual attribution of the Greeks to the Hamian branch of the descent of Noah.

26. Shortly after Jones's death Charles Grant assessed the significance of his work from an Evangelical perspective in his highly influential *Observations on the State of Society among the Asiatic Subjects of Great Britain* (1796).

> He has opened the way into the mythological and scientific arcana of a people, who have for many ages been as remarkable for their adherence to their peculiar institutions, as for arrogating to themselves an unfathomable antiquity, and the possession of a pure and primeval, though carefully concealed system of theology and science; claims which have been as officiously as ignorantly accorded to them by some antichristian philosophers of Europe. He has shown that one of their earliest traces of true history describes an *universal deluge,* in which only *a patriarch,* and *seven* other men (to whom this account gives wives) were saved in an *ark;* and that the whole of their chronology is reconcileable with the Mosaic history.

This reading is wildly unbalanced, missing the solid achievements for which we revere Jones's memory, stressing only the negative side of his assessment of Indian historical traditions. And it is perfectly mistaken when it puts him among the opponents of the idea of the primeval wisdom of the Indians, an idea that is fundamental to his comparative ethnological and linguistic studies. But it is interesting exactly because it is unbalanced and ill informed, showing that the Jonesean rejection of Indian time finds a response in wider circles that are not his own.

27. Trautmann, *Lewis Henry Morgan,* chap. 2 ("Of Time and Ethnology"); and "The Revolution in Ethnological Time" (1992). On the Brixham Cave excavation, the best account is that of Jacob Gruber, "Brixham Cave and the Antiquity of Man" (1965).

28. The connectedness of time and person is the centerpiece of Clifford Geertz's classic essay, "Person, Time and Conduct in Bali" (1973), and it is in my view the most durable part of this much discussed (and criticized) work. It is an important source of inspiration for my analysis of Indian time, though I find the term *authority*

more useful than Geertz's *conduct* as the third point in my interpretative triangle. John Davis's *Times and Identities* (1991) shows how very fruitful the juxtaposition of time and person can be in the examination of many cultures.

29. See, for example, R. C. Majumdar's assessment: "Ideas of History in Sanskrit Literature."

30. Romila Thapar, "Society and Historical Consciousness: The Itihasa-Purana Tradition" (1986). See also the excellent study of royal biographies by V. S. Pathak, *Ancient Historians of India*.

31. Govindaraja, cited in Bühler, introduction to *The Laws of Manu*, xiii. In this paragraph and the next I rely heavily on Bühler's discussion of the authorship problem.

32. Daniel Ingalls has an admirable discussion of this "impersonality" in the introduction to his translation of Vidyakara's *An Anthology of Sanskrit Court Poetry* (pp. 22–27), which nicely evokes the way in which full being attaches to exemplary types rather than to individuals. For example,

> To write of one's patron, say, as Ramapala who has ruled for a few years one of the many regions of India is to make him out a small thing, a human, no more than he might be without the aid of Sanskrit verse. So to select among his qualities and among the vicissitudes of his life as to suggest his identity with Karna or Arjuna or Rama son of Dasaratha [heroes of the great epic] is to magnify him and give him permanence. (pp. 25–26)

33. "In the traditional societies men endeavored consciously and voluntarily to abolish time at periodic intervals, to efface the past and to regenerate time by a series of rituals which in a sense reactualize the cosmogony" (Eliade, "Time and Eternity," 174). This is one of the major themes of *The Myth of the Eternal Return*.

34. I refer to the Mimamsa school of Vedic interpretation, which reconceptualizes the Veda by means of the reincarnation doctrine, thus putting brahminical doctrine on a plane with the anti-Vedic religions.

BIBLIOGRAPHY

Arrian. *Arrian: History of Alexander and Indica*, with trans. by P. A. Brunt and E. Iliff Robson. Loeb Classical Library, 2 vols. Cambridge, Mass.: Harvard University Press; London: William Heinemann, 1976–78.

Asoka. *Inscriptions of Asoka*, ed. E. Hultzsch. Corpus Inscriptionum Indicarum, vol. 1. New ed. Delhi: Indological Book House, 1969.

Augustine. *The City of God*. New York: Modern Library, 1950.

Barnes, R. H. *Kedang: A Study of the Collective Thought of an Eastern Indonesian People*. Oxford: Clarendon, 1974.

al-Biruni, Abu Raihan Muhamad ibn Ahmad. *Alberuni's India*, trans. Edward C. Sachau. 2 vols. in 1. London: K. Paul, Trench, Trübner, 1914.

———. *The Chronology of Ancient Nations*, trans. Edward C. Sachau. London: William H. Allen, 1979.

Bryant, Jacob. *A New System, or, an Analysis of Antient Mythology.* 6 vols. London: J. Walker, 1807. Orig. 3 vols., 1774–76.

Cannon, Garland. *The Life and Mind of Oriental Jones: Sir William Jones, the Father of Modern Linguistics.* Cambridge: Cambridge University Press, 1990.

Cohn, B. S. "The Pasts of an Indian Village." *Comparative Studies in Society and History* 3 (1961): 241–49. Reprinted in *An Anthropologist among the Historians and Other Essays.* Delhi: Oxford University Press, 1990; and in this volume.

Davis, John. *Times and Identities.* Inaugural lecture delivered before the University of Oxford, 1 May 1991. Oxford: Clarendon, 1991.

Dresch, Paul. "Segmentation: Its Roots in Arabia and its Flowering Elsewhere." *Cultural Anthropology* 3 (1988): 50–67.

Eliade, Mircea. "Time and Eternity in Indian Thought." In *Man and Time,* Joseph Campbell, ed., 173–200. Papers from the Eranos Yearbooks, no. 3; Bollingen series, no. 30. New York: Pantheon for the Bollingen Foundation, 1957.

———. *Cosmos and History: The Myth of the Eternal Return,* trans. Willard R. Trask. New York: Harper & Row, 1959. Translation of *Le mythe de l'éternel retour: Archétypes et répétition,* 1949.

Fabian, Johannes. *Time and the Other: How Anthropology Makes its Object.* New York: Columbia University Press, 1983.

Firishtah, Muhammad Qasim Hindu Shah Astarabadi. *The History of Hindostan; From the Earliest Account of Time, to the Death of Akbar,* trans. Alexander Dow. 2 vols. London: T. Becket and P. A. De Hondt, 1768.

Grafton, Anthony T. "Joseph Scaliger and Historical Chronology: The Rise and Fall of a Discipline." *History and Theory* 14 (1975): 156–85.

———. *Joseph Scaliger: A Study in the History of Classical Scholarship.* 2 vols. Oxford: Clarendon, 1983–93.

Geertz, Clifford. "Person, Time and Conduct in Bali." In *The Interpretation of Cultures,* 360–411. New York: Basic Books, 1973.

Grant, Charles. *Observations on the State of Society among the Asiatic Subjects of Great Britain, Particularly with Respect to Morals; and on the Means of Improving it, Written Chiefly in the Year 1792.* First published in 1796. Printed as appendix 1 in *Report from the Select Committee on the Affairs of the East India Company 1831–32,* 3–92. Facsimile rep., Irish University Press Series of British Parliamentary Papers, Colonies: East India, vol. 5. Shannon: Irish University Press, 1970.

Gruber, Jacob W. "Brixham Cave and the Antiquity of Man." In *Context and Meaning in Cultural Anthropology,* Melford E. Spiro, ed., 373–402. New York: Free Press; London: Collier-Macmillan, 1965.

Halhed, Nathaniel Brassey. Translator's preface to *A Code of Gentoo Laws, or, Ordinations of the Pundits, from a Persian Translation, Made from the Original, Written in the Shanscrit Language.* London: [East India Company], 1776.

Hegel, Georg Wilhelm Friedrich. *The Philosophy of History,* trans. J. Sibree. New York: Dover, 1956.
Hutton, J. H. *Caste in India: Its Nature, Function and Origins.* 4th ed. London: Oxford University Press, 1963.
Jaini, P. S. *The Jaina Path of Purification.* Berkeley: University of California Press, 1979.
Jones, William. "On the Chronology of the Hindus." *Asiatick Researches* 2 (1788): 111–47.
Kane, Pandurang Vaman. "Kalivarjya." In *History of Dharmasastra,* 2d ed., vol. 3, chap. 34, 885–968. Poona: Bhandarkar Oriental Research Institute, 1973.
———. "Kalpa, Manvantara, Mahayuga, Yuga." In *History of Dharmasastra,* 2d ed., vol. 5, pt. 1, chap. 19, 686–718. Poona: Bhandarkar Oriental Research Institute, 1974.
Kejariwal, O. P. *The Asiatic Society of Bengal and the Discovery of India's Past, 1784–1838.* Delhi: Oxford University Press, 1988.
La Vallée Poussin, Louis de. "Ages of the World (Buddhist)." In *Encyclopaedia of Religion and Ethics,* vol. 1, J. Hastings, ed., 187–90. Edinburgh: T. & T. Clark, 1908.
Majumdar, R. C. "Ideas of History in Sanskrit Literature." In *Historians of India, Pakistan and Ceylon,* C. H. Philips, ed., 13–28. London: Oxford University Press, 1961.
Manu. *Institutes of Hindu Law; or, the Ordinances of Menu, According to the Gloss of Cullúca,* trans. Sir William Jones. Calcutta: Printed by order of Government, 1794.
———. *The Laws of Manu,* trans. Georg Bühler. Sacred Books of the East, ed. F. Max Müller, vol. 25. Oxford: Clarendon, 1886.
———. *Manusmrtih Medhatithibhasya-samalamkrta.* Gurumandalagranthamala, no. 24. 2 vols. Calcutta: Manasudharaya Mora, 1967–71.
Mill, James. *The History of British India.* 3 vols. New York: Chelsea House, 1968. Photographic reprint of the 5th ed., 1858, with notes and continuation by Horace Hayman Wilson.
Mukherjee, S. N. *Sir William Jones: A Study in Eighteenth Century British Attitudes to India.* Cambridge: Cambridge University Press, 1968.
Pathak, V. S. *Ancient Historians of India.* Bombay: Asia Publishing House, 1966.
Reimann, Luis González. *Tiempo cíclico y eras del mundo en la India.* Mexico City: Centro de Estudios de Asia y Africa, El Colegio de México, 1988.
Schwab, Raymond. *Oriental Renaissance: Europe's Rediscovery of India and the East, 1680–1880.* New York: Columbia University Press, 1984. Translation of *La Renaissance Orientale,* 1950.
Schwanbeck, E. A. *Megasthenes Indica, fragmenta collegit.* Chicago: Argonaut, 1968 (orig. Bonn, 1846).

Thapar, Romila. "Society and Historical Consciousness: The Itihasa-Purana Tradition." In *Situating Indian History: For Sarvepalli Gopal,* S. Bhattacharya and R. Thapar, eds., 353–83. Delhi: Oxford University Press, 1986.
Toulmin, Stephen, and June Goodfield. *The Discovery of Time.* New York: Harper & Row, 1965.
Trautmann, Thomas R. *Lewis Henry Morgan and the Invention of Kinship.* Berkeley: University of California Press, 1987.
———. "The Revolution in Ethnological Time." *Man,* n.s., 27 (1992): 379–97.
Vidyakara. *An Anthology of Sanskrit Court Poetry,* trans. Daniel H. H. Ingalls. Cambridge, Mass.: Harvard University Press, 1965.
Young, Robert. *White Mythologies: Writing History and the West.* London and New York: Routledge, 1990.

Part 3
Time and the Story of the Past

Time and Historical Consciousness: The Case of Ilparakuyo Maasai

Peter Rigby

Iyiolo ening'uaa, kake miyiolo enilo. "You know from whence you come, but not where you are going."

Meisho ilimot; inkulie ebaya. "When events occur, only part of the truth is made known." [All the news is not told; others merely arrive there.]

In a manner somewhat unusual in anthropological discourse, most writers on the Maa-speaking peoples of Tanzania and Kenya include an early section or chapter on what is often called "Maasai traditional history." For example, in his pioneering work, "An Administrative Survey of the Masai Social System," Henry Fosbrooke (1948) devotes a leading and major section to Maasai history, and, in a later publication, he deals explicitly with what he calls a "tribal chronology" of the pastoral Maasai (Fosbrooke 1956a). Similarly, Thomas Beidelman, in an equally pioneering article, pays considerable attention to the historical connections of the Ilparakuyo (Baraguyu) section of the Maasai peoples of Tanzania (1960). It is among the Ilparakuyo that I have carried out most of my work in Maasai studies, and they are the primary focus of the present discussion (Rigby 1977, 1979, 1980, 1981).

There is an "ethnographic" reason for this relatively atypical anthropological obsession with Maasai history, and it arises from the dominance of age-set organization among most of the Maa-speaking peoples. Al Jacobs, who has done extensive fieldwork among the Maasai over a considerable period of time, has noted (1965a:48) that "although the Maasai plan future ceremonial and economic events according to a lunar calendar of thirty days to the month (*olapa*) and twelve months to the year (*olari*), historical events are always reckoned in relation to the age-set (or sets) who were warriors at the time." Fosbrooke (1956a:48) also makes his position clear: "We are fortunate with the Masai," he says, "in having a tribe with an age-group system of regular periodicity, so that it should be possible to get a reasonably accurate dating of an event if it were only known who were the warriors at the time of its occurrence."

Other elements also account for this historical focus in Maasai studies, such as the relative importance of the pastoral Maasai and Ilparakuyo (or "Wakwavi") in the historiography of East Africa over the past 150 to 200 years; but the dominance of the ethnographic fact has also recently attracted a significant number of historians of various persuasions to do research among the Maa-speaking peoples (e.g., Berntsen 1979a, 1979b, 1980; Lawren 1968; Waller 1976, 1978; Vossen 1977a, 1977b).

It is evident that there are fairly sound ethnographic reasons for combining a discussion of Maasai notions and concepts of "time," which should ultimately encompass the entire "culture area" of the Maa-speaking peoples, with a consideration of some elements of the history of a specific social formation within the wider field, that is, the Ilparakuyo. But I have selected a historical approach to the problem of "time" among Ilparakuyo and Maasai for considerations other than merely ethnographic ones; given the theoretical position I adopt, it could scarcely have been any other. It will appear as we proceed that the theoretical problematic I propose has not been employed at the expense of ethnographic detail and historical investigation but arises out of a perception of the latter as being ultimately the same thing.

In order to arrive at a satisfactory theoretical starting point, I first outline briefly what may be called some of the classical forms of anthropological investigation into the time-reckoning concepts and procedures of what are usually referred to as "other cultures," but which I prefer to call "precapitalist" or "noncapitalist" social formations.

I

The various forms of time reckoning manifested in different cultures and through history, and the relationship between conceptualizations of time and space, have been the subject of an enormous amount of inquiry and speculation (see, e.g., Thompson 1967; and Hall 1959). But anthropological studies of the topic may be grouped in four basic (but not exclusive) categories, as follows.

1. The approach of "abstract mentalism," which must not be confused with psychological or psychologistic paradigms.

2. The "abstract empiricist" approach, characteristic of functional analysis.
3. Overlapping closely with item 2, but distinct from it, the approach of "structural totalization."
4. What may be called the "philosophical-theological" approach, characteristic of some recent writings on African (and other "nonwestern") philosophy.

The first category, which I have termed "abstract mentalism" in the strict sense of pertaining to the effects of mental activity, is most clearly foreshadowed in the work of Lucien Lévy-Bruhl, with his literally voluminous exposition of what he called "primitive mentality" or the "mental functions of inferior societies" (1910, 1923, 1938). It may be objected that his work is discredited and is no longer a force in anthropological thinking and interpretation; but Lévy-Bruhl has been restored to an albeit shadowy place in the pantheon of anthropological ancestors by sympathetic and extensive attention in, among others, Edward Evans-Pritchard's *Theories of Primitive Religion* (1965) and, perhaps more significantly, Rodney Needham's *Belief, Language, and Experience* (1972).[1]

In fact, Lévy-Bruhl's thesis, often unrecognized and unbidden, lurks in many anthropological discussions of concepts of time and other "cosmological" notions in nonwestern cultures. Although Evans-Pritchard attempts to clear Lévy-Bruhl of both psychologism and racism, and is successful in the case of the former, he notes (1965:88) that "there is no reputable anthropologist who today accepts [the theory] of two distinct types of mentality," the "civilized" and the "primitive." But despite this, other aspects of Lévy-Bruhl's work apparently still excite the anthropological imagination. As Needham notes (1972:161), "there has . . . persisted in the received ideas of social anthropology what can only be regarded as a travesty of Lévy-Bruhl's theory which seems likely to induce others to continue to ignore or belittle him." And he adds:

> Yet one need do no more than read carefully what he actually wrote in order to see the degree of justice in Mary Douglas's judgement [in *Purity and Danger*, 1966] that "it was he who first posed all the important questions about primitive cultures and their distinctiveness as a class," and to agree with her expostulation that "he has not deserved such neglect."

What *did* Lévy-Bruhl say about concepts of time, and how are his ideas still lurking about, albeit in ghostly and unjustly maligned form? At the end of his book on primitive mentality, he sums up his findings as follows (1923:445–46).

> The primitives' idea of time, which above all is *qualitative*, remains vague; and nearly all primitive languages are as deficient in methods of rendering the relations of time as they are copious in expressing spatial relations. Frequently a future event, if considered certain to happen, and if provocative of great emotion, is felt to be already present.

Although Lévy-Bruhl is clearly trying here to establish an important, socially determined, distinction between his own conception of time and that of his primitives, he cannot possibly account for the causes of it because he (1) adheres to the classical but idealistically erroneous notion that "behavior is a product of thought" (cf. Evans-Pritchard 1965:98), and (2) is unaware that he is making an implicit comparison with another historically and socially determined notion of time: his own. The final part of his statement concerning the conceptualization of future events has, unfortunately, entered into anthropological and philosophical discourse on concepts of time, causing serious errors.

Quite clearly, then, Lévy-Bruhl's conclusion about primitive time involves an unstated epistemological assumption about something else, which we assume is some form of "civilized" time. And, sure enough, here it is (1923: 123–24).

> Primitives do not see, extending indefinitely in imagination, something like a straight line, always homogeneous by nature, upon which events fall into position, a line on which foresight can arrange them in a unilinear and irreversible series, and on which they must of necessity occur one after the other. To the primitive time is not, as it is to us, a kind of intellectualized intuition, an "order of succession." Still less is it a homogeneous quantity. It is felt as a quality, rather than represented.

It is the latter point that is taken up so strongly by Needham. He writes (1972:8) that Lévy-Bruhl was the first in the "anthropological tradition" to inspire in him "as a comparative problem, though with another intent, the question of the line of demarcation between belief and experience." Elsewhere, Needham concludes—and this is important—that Lévy-Bruhl's "crucial advance consisted in concentrating analysis on the most general logical articulations of alien ideologies, rather than their component meanings and their social contexts." It is precisely from this point, Lévy-Bruhl's separation of his so-called alien ideologies from their social contexts, that my critique proceeds, but not in the usual "structural-functional" manner, as I take pains to point out.

In his endeavors to disentangle belief and experience, Needham not only acknowledges the inspiration of Lucien Lévy-Bruhl, among others, but correctly places him in the context of the classical philosophical tradition of the bourgeois enlightenment. His particular exemplars are David Hume and Ludwig Wittgenstein, Hume for his repeated "attentions to the notion of belief, which he could not account for and could not leave alone, [and which] brought out the opacity and the problematical nature of this commonplace category of everyday discourse and psychological report" (Needham 1972:8). Needham explicitly derives his inspiration from Lévy-Bruhl and Wittgenstein precisely because he sees them addressing critical issues in the mainstream of bourgeois philosophy (Needham 1972:161–76) but Lévy-Bruhl also because he was "a founder, together with Durkheim and Mauss, of the comparative and sociological study of forms of classification and modes of ratiocination [and] he effectively inaugurated a comparative epistemology." That Lévy-Bruhl was concerned with some form of a comparative epistemology cannot be countered; that he was the founder of a comparative epistemology, effectively inaugurated by Marx half a century before, must be denied.

There is a reason for this apparent oversight on Needham's part, and it is important for the present discussion to pursue it a little further. Needham himself admits (1972:183) that it was "questionable" for Lévy-Bruhl to lump the "collective representations of non-western societies" into one simple type, even if the ethnographic record was at that time weak, as Marcel Mauss (1923) had already pointed out. Needham also notes that Lévy-Bruhl "recognized" the implication that, by juxtaposing "primitive mentality" and "western thought," he was making assumptions about his own concepts of experience and belief (including, of course, time), but Needham fails to see that this very fact destroys his claims for a Lévy-Bruhlian "comparative theory of knowledge" or epistemology (Needham 1972:184). Despite moments of insight, relegated to footnotes, Lévy-Bruhl obviously assumes, in typical late nineteenth- and early twentieth-century fashion, that his own conceptualizations of the nature of time, space, experience, and belief are self-evidently "rational" and "scientific," representing the "logical unity of the thinking subject."[2] Both Lévy-Bruhl and Needham, following in the footsteps of the same philosophical tradition, fail to apprehend the fundamental epistemological break represented by Marx's much earlier critique of Hegel and Feuerbach, since it uncomfortably involves a self-critique of the historical and sociological grounds for their own ratiocination about time and other "things," a point to which I return.[3]

In other words, by recognizing "differences," the "abstract mentalist" approach (a form of philosophical idealism) does not positively impose the constructs of one culture upon another but "negatively" destroys the basis for understanding by an idealistic "comparison" that ignores the social and historical origins and social production of *both* sets of concepts.

I have characterized the second approach to the anthropological study of time as both abstract and empiricist. Most of the studies we have of African time concepts (with the notable exception of Robert Thornton's 1980 book on the Iraqw of Tanzania) combine elements of this approach with some strands of the former Lévy-Bruhlian one. Such a combination leads to the inclusion of a discussion of concepts in the context of linguistic analysis, on the one hand, and "structural time" in the particular sense of "genealogical time" on the other. A number of these studies reflect with apparent perplexity upon the "dissociation" or contradictions that occur between these two levels of time in the same cultures.

Whatever the theoretical drawbacks of such studies—critiques of functionalism abound, and I mention some of them later—they had the value of focusing attention upon the detailed elements that go to make up a people's cosmological notions of time, from everyday routine to ideas of historical process, if not history. Evans-Pritchard's seminal article on "Nuer Time reckoning" (1939) set a pattern, and two other excellent examples are Beidelman's "Kaguru Time Reckoning" (1963) and Paul Bohannan's "Concepts of Time among the Tiv of Nigeria" (1953).[4]

Commencing with a distinction between "structural time," which is "entirely progressive," and "ecological time," which is "only progressive within an annual cycle," Evans-Pritchard discusses Nuer perceptions about days, months, seasonal cycles, years, genealogical history, and mythology. Despite the promising start of placing Nuer time concepts in their structural context, Evans-Pritchard strays, perhaps unconsciously, into Lévy-Bruhlian territory. In the middle of an incredibly rich ethnographic description, he suddenly states (1939:208): "Though I have spoken of time and units of time, it must be pointed out that, strictly speaking, the Nuer have no concept of time and, consequently, no developed abstract system of time reckoning." Genealogical time is "structural time," not historical time (1939:213), and beyond the constricted limits of genealogical time we arrive at tradition and myth: "Valid history ends a century ago," and "though it astounded me, it is in no way remarkable to the Nuer, that the tree under which mankind came into being was still standing in western Nuerland a few years ago and would still be standing had it not recently been burnt down" (1939:215–16).

During the course of his excellent descriptive study, however, Evans-Pritchard makes two points of considerable theoretical significance to us. First, he writes that "time is not always the same to Nuer at different seasons," the dry season appearing to represent "slow" time, the wet season "fast" time; and, second, there appear to be different "levels" of time reckoning among the Nuer, physical, ecological, and social, each of which has its own "rhythm." He notes (1939:192) that there are "three planes of rhythm: physical rhythm, ecological rhythm based on physical changes, and social rhythm based on ecological changes. Nuer concepts of time are based primarily on the social rhythm," although they obviously "recognize" the others. Although the theoretical integration of these different, and frequently contradictory, "planes of time" is not relevant to the everyday existence of Nuer (or the members of any other culture or social formation for that matter), the general import of Evans-Pritchard's article (one might say its very raison d'être) is that such an integration is possible in certain circumstances relatively unique to the Nuer social formation; but he does not specify what these are. I suspect these circumstances would be manifested in Nuer consciousness of the historical development of the material practice necessary for the reproduction and continual adaptation of their social formation or, more precisely, their political economy.

Beidelman's discussion of Kaguru time reckoning differs from Evans-Pritchard's in two important respects, first in suggesting that "a study of Kaguru time reckoning does not show a dramatic relationship between these concepts and Kaguru social structure and economy," and, second, in establishing that Kaguru have not only one but two terms for the abstract notion of time (Beidelman 1963:10–11). Consequently, his analysis virtually ignores genealogical time and focuses almost exclusively upon conceptual and categorical ideas. It seems to me that these are crucial differences. But while Beidelman notes that if Kaguru (as the Nuer) do not have "chronology as we know it," and although "this is hardly surprising in view of the very different preoccupations of such peoples and the absence of written means to record history," he does not question why differences exist between, say, Kaguru and Nuer, why "chronology" became important to western bourgeois historiography, and why this importance is decreasing, all of which are questions of both theoretical and practical relevance. The lead suggested by the linking of time concepts with "the very different preoccupations" of the Kaguru is not taken up, but Beidelman does make the crucial point that Kaguru notions of time have been radically changed by the impact both of Arabization and of European colonialism, in other words, by the penetration of mercantile and peripheral capitalist relations of production and their vehicles of expression.

In sum, Beidelman and Evans-Pritchard, while not explicitly postulating a theory of "primitive mentality" in the Lévy-Bruhlian sense, and differing in their presentations of "structural time," both offer an analysis of temporal concepts in which comparisons are made with a given notion of "our" time, that is, the time concepts of western philosophical and scientific investigation, but the implications of this comparison are not explored.

Bohannan's study of Tiv time concepts is much along the same lines, except that he does make some interesting additional points. He states, for example, that Tiv place incidents in time, both future and past, by "a direct association of two events," or a "conjuncture" (Bohannan 1967:316). Furthermore, Tiv indicate periods of future time that exceed four or five years either by specific seasonal or cyclical events or by life-expectancy events of a particular person (1967:324). Finally, although Tiv do not associate their fairly deep genealogical reckoning with "myths and legends"—why should they?—they do associate genealogical progression with population trends and territorial expansion (Bohannan 1967:326). Hence, there is quite clearly among the Tiv an awareness of constant historical change, linking together ecological, demographic, and social factors, despite their different temporal "rhythms," into one coherent process.

Before I turn to a discussion of the third approach, it should be noted that these emphases in Bohannan's analysis are consistent with the more strictly structural-functional approach to genealogical time reckoning embodied in the more dynamic studies of descent and kinship such as in Meyer Fortes's work on Tallensi and Ashanti (Fortes 1945, 1949a, 1949b).

The approach I have characterized as "structural totalization" is manifested in a number of recent studies, which emphasize, on the one hand, the symbolic nature of time concepts and, on the other, the use of analogies derived from linguistic analysis such as the abstract notions of synchrony and diachrony (e.g., Lévi-Strauss 1966; Zein 1974; Leach 1961; Rigby 1968). The intellectual debt these studies owe to the second type already discussed is obvious, but some of the latter have gone much further in questioning the epistemological issues involved in discourse about time concepts not only in precapitalist social formations but also in western bourgeois theory and practice. To some extent, then, this approach has shaken the false epistemological certainty of the implicit comparison made in studies of the previous type, and in part this is due to the attempted (if not entirely successful) rapprochement between structuralism and the Marxist problematic in French anthropology and philosophy. If some of these studies ultimately end as further examples of "abstract empiricism," it is not for want of trying to penetrate some of the categories of bourgeois thought.

I begin with the example of Edmund Leach's brief but penetrating, as well as entertaining, studies, "Cronus and Chronos" and "Time and False Noses" (1961:124–36).[5] Leach correctly states that "there is nothing intrinsically geometrical [i.e., linear, cyclical, circular] about time as we actually experience it," and he goes on to demonstrate that the "notion that time is a 'discontinuity of repeated contrasts' is probably the most elementary and primitive of all ways of regarding time" (1961:126, 134; cf. Rigby 1968). But he rests his case upon the epistemological "given" of experience, and one cannot help but wonder if we are not back with Needham's problems with belief and experience, and the shades of Lévy-Bruhl. Leach notes some difficulties here when he comments (1961:132) that "we experience time, but not with our senses. We don't see it, or touch it, or smell it, or taste it, or hear it." How then do we "experience" it? This dilemma is reminiscent of Lenin's remark, following Marx and Engels, that bourgeois philosophers, having set up the abstract idea of reified "objective" time, then want to feel, taste, and see it, that is, *actually* experience it! (Lenin 1962:177ff.; Engels 1940:327, 1975). Leach also states that even historical time is a "fairly simple derivative" of experience (1961:125); but if the experience itself is suspect, where are we?

The problem is that, whereas in the Kachin case Leach emphasizes the time dimension as a product of the history of changing social relations, in the more structuralist analyses under review this historicity is lost. This is at least in part due to his overcommitment to the structuralist dichotomy between synchrony and diachrony and the focus of "mythical time"; "my own explanation," says Leach of the diversity of temporal concepts, "is of a more structural kind" (1961:129).

The Marxist critique of these structuralist tenets is a bridge to the approach I attempt to develop here. But first I will add a brief word on the fourth approach, which I have characterized as the "philosophical-theological." This is the most eclectic of the theoretical positions discussed. John Mbiti, for example, inexcusably (though understandably in view of his terms of reference) generalizing about "African" concepts of time (1969:17), maintains that "according to traditional concepts, time is a two-dimensional phenomenon, with a long *past*, a *present*, and virtually *no future*" (Mbiti's emphasis).[6] His generalization turns out to be based upon his understanding of the tense structures of two Bantu languages, Kikamba and Gikuyu,[7] which in turn is further linguistically expanded into the Kiswahili concepts of *sasa* 'the now period' and *zamani* 'the long past'. The latter period has its own "past," "present," and "future." Mbiti's contention that "African" concepts of time encompass no definite future beyond six months to two years, and no indefinite future at all, is strongly suggestive of Lévy-Bruhlian influence; it has already been refuted for the Tiv by Bohannan (see above) and is insupportable in numerous other cases. Despite the interesting and informative complexity of Mbiti's overall analysis, there is a total lack of historical context in his discussion; and explanatory comparison is further confounded by references to the lack in "African traditional thought" of concepts of history moving "towards a future climax, or towards an end of the world." Why anyone should wish to postulate history as moving toward an end of the world as a condition for a historical conception of the past and the future is a question not examined by Mbiti.

The influence of this approach, proposed as it is by such sympathetic observers as Mbiti, can also be seen in the work of others attempting a more rigorous comparative analysis. In his otherwise excellent study of African religions, for example, Benjamin Ray (1976:40–41) assumes that African concepts of time are relatively "ahistorical," while of course denying that Africa lacks "a sense of history." Ray, like Mbiti, is also led into the implicit epistemological comparisons of bourgeois futurology when he states that "what African thought did not conceive was an indefinite future, stretching beyond the immediate future of the next two or three years" (1976:41). I show later that Ilparakuyo and Maasai, for example, do have such a concept, but it exhibits a specific form commensurate with the nature of their social formations and is not a mere replication of the (also specific) images of the future embodied in capitalist ideology, bolstered by Christian dogma and eschatology.

It can be stated without too much distortion that all of these anthropological and philosophical approaches to the understanding of time and temporal concepts, even when they do make reference to "historical" time and the nature of historical reality, treat the latter as a separate dimension of time, a separate entity, which may, or more often may not, be like "our" conception of history, which is taken as a self-evident truth. History is thus treated as either nonexistent in African (and other so-called primitive) societies or as an analytically distinct conceptual element among other such elements in a "cosmology" of time. There is seldom any attempt to place all these elements or levels of time reckoning, even where their existence is noted, in a total and specific historicity arising from the nature of the specific social formation under consideration, which in turn is itself a product of that specific historical development.[8]

II

What, then, is an alternative problematic? I have noted that the structuralist approach to concepts of time seemed to bring us closest to a comparative and critical epistemological investigation that does not assume the conceptual apparatus of bourgeois society. This achievement is due, at least in part, to structuralism's insistence upon a totalizing theory and upon treating so-called primitive systems of thought as "practico-theoretical logic" (Lévi-Strauss 1966:75 et passim). But this theoretical gain seems to get bogged down in (1) the abstract dichotomy between synchrony and diachrony, in which the latter becomes a contingent element in social reality and hence undermines the very historicity we are pursuing; and, to a lesser extent, (2) the intractable problems of the notion of "experience."[9]

In an illuminating study, "Fulani Penetration into Nupe and Yoruba in the Nineteenth Century," Peter Morton-Williams notes (1968:4) that the

> property of structure permits social anthropologists to join forces with the historians. What separates them is an anthropological formulation of social structure and structural processes that is functionally independent of the dimension of time as the historian conceives it, structural theory employing a notion of generalized time that, like the scientist's time, has different properties.

While taking issue with his assumptions about the nature of historical time (he inserts a disclaimer on epistemological debate), it is relevant to pursue his useful paper a little further. In criticizing the idea of functional stability stemming from a Radcliffe-Brownian position that postulates only "exogenous" sources of social change, Morton-Williams goes on (1968:4–5):

> The historian is concerned at least as much with the problem of changes originating in the society itself, for instance with acts of will intended to modify "the actually existing network of relations". . . . It is merely begging the question to propose a series of synchronic studies of a changing

social structure at a succession of dates in the hope that the result will be a diachronic study illustrating laws of change.

Here again we have a comment upon the unacceptable contingency of the diachronic. Although Morton-Williams achieves a penetrating analysis of a particular case, he does not pursue the matter to its logical conclusion: the further examination, and possible elimination, of the synchrony-diachrony opposition itself. Only then can some coherence be given to the different conceptual levels of time reckoning in the context of the specific historicity of a particular social formation and its development.

Brief mention must be made here of Thornton's important attempt to transcend this structuralist problem in his analysis of the Iraqw social formation (1980). I use the term *social formation* judiciously, for, although Thornton does not, he is at pains to describe Iraqw society and culture not as constituting an ethnic isolate but as the historical product of a complex regional formation involving Iraqw relations with their neighbors, Maasai, Barbaig, and so on. He also attempts to

> develop the argument that the cultural order of Iraqw as we observe it today . . . is the product of historical circumstances; but the structure so constituted is also, at any moment in the process, the producer of a continuing history. Structure and history account for each other. (Thornton 1980:191)

Although this may appear to be little different from Morton-Williams's position, Thornton goes further to argue that Iraqw society cannot be thought of as a given structure, legitimated by oral tradition and history, within which changes are generated; rather, "processes and means [praxis?] are legitimated by the oral tradition, *not* structures and ends" (1980:184). Thornton arrives at this sophisticated position partly through his familiarity with Pierre Bourdieu's work (particularly Bourdieu 1977), but this in itself becomes a limitation upon his advance to a historical materialist problematic, as I show in my conclusion.

Marx's critique of Hegel is concerned basically with the nature of historicity and hence the conditions for the production of concepts of time. Kant was the precursor of the shift in science and philosophy toward a concern with the nature of time and time concepts, but a real critique of the whole idealist philosophical tradition's treatment of time awaited Marx's commentary on Hegel and Feuerbach. As Henri Lefebvre notes (1968:27–28), first to Hegel and then to Marx, the object of investigation and knowledge is time. . . . With Hegel, extensibility in time ("becoming") comes firmly to the fore, takes on primordiality: mankind's life is now [perceived of] in time, is historical, its very consciousness a succession of changing stages and shifting moments.

Marx's critique of Hegel's notion of time was at once an exposition of Hegel's betrayal of "his own finest insight when he gave to understand that his philosophy was the culmination of human thought and the contemporary nation state the end of history" (Lefebvre 1968:28) as well as a critique of Hegel's notion of time and history as the apotheosis of ideological conceptualization.

Appropriately for his own mistaken logic (if somewhat ironically), Hegel declared that "what we understand by Africa, is the Unhistorical, Undeveloped spirit, still involved in the mere conditions of nature" (1900:99; cf. Ray 1976:42), while for Marx, "change becomes truly universal, since both nature and history are now conceived historically. Man and all things human are from now on characterized in temporal terms" (Lefebvre 1968:28). For Hegel, "historical time merely reflects the essence of the social totality of which it *is* the existence," which in turn is "the Spirit"; "Spirit is time," said Hegel in his Jena writings, an assertion later developed in the *Phenomenology*. It is "the essential character of historical time that will lead us, like so many indices, to the peculiar structure of [a] social totality," writes Louis Althusser (1970:93). The distinction between synchrony and diachrony that so pervades contemporary anthropological problematics, structuralist or otherwise, is also based upon a Hegelian essentialist "conception of historical time as continuous and homogeneous and contemporaneous with itself," the idealist historian's "time present" (Althusser 1970:96).[10] Althusser continues:

> The synchronic therefore presupposed the ideological conception of a continuous-homogeneous time. It follows that the diachronic is merely the development of this present in the sequence of a temporal continuity in which the "events" to which "history" in the *strict sense* can be reduced (cf. Lévi-Strauss) are merely successive contingent presents in the time continuum [cf. Morton-Williams's critique previously mentioned]. Like the synchronic, which is the primary concept, the diachronic therefore presupposes both of the very two characteristics I have isolated in the Hegelian conception of time: an *ideological* conception of historical time.

Structuralism, therefore, far from having fulfilled its promise of liberating us from an ideological epistemology of all time concepts such as bedevil Lévy-Bruhl, his followers, and the other anthropological approaches I have described, perpetuates the very error we are trying to escape. And I can agree with Althusser when he concludes (1970:107):

> If what I have said has any objective meaning, it is clear that the synchrony/diachrony opposition is the site for a misconception, since to take it for a knowledge would be to remain in an epistemological vacuum, i.e.— ideology abhorring a vacuum—in an ideological fullness, precisely in the fullness of the ideological conception of history whose time is continuous-homogeneous/self-contemporaneous.

How, then, can we lay a foundation for the understanding of historical time, which is at once the source of all levels of the conceptions of time in a social formation as well as their embodiment? In order to avoid an ideological conception of history, and hence the implicit ideological distortion of any discussion of time concepts,

> it needs to be said that, just as there is no production in general, there is no history in general, but only specific structures of historicity which, since they are merely the existence of determinate social formations (arising from specific *modes of production*), articulated as social wholes, have no meaning except as a function of the essence of these totalities, i.e., of the essence of their *peculiar complexity*. (Althusser 1970:108–9, emphasis added)

This "essence," therefore, lies not in an abstract essence of conceptual time (Lévy-Bruhl's "scientific," "civilized" time) with which other concepts are compared but in the temporal embodiment of social formations and their modes of production, which give rise to those concepts themselves. The concept of historical time, together with the different conceptual levels and "rhythms" that make it up,

> can only be based on the complex and differentially articulated structure in dominance of the social totality that constitutes the social formation arising from a determinate mode of production, it can only be assigned a content as a function of the structure of that totality, considered either as wholes, or in its different levels. (Althusser 1970:108)

I now turn to the Ilparakuyo and Massai material, commencing with a selection of conceptual issues and various notions of temporal units, days, months, seasons, years, etcetera, looking finally at their specific embodiment in the historical social formation of the Ilparakuyo section of the Tanzanian Massai.

III

There is a term in Maa, the Maasai language, which can generally be translated into the English word "time," as long as the latter is freed as much as possible from the epistemological assumptions discussed earlier. This Maasai word, *enkata,* has, as does its English counterpart, a plural form, *inkaitin* (*enkatitin* in some dialects). It may, for example, be used to refer to the two major seasonal units of the year, *enkata o'lari* 'the rainy season' and *enkata o'lameyu* 'the time of dryness, or the dry season'. The word *erishata* (pl. *irishat*) can be used to refer to varying units of time and also to an area of space (from *arish* 'to separate, divide').[11]

Enkata may also refer to specific points or periods in time, as in *ti ai kata* 'at another time', *te ena kata* 'at this moment', or *nabo kata* 'one time' (cf. Mol 1978:159–60). Similarly, one may say *te nkiti rishata* 'in a short time'. Then again, *enkata* may refer to times with a particular *quality* such as *enkata e'leng'on,* a time when everything is green, there is food for herds, the cattle are vigorous, and everything is healthy, a period that in other contexts may be referred to by other terms such as those for particular months.

Although it is not usually expressed by a verbal tense (most Maasai verbs combine present habitual and future in one form, except for a few that take an extra terminal vowel, *-u,* in the future), Maasai do have a very definite concept of an "indefinite" future, expressed adverbially. For this the word *kenya,* or *akenya,* may be used; it refers to any time from the present to an unspecified future day, season, epoch, generation, or condition. Or one may say *inkolong'i naapuonu* 'the days which will come'.

Other, fairly specific temporal concepts are *te ina kata* 'at that time', *taata* 'today', *oshi* 'usually', *oshi taata* 'nowadays', *oshi det* 'recently', *ade* 'later today', *taaisere* 'tomorrow', *metabaiki* 'tomorrow all being well' (the "possibilities of tomorrow," lit. 'may it be'), and so on. Referring to the past, there is *naaji* 'a little time ago', *duoo* 'a short while ago' or 'this morning', *ng'ole* 'yesterday', *naarri* 'some time ago' (longer than recently), and *opa* 'formerly' (also *apa* 'long ago'). This last concept, *apa,* may refer to any relatively remote time in the past up to "mythical time," the origins of man and the universe, although, as we shall see, these are not topics of particular concern to Ilparakuyo and Maasai. It may be specified as "long ago" by the addition of *moitie,* and in this sense refers to the period of the origin of Maasai and, more specifically, the beginning of such sections as the Ilparakuyo. These are among the few true adverbs in the Maa language, functioning as demonstratives of time; others refer to manner.

Maasai divide the year (*olari,* pl. *ilarin*) into twelve months (*ilapaitin,* sing. *olapa*) of thirty days (*inkolong'i,* sing. *enkolong'*). *Olari* also refers to the rainy season, *olapa* means "moon," and *enkolong'* "the sun," although "day" or "daylight time" can also be rendered by *endama* (pl. *indamaritin*), *dama* implying "during the day." These terms may, as we have seen, be used to refer to specific periods or times, as in *inkolong'i naapuonu* 'the days which will come' (i.e., "the future"), *ai olong'* 'another day', and so on.

The month is divided very specifically into thirty days according to the position of the moon, grouped into fifteen "bright" days, *enkiborra* (from *aibor* 'to be white'), and fifteen "dark" days, *enaimin* (from *enaimin* 'darkness'). Daylight time and the night (*enkewarie,* pl. *inkewarieitin*) are subdivided into a number of periods, some relating to activities concerning the herd; *eata Ilmaasai inkataitin enye enkolong' o enkewarie* 'the Maasai have various time divisions of day and night'; but the details need not concern us here (cf. Hollis 1905:332; Mol 1978:52, 112).

The complex progression of the thirty days of each month are described in great detail by Mol (1978:105). It is sufficient here to make two points: (1) as Mol states, "it is better to speak of nights or evenings of the lunar cycle than of days," and (2) knowledge of this complex classificatory structure is now highly specialized, and Kiswahili concepts have practically taken over in everyday usage, except where elders, warriors, or women gather for ritual occasions. We may also note that in Maasai and Ilparakuyo folklore there is a symbolic association between the night and men, and the daylight and women, the reason for this being, say Maasai, that "men, who are strong, go to fight [an enemy] at night, while women can work only during the day" (Hollis 1905:279).

Since the yearly cycle of twelve lunar months is based upon seasonal activities and the various Maasai sections inhabit a considerable area and variety of ecological and rainfall regimes, the exact relationship between particular months, seasonal divisions, and adjustments to discrepancies in the lunar year varies a great deal from one section to another. A. C. Hollis (1905:333–34), M. Merker (1910:160), S. S. ole Sankan (1971:65–66), and Frans Mol (1978:105–6) give various versions and, although the Hollis material is partially Ilparakuyo in origin (his main informant and assistant, Sameni ole Kivasis—or Justin Lemenye—was Olparakuoni), the versions I have recorded differ slightly from all of these. That such variations are true for the Nuer as well (Evans-Pritchard 1939:202–3) implies that they can be expected to occur in such widespread pastoral and semipastoral social formations.

IV

Ilparakuyo and Maasai myths and legends are not generally concerned with the ultimate origins of mankind, and even the few that do make some reference to beginnings attempt to explain the differentiation of Maasai from others, such as the Iltorrobo hunters and gatherers or their Bantu-speaking neighbors (cf. Jacobs 1965a:20–29; Hollis 1905:264–73; Galaty 1977, 1978a, 1978b; Beidelman 1960:245–49), rather than any overall "human origin." In this, Ilparakuyo are similar to the Iraqw (Thornton 1980:181, 183–84). But, unlike the Iraqw whose lack of myths of origin results in their also being without a "zero point against which to set a regularized chronology," and hence a history "with a backbone" (Thornton 1980:186), the Ilparakuyo and most other Maa-speaking peoples have a history with a sequence as well as a form of chronology, the latter provided by the overriding succession of age-sets, generation-sets, and the culturally attributable uniqueness and peculiarity of each.[12] Of course, Ilparakuyo and Maasai do not postulate a fixed zero point such as the B.C.-A.D. distinction of the Gregorian calendar, which, despite its relatively scientific adjustability to cosmic processes, is, after all, tied to the prevailing religious beliefs as well as the nonreligious ideology of western capitalist society (cf. Thornton 1980:183). But the sequence of age-sets provides Ilparakuyo and Maasai with a conception of time past that can be translated into the system with which we are stuck, though not without some difficulty.

Ilparakuyo, as well as other pastoral Maasai, explicitly link the (albeit oral) historiography of age-sets to the genealogy of the most famous line of religious leaders and prophets—*iloibonok kituaak* 'the great diviners'—who belong to a particular subclan of the Ilaiser clan, the Inkidong'i, founded by Kidong'oi, the first, or in some versions second, *oloiboni kitok* 'great diviner' (cf. Hollis 1905:324–30; Fosbrooke 1948:13 et passim; Jacobs 1965a:48–54, 1968:10–31; Berntsen 1979a:134–35 et passim; Huntingford 1953:103, 121–22; Merker 1910:18–22).[13] The main sections of the pastoral Maasai trace this genealogy of the *iloibonok* some nine to ten generations back from the present, taking us back to the middle or late seventeenth century. Writing in the 1940s, Fosbrooke gives a date of 1640 for the "finding" of the first *oloiboni kitok* (1948:3, 12; cf. Hollis 1905:326) and notes that "this agrees with all my informants, who state that [it] took place before the earliest remembered age-grade [*sic*], i.e., before the beginning of the 18th century."[14]

The details of Fosbrooke's original correlation between age-sets and the genealogy of the great prophets, his own revisions, and Jacob's comments and revisions have been published (Fosbrooke 1948:11–12, 1956a:193; Jacobs 1968a:14–18) and need not be repeated. The progressive increase in the historical "accuracy" of these chronologies has certainly been established, and they provide a reasonable framework for the translation of the past two hundred years of Maasai history, although this could be extended by further analysis of the actual relationships of the *iloibonok* to Maasai society through a genealogy of some three hundred years (cf. Berntsen 1979a, 1979b).

My purpose at present, however, is not to unravel the details of Maasai historiography but to develop a notion of Ilparakuyo consciousness and apprehension of history and hence the theoretical context for the development of Ilparakuyo (and Maasai) views of time. This may be achieved by limiting the discussion to two further aspects of the Ilparakuyo and Maasai social formations: (1) the conceptual categories used in distinguishing between myths, legends, and stories about the past; and (2) the tracing of an outline history of Ilparakuyo separation from the main body of pastoral Maasai, using a combination of age-set chronology and the genealogies of the major Ilparakuyo "great ritual leaders" since that separation, in conjunction with written sources.

Jacobs notes that Maasai make a clear distinction between what they call *enkatinyi*, "history and oral tradition" (from *enkata* 'time', pl. *inkatitin* [see above]), on the one hand, and *enkiterunoto* "myths" (lit. 'beginnings', from *aiteru* 'to start', 'to begin') on the other (Jacobs 1968a:14–15). Concepts that might include "history," "myth," and even "pedagogy," depending upon the context, are *inkopa* 'things (matters, affairs) of long ago' and the word *inkoon* 'advice', 'history', derived from *aikok* 'to instruct, advise, warn', the last described by Mol (1978:82) as "such instructions e.g. are given by the *olpiron* ['firestick'] elders to the warriors on various occasions, [and] such instructions will also include bits and pieces of old customs, old happenings, etc."

Although very similar distinctions could apply, mutatis mutandis, to the Ilparakuyo conceptualization of the past, the differences between the two main types of narrative are not always clear-cut. Ilparakuyo, as with other pastoral Maasai, do recount historical events in relation to the succession of age-sets and the genealogy of the *iloibonok,* but the order is often confused beyond the third or fourth age-set prior to the present senior elders. This is due in part to the somewhat difficult categories of historical discourse available to the investigator. For example, "left hand" and "right hand" sets (*emurata e kedyenye* and *emurata e tatene*) may have different "nicknames" before the *eunoto* ceremony that promotes them to senior warriorhood, and different names may be used for the same age-set according to whether it is the *eunoto* name or that given at *olng'eher,* the ceremony that confirms the status of elderhood. These problems in themselves are relatively amenable to a method of systematic cross-reference; but the fact remains that elders give in different contexts varying versions of the names of earlier age-sets and the genealogy of the *iloibonok.* Furthermore, there is as yet no comprehensive published account of the actual operation of Ilparakuyo age-set organization, which differs in some important respects from that of pastoral Maasai.[15]

I have described elsewhere (Rigby 1979, 1980, 1981) certain aspects of Ilparakuyo age organization and age-sets functioning as relations of production not only within the historical development of Ilparakuyo society itself but also in the context of the wider social formations of neighboring peoples and the modern state of Tanzania (cf. also Beidelman 1961a, 1968, 1960:262–67). It is through the succession of age-sets, the attribution of particular qualities and practices to them as distinct historical units (remembered very often in songs and epic recitations), and their continuing role in the social (as well as symbolic) reproduction of the Ilparakuyo social formation that the other elements of social structure such as descent, kinship, and generational relationships are articulated with wider historical developments.

With these points in mind, I present a tentative and diagrammatic sketch of historical developments since Ilparakuyo separation as a distinct social formation from other pastoral Maasai, concluding with a discussion of its implications for historical consciousness and time reckoning.[16] I have not distinguished in the age-set chronology between right-hand and left-hand groups, nor have I distinguished them from "generation-sets" (cf. Hollis 1905:262–63; Fosbrooke 1956a; Jacobs 1965a, 1968a). The dates given in table 1 are very approximate times for the *opening* of new circumcision periods for what later became complete age-sets (*ilajijik,* sing. *olaji*) at *eunoto* ceremonies, including all subdivisions and all three subsets (Ilcang'enopir, Ilparing'otua, and Ilkerimbuot) of each age-set (see Rigby 1979:339). They were established upon the basis of oral evidence, textual corroboration, and the general principle that, although "new age-sets among the Maasai [and Ilparakuyo] tend to be formed at regular intervals of about fifteen years, each age-set serves in the warrior grade for an average of twenty years," depending upon the political,

TABLE 1. Ilparakuyo Age-set Chronology and Genealogy of the Great Diviners/Prophets (Iloibonok kituaak)

Age-set Name[a]	Year (ca.) Opening Circumcision and Events		Genealogy of Iliobonok Kituaak and Events	Date (ca.) of Succession
?	1795	Beginning of the conflict with the other Maasai sections	Teliang' (?)	?
?	1815		Lengunat	1805
Ilmeriho I	1832	Final break with Kinsongo Maasai and beginning of movement south into present Bagamoyo and Morogoro Districts[b]	Mtango	1835
Ilkishomu	1854		Kirigo	1855
Ilkenyeiyie/Ilsujita	1874[c]		⎯ Kirkong'	
Ilpariho/Isiyiapai (Iltapali)	1890	Movement westward and south into Ugogo and Uhehe[e]	Maitei Resigned as "official representative" of Ilparakuyo to German administration, 1918[d]	1885
Ilkijaro/Ilmetimpot(?)	1907	Beginning of movement into Mbeya District	Moreto	1920

Iseeta (*eunoto*)/Iltwati (*olng'eher*)(Ilkisalie)	1924	Current senior elders (1980)
Ilkidotu (*eunoto*)/Ilmeriho II (*olng'eher*)	1942	Current junior elders (1980)
Ilmedoti (*eunoto*)/ "Ildobola"/"Ilmesokile"	1957	Current senior warriors
Ilkipone (*eunoto*)/Iltareto/ "Ilmakaa"	1972	*Eunoto* ceremony, 1980, but just moving out of junior warriorhood

```
           Mutari      Senteu
                       1950
```

^aWhere available, I have given both names received, first at *eunoto* ceremonies (promoting junior warriors to senior warriors), and second at *olng'eher* rituals (promoting senior warriors to junior elderhood). Nicknames of age-sets given before *eunoto* (there are no proper names before this) are in quotation marks.

^bSee Lemenye 1956:52, n.11 (by Fosbrooke), 57; and Thomson 1885:240–42.

^cIn some versions of oral history, Iiparakuyo claim that the Ilkenyeiyie age-set fought the Germans at Handeni. If they were junior warriors (*ilmurran*) at the time, this would have been possible only if they had come into conflict with the forces deployed by the Society for German Colonization, which sent an expedition in 1884 to "sign treaties with native chiefs" in the Pangani River area and hinterland, or those of the German East Africa Company, incorporated in 1887–88. Or they may have fought the Germans as senior warriors, after *eunoto*, whose task it is to direct operations carried out by the junior age-set.

^dSee Rigby 1981:115.

^eBeidelman 1960:248–53; 1962.

ecological, and ritual conditions at the time. Similarly, the genealogy of the Ilparakuyo *iloibonok kituaak* rests on oral evidence, textual corroboration, and a generation span of about thirty years.

It seems fairly certain that Ilparakuyo finally established their political and ritual "independence" from Kisongo and other pastoral Maasai either just prior to or during the incumbency of Mtango as *oloiboni kitok,* or great prophet, and many elders cite him as the first in the line of Ilparakuyo prophets. But there is evidence that there were moves to gain ritual control of the Ilparakuyo age-set system prior to this, as suggested in the "composite" genealogy represented in table 1. Two further points must be made: first, the genealogy given for the *iloibonok kituaak* refers only to the lineage of the Inkidong'i subclan of the Ilwarakishu (Ilaiser) clan, which provides the ritual center for northern and eastern Ilparakuyo, but their exact relationship with the Ilaiser Inkidong'i of the other pastoral Maasai is not known; and, second, southern and southwestern Ilparakuyo communities have another ritual center with its own lineage of *iloibonok kituaak* (Rigby 1979:334; Beidelman 1960:264–65).

Before proceeding to draw some conclusions from this brief outline of Ilparakuyo age-set chronology and separation from other pastoral Maasai sections, I hasten to state that this is not the place to join the debate upon the frequent identification of Ilparakuyo with "agricultural Maasai," also called "Wakwavi" and "Iloikop," and their presumed hostility and opposition to pastoral Maasai (e.g., Jacobs 1965a:39–48, 1968b:21–28, 1975, 1979:36 et passim; Krapf 1854, 1968; Galaty 1977, 1978b; cf. Berntsen 1979b, 1980). Certainly, Ilparakuyo have always attempted to maintain as "pastoral" a system of production and reproduction of their social formation as they can, eschewing agriculture, although, as with most pastoralists, they frequently depend to a lesser or greater extent upon the agricultural produce of their cultivating neighbors, for which they trade. The arguments on this issue and the literature relating to the debate have been admirably surveyed by Berntsen (1980).

Ilparakuyo historical traditions describe hostile relations with other Maasai sections, particularly during the period of some forty years until about 1832, when they finally broke away and began moving into their present areas, dotted about among primarily Bantu-speaking cultivators; but their general drift south and west continues to the present time. It is also true that Bantu speakers on the coast currently refer to Ilparakuyo as "Wakwavi"; the Gogo, Kaguru, Kimbu, and other Bantu-speaking peoples of central Tanzania, however, among whom large groups of Ilparakuyo live, refer to all Maa-speaking peoples (except the Ilarusa, "Arusha") as "Wahumha," or variations thereof (from "Humba," "Ilumbwa"). Several Ilparakuyo elders claim historical connections with Ilumbwa whom they currently identify as the Kalenjin-speaking Kipsikis who are frequently (but erroneously) known by that name, and who are also called "Ilkakisang'" (sing, Olkakisang'i) by pastoral Maasai (cf. Huntingford 1953:10; Hollis 1905:280–81); and some other pastoral Maasai sections call Ilparakuyo "Ilumbwa" (cf. Beckwith and Saitoti 1980:18). That at least some elements of Ilparakuyo have had an identity distinct from contemporary pastoral Maasai sections dating from a period much earlier than outlined, including perhaps ritual independence, is suggested by evidence in various sources (e.g., Thomson 1885; Johnston 1886). And linguistic evidence linking the Ilparakuyo dialect of Maa with that of the Isampur (Samburu) at the extreme northern end of the Maa-speaking diaspora offers the possibility of close contact with Kalenjin speakers on their movement south (Vossen 1977a:10–13), as well as earlier separation. This does not, however, alter materially the connections and conflicts with Kisongo Maasai at the later period of the eighteenth and nineteenth centuries as outlined.

The term *Iloikop* is even more ambiguous as an appellation for Ilparakuyo than Ilumbwa. The fact that Samburu call themselves Iloikop may lend some weight to the identification of Ilparakuyo as Iloikop. But Jacobs's insistence that the term is related to one of the Maasai words for "murder" (e.g., someone who has committed serious assault or murder is in a state of "taboo" and is said "to have *iloikop*," "*eeta iloikop*") is on shaky ground, at least in this context, when we note that Samburu deny any association between their own term for themselves and the word for the contamination of murder (Spencer 1973:109, n.7). As Berntsen correctly maintains, these usages and concepts have not remained historically consistent, and there is strong evidence to suggest that there were several groups of Maa speakers referred to as Wakwavi (or its variations) in the nineteenth century that have little or no connection with the contemporary Ilparakuyo apart from the common one of language.

It is clear that Ilparakuyo historical consciousness is expressed not as a "mythical charter" for a unique and fixed identity as such but as a narration of past spatial movements associated with other peoples, both Maa and non-Maa, and adaptation to constantly changing circumstances, political, economic, ecological, and ritual. At all of these times, specific age-sets acting as warriors (and later as elders directing and advising new warrior groups) are conceived of as playing decisive roles of a historically unique kind. Ilparakuyo history does have elements of what can be called a "topological" chronology, but it also has much more. The spatial elements inseparable from Ilparakuyo time and historical concepts do not establish "exclusive spaces," since Ilparakuyo appropriation of their environment is temporary in the extreme, and subject to ever-changing politicoeconomic circumstances, more so than pastoral formations that do have contiguous territories, and much more so than cultivators (Rigby 1977, 1979, 1981, but particularly 1980:47–48; Marx 1973:472, 474; Beidelman 1960:259). Ilparakuyo do have a very explicit idea of the territorial segments of their social formation; these are not, however, seen as fixed, bounded units but as constantly changing through loss, and accretion, and migration, a process stretching back through the whole period of the known historical development of the social formation itself.

In illustration of these condensed statements, one example must suffice. It concerns the continuity of descent groups, which, as noted, is conceptually inseparable from the "complementary" reproduction of society as a whole through the succession of age-sets. Prior to the actual circumcision of a man's first child, a special ceremony, called *olkiteng' loo 'lbaa* 'the ox of the wounds', must be performed by the father. Although this is a "private," kinship ceremony, the involvement of all kin and affines is combined with a *restatement* of the specific relationships among the father's age-mates, and hence between their age-set and others, within that specific neighborhood at that specific time. This includes not only a reappraisal and reinterpretation of the history of the local community in relation both to its own past and to other Ilparakuyo and Maasai localities, but also the past, future, and current conditions that may exist within the context of other, non-Ilparakuyo neighbors, as well as political and economic events beyond local control. Thus, prior to commencing the critical series of ceremonies that integrate an individual into the community at large, a context for the historical reappraisal of, and adjustment to, "global" conditions—social, political, and economic—is created.

Finally, age-set chronology and the genealogy of the senior Ilparakuyo religious experts place the most significant conflicts and final separation or disengagement from other pastoral Maasai sections in the period of about forty years prior to 1832, precisely when other historical evidence indicates them to have occurred.[17] Since this period, and probably (but less significantly) even before, Ilparakuyo and other Maasai age-sets have had different though sometimes overlapping names, just as they have their own branch of the clan and subclan of the great religious leaders. This historicity and the territorial and spatial connotations of Ilparakuyo and Maasai historical consciousness have been admirably expanded by Rainer Vossen on the basis of linguistic evidence, and his conclusion, with the exception of the final period, which I would change to "early nineteenth century," is in no way contradictory or incompatible with the interpretation presented in this essay (Vossen 1977a:13):

> In summary, the following hypothesis is set up. The Baraguyu [Ilparakuyu] are either close relatives or at least former neighbors of the Sampur. Their separation might be dated back to the end of the 15th century, slightly before the southward migration of the Baraguyu. Coming through the rift valley, they settled down in what is now the L-Oitai region [now the western part of Kenya Maasailand bordering Tanzania]. There they lived side-by-side with L-Oitai from around 1700. By the end of the 18th century, both groups got into quarrels which resulted in the defeat and expulsion of the Baraguyu. In the environs of Pangani river they formed a new political [and ritual?] center which was gradually destroyed by the Kisongo during the mid-19th century and ultimately ended up with the scattering of Baraguyu to their present settlement areas.

V

Although in my previous article on Ilparakuyo meat-feast ceremonies (*ilpuli*) and the notion of *entoroj* (also spelled *enturuj*) I emphasized the real and symbolic role of age-sets in social reproduction (Rigby 1979:340–43), it should be stressed here that the succession of age-sets through the age-grade system should not be thought of as merely a repetitious cycle, the life cycle writ large; the Ilparakuyo certainly do not think of age-sets as such. Each age-set, the circumstances of its formation, and the creativity and originality of its actions, are emphasized, making it unique in the annals of Ilparakuyo historical consciousness as well as a referent for historical interpretation and, ultimately, the future of the social formation itself. If Ilparakuyo have any "identity" as a social formation, it lies not merely in an endless repetition of the past (despite a strong emphasis upon adherence to certain basic social practices and their associated values) but in a consciousness of a developing entity having a definite historical beginning and moving in a specific historical manner. This movement is not toward a destiny but embodies both the relations of past and present in the creation of a future.

In this context, the specific temporal concepts of life cycles, days, months, years, generations, and age-sets are incorporated in a very specific historical manner in a social formation that has managed to retain its singular identity *not* through conservatism and the mere reproduction of the past but in the creative use of the past in the praxis required for adaptation to an always uncertain future.

What I am arguing here is that the specific elements of the pastoral Ilparakuyo and Maasai social formations and pastoral praxis permit the development of a theory of historical time, closely linked to the spatial relations of migration, conflict, and the temporary appropriation of nature, which has a specific meaning for pastoral praxis itself. Any changes in spatial relationships, through loss of access to grazing, water, etcetera, and movement subsequent thereto must inevitably reflect upon, and be reflected in, views of time and history. The very fluidity of Ilparakuyo concepts dealing with time and history are therefore a condition of continuing and successful pastoral praxis.[18] History is not merely "a rationalization of customary behavior" (Thornton 1980:187), although it is that on one level; rather, it is a theoretical reaffirmation of material practice geared to the pastoral appropriation of nature, in the context of—one might say "in face of"—other forms of such appropriation, either by agriculture or hunting and gathering, accompanied by its ritual and symbolic elements (Rigby 1979). It is only in this total setting, which includes Ilparakuyo historical relationships with other Maasai and the numerous non-Maasai among whom they live, as well as the modern state of Tanzania and its development policies and practices, that we can understand the other temporal rhythms of days, months, years, etcetera, which Ilparakuyo and Maasai conceptualize.

Finally, while I am in general agreement with Bourdieu's sensitive analysis of time and practice of the Kabyle (Bourdieu 1963; 1977:171–79), his elaboration of the idea of "symbolic capital" and his insistence upon the "interconvertibility" of symbolic and other forms of capital by broadening the definition of "economic interest" are ultimately reductionist in the senses already discussed for other approaches to concepts of time and cannot be applied to the Ilparakuyo and Maasai case. To assume that practice in the latter social formations "never ceases to conform to economic calculation even when it gives every appearance of disinterestedness by departing from interested calculation (in the narrow sense) and playing for stakes that are nonmaterial and not easily quantified" (Bourdieu 1977:177) would be to obscure the very historical process we are trying to grasp. I have attempted to propose in this essay that the apprehension of temporal categories can be understood only in the context of particular historical transformations and the specific forms of historical consciousness generated by them. Any notion of a *universal* "symbolic commoditization" of time, as suggested by Bourdieu, runs counter to this proposal and hinders our apprehension of the conditions under which "commoditization" itself occurs, that is, the penetration of capitalist relations of production. As E. P. Thompson ably demonstrates for a very different time and place, during the period 1300 to 1650 the "intellectual culture of western Europe [underwent] important changes in the apprehension of time" (1967:56 et passim), and the backward capitalist countries of Europe (such as Greece) are still undergoing this transformation toward the accumulation, use, and "budgeting" of time, as are peasant social formations elsewhere, in short, its commoditization (cf. Mead 1955:70–72; Taussig 1980:5 et passim).

The formation of classes and the penetration of capitalist relations of production and ideology have been more fully explored for Maasai in Kenya, where these processes are further advanced than in Tanzania (e.g., Bonte 1975; Hedlund 1980; Rigby 1981; ole Sena 1981). While these transformations need much further elaboration for the Ilparakuyo and Maasai social formations of Tanzania, I have attempted in this paper to provide a theoretical and historical background for such an account by locating Ilparakuyo notions of time and change in the specific historicity of this social formation at a particular historical conjuncture.[19]

NOTES

1. Evans-Pritchard also considered in some detail Lévy-Bruhl's ideas while preparing, in 1934, *Witchcraft, Oracles, and Magic among the Azande* (1937).

2. At some points in his argument, Lévy-Bruhl comes close to recognizing the social and economic conditions that engender the production of various temporal concepts, but his false epistemological position prevents him from grasping their significance. Commenting upon the issue that, as far as he knew, "primitive" peoples did not see a future event as "situated at any point on the line of futurity," he notes (1923:124, n.1), "This is one of the principal reasons for that 'lack of foresight' so often observed and deplored by those who study uncivilized peoples. It is *undoubtedly also due to other causes* of a *social and economic order;* but it proceeds mainly from the mental habits of primitives" (emphasis added).

3. That Clifford Geertz, despite the richness of the analyses he achieves, makes the same error in this respect is evident throughout his work (e.g., see Geertz 1973:361).

4. In an often neglected but frequently perceptive paper on what he calls "temporal orientations" among the Berens River Saulteaux, an Ojibwa-speaking group of hunters and fishermen, A. I. Hallowell makes explicit his comparison with "western time concepts," commenting upon the "relativity and provinciality" of the latter. He notes

the reification and commoditization of time in capitalist society, quoting from Lewis Mumford (1934): "Under capitalism, time-keeping is not merely a means of coordinating and inter-relating complicated functions; it is also like money an independent commodity with a value of its own" (Hallowell 1937:649, n.9).

5. Leach's study of Kachin social structure (1954) is an excellent and in many ways pioneering analysis of the "time dimension" in social structure (cf. Friedman 1975).

6. It is pertinent here to note Robin Horton's much more accurate contention (1967:176) that, "in traditional Africa, methods of time reckoning vary greatly from culture to culture," an observation in keeping with the analysis presented here.

7. Somewhat similar, but less "theological," versions of time reckoning among the Kamba are given by Kivuto Ndeti (1972:178–85) and Gerhard Lindblom (1920:338–42).

8. In an extremely penetrating and often ignored analysis of Iatmul ritual, ideology, and social structure, Gregory Bateson (1958:111–14 et passim) emphasizes the importance of a historical approach to the study of precapitalist societies and their conceptual systems.

9. Lévi-Strauss (1963:21) maintains that "even the analysis of synchronic structures . . . requires constant recourse to history. By showing institutions in the process of transformation, history alone makes it possible to abstract the structure which underlies the many manifestations and remains permanent throughout a succession of events" (cf. also Lévi-Strauss 1966:66–74, 245–69). I show later why this position entails an unacceptable notion of "history," but we may note here that the strong Hegelian influence underlying structuralist theory accounts for the reification of time concepts and hence the emphasis upon "experience" (cf. Dunayevskaya 1973:15–16 et passim).

10. Jean Hyppolite's contention that essentialism is "displaced forever" is erroneous in that essentialism is in fact embodied in the very structuralist problematic he sees as displacing the "tide of neo-Hegelian existentialism" (Hyppolite 1969:viii, 13).

11. That most concepts of time are inextricable from concepts of space is admirably documented for the Iraqw by Thornton (1980); this is, of course, a universal condition.

12. The implications of age- and generation-set systems for chronology and history are explored for a number of societies in P.T.W. Baxter and Uri Almagor (1978) and in such monographs on particular societies as N. Dyson-Hudson (1966:258–70 et passim) and J. Lamphear (1976).

13. Some versions give Kidong'oi as the son of "ole Mweiya," but, as Jacobs points out (1965a:321), this is not a proper name but an appellation given to "the first one" in any story or legend when the true name is not known. In his later article, Jacobs (1968a:20) suggests that Sitonik was the "first ritual expert," the third on Hollis's list.

14. According to Maasai and Ilparakuyo legend, a warrior of the Ilaiser clan "found" ole Mweiya (Kidong'oi or Sigiraishi) on the mountain called Enkong'u e 'Mbakasi, "the source (lit. the 'eye') of the Mbakasi (Athi) River." The mountain is

also called Oldoinyo loo 'Laiser 'of the Ilaiser clan' or Oldoinyo lo 'leMweiya 'of Mweiya's place' and is known as the Ngong Hills by non-Maasai.

15. Melkiori Matwi, of the Tanzanian Ministry of Information and Culture, has prepared such an account, but access to it has not yet been made possible.

16. I must emphasize the extremely tentative nature of this provisional outline, which is subject to further verification from oral sources as well as comparative research. The version of Ilparakuyo historical relations presented here is gleaned from cross-referencing my own research with materials from Beidelman (1960, 1961a, 1961b, 1962, 1968); Berntsen (1979a, 1980); Jacobs (1965a, 1968a, 1968b, 1975, 1979); Fosbrooke (1948, 1956a, 1956b); Lemenye (1956); Thomson (1885); Krapf (1968); and Berntsen (personal communication). Other historical references are cited in the text.

17. In commenting upon a story told to Justin Lemenye by an Olparakuoni elder in 1893, Fosbrooke places the final conflict between "Kwavi" and Maasai and their final separation in the period 1825–39 (Lemenye 1956:51, n.11). The evidence from such writers as O. Baumann (1894), Joseph Thomson (1885), and others is summarized by Beidelman (1960:245–50) and roughly confirms this period. There have been Ilparakuyo communities on and off since at least 1837 in the areas of West Bagamoyo District where I have lived with them.

18. I am in fact suggesting that Ilparakuyo and Maasai consciousness of "time in historicity" is dialectical, and to that extent it is antithetical to the "metaphysical mode of thinking" characteristic of bourgeois philosophy. It is similar, on the other hand, to what Engels identified as conceptions of history and time in "this primitive, naïve, but intrinsically correct conception of the world" characteristic of ancient Greek philosophy, "first clearly formulated by Heraclitus: everything is and is not, for everything is fluid, is constantly changing, constantly coming into being and passing away" (Engels 1975:27–28; cf. Galaty 1977, 1978b, for Kenya Maasai).

19. A somewhat different interpretation of the data on relations of production and their transformation in the Ilparakuyo social formation is given by Bryceson and Mbilinyi (1980:95–98 et passim).

REFERENCES

Althusser, Louis. 1970. "The Errors of Classical Economics: An Outline for a Concept of Historical Time." In *Reading Capital,* Louis Althusser and Étienne Balibar, eds. London: New Left.
Bateson, Gregory. 1958 [1936]. *Naven.* Stanford: Stanford University Press.
Baumann, O. 1894. *Durch Maasailand zur Nilquelle.* Berlin: Reimer.
Baxter, P. T. W., and Uri Almagor, eds. 1978. *Age, Generation and Time.* New York: St. Martin's.
Beckwith, C., and T. ole Saitoti. 1980. *Maasai.* New York: Abrams.

Beidelman, Thomas O. 1960. "The Baraguyu." *Tanganyika Notes and Records* 55:245–78.

———. 1961a. "A Note on Baraguyu Housetypes and Economy." *Tanganyika Notes and Records* 57:56–66.

———. 1961b. "Beer Drinking and Cattle Theft in Ukaguru." *American Anthropologist* 63:3, 524–49.

———. 1962. "A Demographic Map of the Baraguyu." *Tanganyika Notes and Records* 58/59, 8–10.

———. 1963. "Kaguru Time Reckoning: An Aspect of the Cosmology of an East African People." *Southwestern Journal of Anthropology* 19 (Spring): 9–20.

———. 1968. "Some Hypotheses Regarding Nilo-Hamitic Symbolism and Social Structure." *Anthropological Quarterly* 41:2, 78–89.

Berntsen, John L. 1979a. "Massai Age-sets and Prophetic Leadership: 1850–1910." *Africa* 49:2, 134–46.

———. 1979b. "Pastoralism, Raiding, and Prophets: Maasailand in the Nineteenth Century." Ph.D. diss., University of Wisconsin.

———. 1980. "The Enemy is Us: Eponymy in the Historiography of the Maasai."

Bohannan, Paul. 1967 [1953]. "Concepts of Time among the Tiv of Nigeria." *Southwestern Journal of Anthropology,* 9:3, 251–62. Reprinted in *Myth and Cosmos,* John Middleton, ed. New York: Natural History Press, 1967.

Bonte, Pierre. 1975. "Cattle for God: An Attempt at a Marxist Analysis of the Religion of East African Herdsmen." *Social Compass* 22:3/4, 381–96.

Bourdieu, Pierre. 1963. "The Attitude of the Algerian Peasant toward Time." In *Mediterranean Countrymen,* J. Pitt-Rivers, ed. Paris: Mouton.

———. 1977. *Outline of a Theory of Practice.* Cambridge: Cambridge University Press.

Bryceson, D. F., and Mbilinyi, M. 1980. "The Changing Role of Tanzanian Women in Production." In *Jipemoyo 2/1980,* A. O. Anacleti, ed. Uppsala: Scandinavian Institute of African Studies for Ministry of National Culture and Youth, Dar es Salaam, and the Academy of Finland.

Douglas, Mary. 1966. *Purity and Danger.* London: Routledge and Kegan Paul.

Dunayevskaya, Raya. 1973. *Philosophy and Revolution.* New York: Dell.

Dyson-Hudson, N. 1966. *Karimojong Politics.* Oxford: Clarendon.

Engels, Frederick. 1940 [1878]. *Dialectics of Nature.* New York: International Publishers.

———. 1975 [1924]. *Anti-Dühring.* Moscow: Progress Publishers.

Evans-Pritchard, E. E. 1937. *Witchcraft, Oracles, and Magic among the Azande.* Oxford: Clarendon.

———. 1939. "Nuer Time Reckoning." *Africa* 12:2, 189–216.

———. 1940. *The Nuer.* Oxford: Clarendon.

———. 1965. *Theories of Primitive Religion.* Oxford: Clarendon.

Fortes, Meyer. 1945. *The Dynamics of Clanship among the Tallensi.* London: Oxford University Press.

——. 1949a. "Time and Social Structure: An Ashanti Case Study." In *Social Structure: Studies Presented to A. R. Radcliffe-Brown*, M. Fortes, ed. Oxford: Clarendon.

——. 1949b. *The Web of Kinship among the Tallensi*. London: Oxford University Press.

Fosbrooke, Henry A. 1948. "An Administrative Survey of the Masai Social System." *Tanganyika Notes and Records* 26, 1–50.

——. 1956a. "The Masai Age-group System as a Guide to Tribal Chronology." *African Studies* 15:4, 188–206.

——. 1956b. "Introduction and Annotations." In Justin Lemenye, *"The Life of Justin:* An African Autobiography." *Tanganyika Notes and Records* 41/42:31–57, 19–30.

Friedman, Jonathan. 1975. "Tribes, States, and Transformations." In *Marxist Analyses and Social Anthropology*, M. Bloch, ed. London: Malaby.

Galaty, John G. 1977. "In the Pastoral Image: the Dialectic of Maasai Identity." Ph.D. diss., University of Chicago.

——. 1978a. "Pollution and Pastoral Anti-Praxis: The Issue of Maasai Inequality." Manuscript.

——. 1978b. "Being 'Maasai'; Being 'Under Cows': Ethnic Shifters in East Africa." Paper presented to the 77th Annual Meeting of the American Anthropological Association, Los Angeles.

Geertz, Clifford. 1973. *The Interpretation of Cultures*. New York: Basic Books.

Hall, Edward T. 1959. *The Silent Language*. New York: Doubleday.

Hallowell, A. I. 1937. "Temporal Orientations in Western Civilization and in a Preliterate Society." *American Anthropologist* 39:4, 647–70.

Hedlund, Hans. 1980. "Contradictions in the Peripheralization of a Pastoral Society: The Maasai." *Revue of African Political Economy* 15/16, 15–34.

Hegel, G. W. F. 1900. *Philosophy of History*. New York: Colonial Press.

Henderson, W. O. 1965. "German East Africa, 1884–1918." In *History of East Africa*, V. Harlow, E. M. Chilver, and Alison Smith, eds., vol. 2. Oxford: Clarendon.

Hollis, A. C. 1905. *The Masai: Their Language and Folklore*. Rept. 1970. Westport, Conn.: Negro Universities Press.

Horton, Robin. 1967. "African Traditional Thought and Western Science." *Africa* 37:50–71, 155–87.

Huntingford, G. W. B. 1953. *The Southern Nilo-Hamites*. London: International African Institute.

Hyppolite, Jean. 1969. *Studies on Marx and Hegel*. New York: Basic Books.

Jacobs, Alan H. 1965a. *The Traditional Political Organization of the Pastoral Masai*. D. Phil. thesis, Oxford University.

——. 1965b. "African Pastoralists: Some General Remarks." *Anthropological Quarterly* 38:3, 144–54.

———. 1968a. "A Chronology of the Pastoral Maasai." In *Hadith I*, B. A. Ogot, ed. Nairobi: Historical Association of Kenya.

———. 1968b. "The Irrigation, Agricultural Maasai of Pagasi." In *Proceedings of the Annual East African Social Science Conference*, Dar es Salaam.

———. 1975. "Maasai Pastoralism in Historical Perspective." In *Pastoralism in Tropical Africa*, T. Monod, ed. London: Oxford University Press.

———. 1979. "Maasai Inter-Tribal Relations: Belligerent Herdsmen or Peaceable Pastoralists?" In *Warfare among East African Herders*, K. Fukui and D. Turton, eds. Senri Ethnological Studies, no. 3. Osaka: National Museum of Ethnology.

Johnston, H. H. 1886. *The Kilima-Njaro Expedition*. London: Kegan Paul, Trench.

Krapf, J. L. 1854. *Vocabulary of the Enkutuk Eloikob: Or, the Language of the Wakuafi-Nation in the Interior of Equatorial Africa*. Tübingen: Fues.

———. 1968 [1860]. *Travels, Researches, and Missionary Labours during Eighteen Years Residence in East Africa*. 2d ed. London: Frank Cass.

Lamphear, J. 1976. *Traditional History of the Jie of Uganda*. London: Oxford University Press.

Lawren, W. L. 1968. "Masai and Kikuyu: An Historical Analysis of Culture Transmission." *Journal of African History* 9:4, 571–83.

Leach, Edmund. 1954. *Political Systems of Highland Burma*. London: Bell.

———. 1961. *Rethinking Anthropology*. London: Athlone.

Lefebvre, Henri. 1968. *The Sociology of Marx*. London: Penguin.

Lemenye, Justin. 1956. *"The Life of Justin:* An African Autobiography," H. A. Fosbrooke, trans., ed., and ann. *Tanganyika Notes and Records* 41/42, 31–57, 19–30.

Lenin, V. I. 1962 [1908]. *Materialism and Empirio-Criticism*. Moscow: Progress Publishers.

Lévi-Strauss, Claude. 1963. *Structural Anthropology*. New York: Basic Books.

———. 1966. *The Savage Mind*. London: Weidenfeld and Nicholson.

Lévy-Bruhl, Lucien. 1910. *Les fonctions mentales dans les sociétés inférieures*. Paris: Alcan.

———. 1923. *Primitive Mentality*. Boston: Beacon.

———. 1938. *L'Expérience mystique et les symboles chez les primitifs*. Paris: Alcan.

Lindblom, Gerhard. 1920. *The Akamba*. Uppsala: Appelbergs.

Marx, Karl. 1973 [1857–58]. *Grundrisse*, Trans., fwd., and ann. by Martin Nicholaus. Harmondsworth: Penguin.

Mauss, Marcel. 1923. "Comments on an Address by Lucien Lévy-Bruhl on 'La Mentalité primitive.'" *Bulletin de la Société française de philosophie* 18:2, 24–28.

Mbiti, John S. 1969. *African Religions and Philosophy*. London: Heinemann.

Mead, Margaret, ed. 1955. *Cultural Patterns and Technical Change*. New York: Mentor for UNESCO.

Merker, M. 1910. *Die Masai*. Berlin: Reimer. Reprinted, Johnson Reprint Corporation, New York, 1968.

Mol, Frans. 1978. *Maa: A Dictionary of the Maasai Language and Folklore.* Nairobi: Marketing and Publishing, Ltd.
Morton-Williams, Peter. 1968. "The Fulani Penetration into Nupe and Yoruba in the Nineteenth Century." In *History and Social Anthropology,* I. M. Lewis, ed. London: Tavistock.
Mumford, Lewis. 1934. *Technics and Civilization.* New York: Harcourt, Brace.
Ndeti, Kivuto. 1972. *Elements of Akamba Life.* Nairobi: East African Publishing House.
Needham, Rodney. 1972. *Belief, Language, and Experience.* Oxford: Blackwell.
Ray, Benjamin. 1976. *African Religions.* Englewood Cliffs, N.J.: Prentice Hall.
Rigby, Peter. 1968. "Some Gogo Rituals of 'Purification': An Essay on Social and Moral Categories." In *Dialectic in Practical Religion,* Edmund Leach, ed. Cambridge: Cambridge University Press.
———. 1977. "Critical Participation, Mere Observation, or Alienation: Notes on Research among the Baraguyu Maasai." In *Jipemoyo 1/1977.* Helsinki: University of Helsinki for the Department of Research and Planning, Ministry of National Culture and Youth, Dar es Salaam, and Academy of Finland.
———. 1979. "Olpul and Entoroj: The Economy of Sharing among the Pastoral Baraguyu of Tanzania." In *Pastoral Production and Society,* Equipe écologie et anthropologie des sociétés pastorales, eds. Cambridge: Cambridge University Press.
———. 1980. "Pastoralist Production and Socialist Transformation in Tanzania." In *Jipemoyo 2/1980,* A. O. Anacleti, ed. Uppsala: Scandinavian Institute for African Studies for the Ministry of National Culture and Youth, Dar es Salaam, and the Academy of Finland. To be reprinted in *Nomadic Peoples in a Changing World,* P. C. Salzman and J. G. Galaty, eds. Philadelphia: ISHI.
———. 1981. "Pastors and Pastoralists: The Differential Penetration of Christianity among East African Cattle Herders." *Comparative Studies in Society and History* 23:1, 96–129.
Sankan, S. S. ole. 1971. *The Massai.* Nairobi: East African Literature Bureau.
Sena, Sarone ole. 1981. "Schemes and Schools: Two Agents of Change among the Maasai of Kenya." Paper presented to joint session of the Temple University African Studies Committee and Department of Anthropology, April.
Spencer, Paul. 1973. *Nomads in Alliance.* London: Oxford University Press.
Taussig, Michael T. 1980. *The Devil and Commodity Fetishism in South America.* Chapel Hill: University of North Carolina Press.
Thompson, E. P. 1967. "Time, Work-Discipline, and Industrial Capital." *Past and Present,* no. 38, 56–97.
Thomson, Joseph. 1885. *Through Masai Land.* London: Sampson, Low, Marston. 3d ed. London: Frank Cass, 1968.
Thornton, Robert. 1980. *Space, Time, and Culture among the Iraqw of Tanzania.* New York: Academic.

Vossen, Rainer. 1977a. "Notes on the Territorial History of the Maa-speaking Peoples: Some Preliminary Remarks." University of Nairobi, Department of History, Staff Seminar Paper no. 8.

———. 1977b. "Eine wort geographische Untersuchung zur Territorialgeschichte der Maa-Sprechenden Bevölkerung Ostafrikas." M.A. thesis, University of Cologne.

Waller, Richard. 1976. "The Maasai and the British." *Journal of African History* 17:4, 529–53.

———. 1978. "The Lords of East Africa: The Maasai in the Mid-nineteenth Century (c. 1840–1885)." Ph.D. diss., University of Cambridge.

Zein, Abdul Hamid M. el. 1974. *The Sacred Meadows*. Evanston: Northwestern University Press.

Postscript

This essay has been left in its original form as published in *Comparative Studies in Society and History* in 1983. It represents an interpretation produced at a specific time, both in my theoretical position and my appreciation and knowledge of Ilparakuyo Maasai history, society, and culture. What follows, then, is a response to some critical assessments of this earlier interpretation as well as an attempt to develop some of the major issues raised in the essay itself and by some reactions to it.

The most important of these is the dialogue between Prof. Valentin Mudimbe and myself, initiated by his critique of a version of this essay, which was republished as chapter 4 in my 1985 book, *Persistent Pastoralists: Nomadic Societies in Transition*. Mudimbe's criticisms are all the more valuable because they are part of an overall appraisal of broader aspects of my work among Ilparakuyo Maasai as represented in that book. These in turn involve a particular version of a Marxist problematic in anthropology, which I have been developing for a number of years, and its relation to recent issues in African philosophy.

Mudimbe begins his commentary on my representation of Ilparakuyo Maasai notions of history and time as follows (Mudimbe 1991:184).

> Rigby's critique of idealist models that have dominated Africanist discourses underscores the productivity of a Marxist evaluation of the history of anthropology. . . . They seem to represent anti-Marxist stances and very subtly contribute to the controversial thesis of the ahistoricity of African experience of time. Rigby's analysis of these trends is masterful.

On the other hand, Mudimbe questions my "romantically beautiful Marxism" yet concludes (p. 191) that the

> anthropologist's action and confession [that we consider scientific knowledge] spatializes a place. If there is in it any truth, this should be accounted for by its isomorphism with a basic construct made of dreams and traces, functioning in the materiality of a locus.

That I could even have approached such an achievement in the opinion of one of the most distinguished of contemporary African scholars is, indeed, most gratifying, but its implications are more important. These are concisely summarized by Mudimbe (1991:191) as follows, and I quote in extenso.

> The awareness that Rigby's book brings about imposes itself as an epistemological demand. Most African Marxist projects ignore the complexity of their own epistemological roots and thus erase the paradoxes of their own discourse and practice. . . . Yet he is fundamentally right in insisting on the fact that there is a relation of necessity between the practice of social science and politics. One might oppose his political deductions, but there is no way of ignoring their significance and the evidence they unveil: the cost (or price) of social mythologies (development, modernization, etc.) invented by functionalism, applied anthropology, colonialism is such that a redefinition of the anthropological practice should be isomorphic with that of our political expectations and actions.

This brings us to the substantive issues upon which I have been reflecting since, in my dialogue with Professor Mudimbe and in more recent work on Ilparakuyo and Maasai social formations (Rigby 1991, 1992).

Most of these subsequent reflections concern what may be called the anthropological misuses of time and their impact upon both the authenticity of anthropological representations of "other" societies, as well as upon the political dimensions of anthropological praxis. These, in turn, demand an inevitable and dialectical union of *theory* and practice from Ilparakuyo Maasai and Marxist points of view (Rigby 1992:166–96 et passim). Common to each level of discourse are the relations of power generated by all forms of the production and appropriation of knowledge and its potential for either domination or liberation.

The first of these arguments is derived from Johannes Fabian's brilliant exposition of the way in which the misuses of time are involved in "how anthropology makes its object" (Fabian 1983). The ramifications of this short yet complex and profound book are seemingly inexhaustible, and I can deal with only a couple of them here.

Classical anthropological practices, including fieldwork, were not only thought to be based upon temporal and spatial forms of distancing but they were *demanded* as methodological and epistemological necessities.[1] All forms of distancing, argues Fabian, are functions of the operation of power relations, domination, and exploitation. The latter can only be overcome by establishing "coevalness" in terms of time and space. This entails the *production* of an intersubjective time in anthropological practice (Fabian 1983:30–31).

> To recognize *Intersubjective Time* would seem to preclude any sort of distancing almost by definition. After all, phenomenologists tried to demonstrate with their analyses that social interaction presupposes intersubjectivity, which in turn is inconceivable without assuming that the participants involved are coeval, i.e. share the same Time. In fact, further conclusions can be drawn from this basic postulate to the point of realizing that for human communication to occur, coevalness has to be *created*. Communication is, ultimately, about creating shared Time. (original emphasis)

If distancing is no longer an acceptable methodological and epistemological condition for anthropological practice, its elimination is not easy. The communication models encouraged by structuralism (Leach 1976), the ethnography of speaking (Moerman 1988), and even ethnomethodology promote temporal distancing. Fabian comments (1983:31),

> Beneath their bewildering variety, the distancing devices we can identify produce a global result. I will call it *denial of coevalness*. By this I mean *a persistent and systematic tendency to place the referent(s) of anthropology in a Time other than the present of the producer of anthropological discourse.*

To this we may add, with Mudimbe (1988, 1991) and de Certeau (1969, 1984), the tendency to place referents in another epistemological "Space." As MacCannell has phrased it (1992:290):

> One way classical theoretical understanding is flawed stems from the fact that "primitive" and "modern" types of culture were conceived to be in

mutual isolation when actually they exist in a determinate historical relation, empirically manifested at the community level in ways that are not specified by [classical] theory. . . . Even as they were put on a pedestal as "original" and "authentic," traditional communities were specifically denied all known paths to development, exploited for cheap labor and raw products, their local industries were destroyed and they were forced to consume western things from pop culture to Fast Food.

But the instantiation of coevalness is necessary not only for establishing the conditions for the production of authentic anthropological knowledge; it is also essential for the liberating potential of anthropological praxis. Coevalness recognizes both a common history *and* the creation of dialectically linked historicities in which both the anthropologist and the people and culture she purports to represent are embedded (Fabian 1983:154–55).

As it is understood in these essays, coevalness aims at recognizing cotemporality as the condition for truly dialectical confrontation between persons as well as societies. It militates against false conceptions of dialectics—all those watered-down binary abstractions which are passed off as oppositions: left vs. right, past vs. present, primitive vs. modern. Tradition and modernity are not "opposed" (except semiotically), nor are they in "conflict." All this is (bad) metaphorical talk. What are opposed, in conflict, locked in antagonistic struggle, are not the same societies at different stages of development, but different societies facing each other at the same Time.

This dialectic of historicities is not merely hermeneutic, as Ricoeur informs us (1974:46–47). It *must* also be *political* and *material*. In his discussion of Benjamin's theses on the philosophy of history (Benjamin 1969) in the context of "modernity's consciousness of time," Habermas (1987:14–15) reveals that:

What Benjamin had in mind is the supremely profane insight that ethical [and political] universalism also has to take seriously the injustice that has already happened and that is seemingly irreversible; that there exists a solidarity of those born later with those who have preceded them, with all those whose bodily or personal integrity has been violated at the hands of other human beings; and that this solidarity can only be engendered and made effective by remembering. Here the liberating power of memory is supposed not to foster a dissolution of the power of the past over the present, as it was from Hegel to Freud, but to contribute to the dissolution of a guilt on the part of the present with respect to the past: "For every image of the past that is not recognized by the present as one of its own concerns threatens to disappear irretrievably" (Thesis V).

This political dimension of a dialectic of historicities and its accompanying coevality in time also provides the basis of my later work on the necessity of a *rapprochement* between a historical-materialist anthropology and African philosophy (Rigby 1992:185).

If my argument on the close analogy or, more strongly, a direct parallel between the epistemological needs and contributions of African philosophy and African history is accepted, then neither philosophy nor history can merely be "Africanized"; they must go through an epistemic transformation. The way in which both may be achieved, I submit, is through a responsive and constructive historical materialism. . . . The central role of a theorized historicity in the revitalization of *all* African social sciences has been expressed by Mudimbe (1988:177).

My discussion so far is clearly remote from current intellectual concerns about what is "post" about postmodernism, a region in which all questions about authenticity, universalism, and, in effect, theory, vanish. Indeed, my argument about the imperative interrelatedness of modes of production and the materiality of history and notions of time among Ilparakuyo Maasai, even if it was based on Althusser's misreading of Hegel (an issue in which my concern is with Althusser not Hegel), is enhanced by David Harvey's recent critical assessment of experiences of space and time in "the condition of postmodernity" (Harvey 1990:203–4).

> I think it important to challenge the idea of a single and objective sense of time and space, against which we can measure the diversity of human conceptions and perceptions. I shall not argue for a total dissolution of the objective-subjective distinction, but insist, rather, that we recognize the multiplicity of the objective qualities which space and time can express, and the role of human practices in their construction. . . . From this materialist perspective we can then argue that objective conceptions of time and space are necessarily created through material practices and processes which serve to reproduce social life. The Plains Indians or the African

Nuer objectify qualities of time and space that are separate from each other as they are distant from those ingrained within a capitalist mode of production. The objectivity of time and space is given in each case by the material practices of social reproduction, and to the degree that these latter vary geographically and historically, so we find that social time and social space are differentially constructed. Each distinctive mode of production or social formation will, in short, embody a distinctive bundle of time and space practices and concepts.

I preface my conclusion, then, with a statement I made in response to Mudimbe's stimulating critique (Rigby 1991:203): "I think also that my discussion of time and history would not have been possible without the resolution I derived from Althusser-Balibar on the conditions for the production of specific forms of history." While Fabian is correct in his critique of some forms of Marxist anthropology, which he sees as "little more than theoretical exercises in the style of Marx and Engels" and which "share with bourgeois positivist anthropology certain fundamental assumptions concerning the nature of ethnographic data and the use of 'objective' methodology" (cf. Rigby 1985:30 et passim), the essay reproduced above is not in contradiction with a "Marxist praxis on the level of the production of ethnographic knowledge" (Fabian 1983:155). This is a goal to which I have devoted a considerable amount of my subsequent work; perhaps I cannot do better than end with a statement of my efforts in this direction (Rigby 1992:172).

What is the relation between African philosophy and Marxism? For my present purposes, the construction of an epistemology grounded in Ilparakuyo history and experience as well as in historical materialism, this question has two levels. The first (and most obvious) concerns the way in which various African philosophers read and use (or reject) Marx; the second (and more important) addresses the issue of the *dialectic* that may be established between the "logic" of Ilparakuyo notions of history and culture and their apprehension of the *transparency* of socio-political relations in this social formation, both past and present (Rigby 1985), on the one hand, and what I have called a "specific form of Marxist phenomenology," on the other. (original emphasis)

However, it is ultimately my relationship with Ilparakuyo Maasai individuals, and the knowledge *they revealed to me* about a community's culture and historical consciousness and conceptions of time (which assimilates me,

"Pita," as an inquiring subject), that have allowed me to grasp my own position. This includes my apprehension of the capacities of anthropological and Marxist discourses. While I have been trying to put anthropology in its place, Ilparakuyo have put me in mine.

In final illustration of this, in exploring Ilparakuyo views of peripheral capitalism, I wrote (Rigby 1992:128) that:

> The "Time" expressed in . . . Ilparakuyo and Maasai discourses is, in a sense, "transcendental," expressing the continuity of Ilparakuyo and Maasai history; but this is a continuity punctuated by the temporal dimensions of "significant" events that constitute an irreversible historicity. The concept of the irreversibility of events is neither new nor foreign to Ilparakuyo and Maasai [as I have indicated above]. But the temporality of these events is controlled through the dialectical relations between historical discourses about them on the one hand and the continuing transcendental unity of pastoral praxis and discourse about it on the other, itself expressed as a historicity.

NOTES

1. How such distancing is achieved by various forms of *writing* is elaborated in Geertz's interesting reflections on some major anthropological ancestors and contemporaries (Geertz 1988).

REFERENCES

Benjamin, Walter. 1969. "Theses on the Philosophy of History." In *Illuminations,* ed. Hannah Arendt. New York: Schocken.
Certeau, Michel de. 1969. *L'Etranger: Ou L'Union dans la différence.* Paris: Desclée de Brouwer.
———. 1984. *The Practice of Everyday Life.* Berkeley: University of California Press.
Fabian, Johannes. 1983. *Time and the Other: How Anthropology Makes Its Object.* New York: Columbia University Press.
Geertz, Clifford. 1988. *Works and Lives: The Anthropologist as Author.* Stanford: Stanford University Press.
Habermas, Jurgen. 1987. *The Philosophical Discourse of Modernity.* Oxford: Polity.
Harvey, David. 1990. *The Condition of Postmodernity: An Enquiry into the Origins of Cultural Change.* Oxford: Basil Blackwell.

Leach, Edmund. 1976. *Culture and Communication: The Logic by Which Symbols are Connected.* Cambridge: Cambridge University Press.

MacCannell, Dean. 1992. *Empty Meeting Grounds: The Tourist Papers.* New York: Routledge.

Moermen, Michael. 1988. *Talking Culture: Ethnography and Conversation Analysis.* Philadelphia: University of Pennsylvania Press.

Mudimbe, Valentin Y. 1988. *The Invention of Africa: Gnosis, Philosophy, and the Order of Knowledge.* Bloomington: Indiana University Press.

———. 1991. *Parables and Fables: Exegesis, Textuality, and Politics in Central Africa.* Madison: University of Wisconsin Press.

Ricoeur, Paul. 1974. *The Conflict of Interpretations: Essays in Hermeneutics.* Evanston, Ill.: Northwestern University Press.

Rigby, Peter. 1985. *Persistent Pastoralists: Nomadic Societies in Transition.* London: Zed.

———. 1991. "Response to 'Anthropology and Marxist Discourse,'" in Valentin Y. Mudimbe, *Parables and Fables.* Madison: University of Wisconsin Press.

———. 1992. *Cattle, Capitalism, and Class: Ilparakuyo Maasai Transformations.* Philadelphia: Temple University Press.

Time and the Sense of History in an Indonesian Community: Oral Tradition in a Recently Literate Culture

R. H. Barnes

Les événements des mythes, il est vrai, se passent, semble-t-il, hors du temps ou, ce qui revient au même, dans l'étendue totale du temps, puisque, comme le montre en particulier leur répétition dans les fêtes, ils réussissent à être également contemporains de dates espacées dans le temps normal. Cependant toutes les mythologies ont fait effort pour situer cette éternité dans la série chronologique.

—H. Hubert

Lamalera, Lembata, Indonesia, is a Lamaholot-speaking community of eastern Indonesia, which stands apart from its neighbors in many ways, among them in assuming a lead in acquiring literacy and in conversion to Christianity. Elsewhere (Barnes 1986) I have written about their many successes and difficulties in adapting to a modern world. Sons and daughters of the village have moved into modern occupations, including business and the learned professions, while others have become nuns and priests. At the same time their relative success has led to adverse demographic effects on the local community and placed its subsistence economy under strain. Catholicism has introduced a new ritual regimen and radically transformed the older ceremonies. Nevertheless, I once witnessed a spectacular ceremony of expiation on the outskirts of the village, conducted by impersonated pagan spirits, in which several hundred people participated, every one of whom was Catholic and many of whom were in the employment of the government as teachers or clerks (Barnes 1989). I have more than once been surprised by similar unexpected manifestations of the past, demonstrating the ability of myth and legend to exert a practical influence on daily life.

History and Current Events in Written Form

Others have written about the plurality of pasts in local communities (Cohn 1961, Appadurai 1981, and Davis 1989, for example). Lamalera certainly has many pasts, and these pasts are various in their nature. News items and articles touching on village events and culture, for example, appear in the regional monthly *Dian* (e.g., Wignyanta 1981), the managing editor of which is a son of the village. In 1986, the village celebrated "100 Years of Religion," marking the century since pastors Cornelius ten Brink and R. P. J. de Vries, accompanied by the Crown Prince of Larantuka, Flores, stopped on 8 June 1886 at Lamalera (Heslinga 1891:68–70) and baptized more than two hundred children. Eventually, using Lamalera as a base, the mission converted wide areas of south Lembata and established schools that opened new opportunities, which many eagerly seized. For this anniversary Alex Beding prepared a pamphlet (Beding 1986), complete with tables and photographs, setting out the history of the Catholic Church on Lembata.

History in its written form is familiar to local experience, and I do not mean that this experience is limited only to village history. Local schooling is available through junior high school, and current affairs are followed through radio news broadcasts and the occasional copy of a newspaper. Issues of *Intisari,* the Indonesian *Readers' Digest,* sometimes circulate in the village.

Villagers also use their literate skills to produce documents for private use. My collection of local writing that I have been given or allowed to copy includes a diary of significant events, a family genealogy, a clan history, an extended village history written by a local schoolteacher, a history of the core clans of Lamalera written especially for me by the former head of Lamalera district, Kakang Petrus Bau Dasion, and a young girl's school report on the history of the village. One of the most illustrious products of the village, Professor Gregorius Keraf,[1] has published his doctoral dissertation on the grammar of the local dialect (Keraf 1978). This grammar reproduces in an appendix a dialogue between himself and a relative about village history.[2] These written versions of village and regional history typically put to paper oral accounts of history and legend, which continue to be told and commented on in conversation. This essay will draw on the histories written by Kakang Petrus Bau Dasion (henceforth Dasion) and Guru Josef Bura Bata Onā (henceforth Bata Onā).

Experiences and Representations of Time

Just as their access to current affairs and to means of representing the historical past are diverse, so, too, are their experiences of the passing of time. The schedule of church events, the tolling of the church bell, western-style calendars in schools and the houses of village officers, radio programs, the constant use of the months and days of the week of the western calendar, and so on, provide for a familiar understanding of time in an international sense. People generally know the dates of their birth, confirmation, marriage, and the birth of their children.

Also important, however, are more traditional ways of viewing time and the past. For example, productive activities throughout the year follow a shifting rhythm associated with what anthropologists call time marks. Keraf (1978:240–45) includes as appendix 2 in his grammar a description of the pursuit of livelihood as described by Gabriel Blido Keraf, master builder of the boat *Nara Ténā*. This description, which I here condense and paraphrase, includes an account of the working calendar.[3]

In a year, he says, there are several seasons: *lerā*,[4] *barafāi, keronā,* and *temakataka. Lerā* lasts around five months, beginning in May and continuing until September. At that time the sun is very bright and rain never falls. *Barafāi* lasts one or two months during September and October. During *barafāi* the sun is extraordinarily hot, the trees all lose their leaves, palm wine flows abundantly, and there is no wind. Following *barafāi*, we pass to *keronā*, which begins with November and ends in February. In *keronā*, much rain comes down. We further divide *keronā* into the planting season, mid-*keronā*, and bad sea worms.[5] March is the beginning of *temakataka* which is also called good sea worms.[6] *Temakataka* ends with April.[7] Thus they [*sic*] reckon the seasons in a year in relation to rain and sunshine. When *temakataka* begins to change to *lerā* all the elders of the village assemble the population of the two villages and the lords of the land in order to discuss fishing, [i.e., to hold the annual village ceremony at the start of the main fishing season, now conventionally assigned to 1 May].

There are, of course, further detailed time marks that could be included in a more thorough description of this calendar, and mention would have to be made of the characteristic economic activities. House building and boat building have their own periods, and these differ from the period in which the large boats routinely put to sea. The bad and good sea worms periods refer to the nature of the weather at these times. Coastal peoples in Nusa Tenggara Timur commonly include in their time systems two occasions when sea worms release their sexual parts, which can be gathered in large quantities and cooked (see Fox 1979a). These events occur first during the height of the storms of the rainy season, in February, and second in the calm, sunny weather of March. *Good* and *bad* refer to the prevailing weather conditions, *not* to the worms.

Such annually recurring patterns have given rise to discussions of cyclical conceptions of time. As is well known, Leach (1961:126) questioned the use of the word *cycle* when discussing "primitive" communities. Nevertheless, for Leach (1961:125, 133), primitive time concepts emphasize the fact that certain phenomena of nature repeat themselves, while modern concepts dwell on the irreversibility of life change. According to Leach (p. 126), in some "primitive" societies, there is no sense of time going on and on in the same direction or round and round the same wheel. Instead, time is experienced as a sequence of oscillations between polar opposites such as "night and day, winter and summer, drought and flood, age and youth, life and death." In discussing time among another Lembata culture, in Kédang, I questioned this assertion and argued that the people of Kédang consider time as repetition resulting from irreversibility (Barnes 1974a:127–28). In other words, I placed emphasis on the fact that there is a *sequence* to the seasonal oscillations. Spring returns with a fresh face, but it is not the same spring as last year's. Today, I would take more care to distinguish between time the medium or dimension, on the one hand, and the ordering of events in our memories on the other (cf. Barnes 1987a:127–29). Nevertheless, my attempt to break down the opposition between linear and cycle time concepts was taken up by Howe (1981:231) in his influential discussion of Bloch's and Geertz's interpretations of Balinese time. He argued there that Balinese representations of duration exhibit simultaneously properties of cyclicity and linearity: "The Balinese possess only one notion of time and . . . this incorporates both cyclical and linear features." Farriss (1983:572) in turn took up Howe's formulation and universalized it.

In one sense all conceptions of time incorporate both linear and cyclical features, as proposed by Leopold Howe in his study of Balinese time (1981). In a cyclical conception there is linear progression within each cycle. Conversely, cycles (days, months, years, centuries) can be added up to produce a linear sequence.

However she also states (p. 580) that "linear and cyclical conceptions of time can coexist within a single code or system only if one is subordinated to the other." This stance suggests, perhaps inadvertently or unintentionally, that linear and cyclical conceptions can nevertheless exist separately, although I think that all she really wants to say is that there will be a subordination. By comparison, after reviewing some of the work of his new colleagues in his inaugural lecture for the Oxford Chair of Social Anthropology, Davis (1991:12) concludes that the "Uduk construed duration as a series of alternations; Kédang added to that a sense of direction, and hence of cycles of renewal." So here we are back again to the notion that linearity and cyclicity are alternative and contrasted cultural options. I would prefer to say that how a culture, or private memory, chooses to organize the relationships between a selected number of events may be open to such variations as alternations, lines, cycles, and reversals, but that the underlying experience of duration is irreversible. Even Hawkings (1988:150–51), who toyed with the idea of a universe in which the direction of time could be reversed, has reverted to the position that time's arrow does not change.[8]

Given my position on Kédang time and the close cultural and geographical ties between them and their Lamaholot-speaking neighbors, there is *no* room in my interpretation for associating the "modern" and exogenous features of Lamalera time concepts exclusively or primarily with linear aspects and the indigenous with cyclical ones. Hubert's essay on the representation of time in religion and magic, which so strongly influenced Leach, was not, as might be expected, drawn from a study of simple societies but was in large part based on his understanding of Christian ritual. There he denied that the religious calendar is a system of measuring time. Instead, it is a system devised specifically for regulating the periodicity of religious acts: "For religion and magic, the purpose of the calendar is not to measure, but to put rhythm into time" (1905:6–7). For Hubert the relevant distinction was not between primitive and modern but between religious and "normal" contexts, a contrast that applies everywhere. However, I do not accept that the religious calendar is irrelevant to measuring the passage of time, or that religion, especially Christianity or Islam, is uninterested in measuring time. I will not try to impose Hubert's distinction on Lamalera because I think it is flawed, even in his cautious treatment of it.[9]

What Christianity and the twentieth century did bring to Lamalera was a long-term chronology. Chronology can, of course, make use of cycles, as is the case with those of the Maya and the Balinese, but the Christian chronology, as opposed to its ritual calendar, has a unique beginning and aims at a unique end (Farriss 1983:572–73).[10]

Oral Accounts of the Past and the Effects of Legend on the Present

Hubert wrote that the events in myth happen outside time or in the total expanse of time and that they are contemporary with widely different dates in normal time through their repetition in festivities. This observation helps in some ways in formulating an aspect of Lamalera life that has often struck me. Events in village history that are confined to a legendary or even mythical past sometimes have striking effects on what happens in the village. Three examples may be selected to illustrate ways in which the past may become active in the present. The first involves the reasons why the son of a prominent member of Lefo Tukan clan was unable to fulfill his desire to marry a woman of Léla Onā due to an ancient dispute, despite the best efforts of the missionary Bernardus Bode to bring about reconciliation.[11] The second concerns an outsider's futile attempt to restore ties to the community that had been ruptured long ago. The third has to do with the way legend explains, or fails to explain, the anomalous position of the principal groups of the village who have claims to leading rank in the region but lack autochthonous status.

Why Léla Onā and Lefo Tukan Cannot Marry; Boli Léla Escapes, and Pati Mangun Thwarts the Missionary

According to the village history written down from oral sources by Bata Onā, the ancestors of Léla Onā clan accompanied the village founder Korohama when his group moved to Lamalera from the village Nualéla farther to the west. At Nualéla, Korohama had taken on the rank of Kakang (regional ruler) from the people of Nualéla. The Léla Onā family came from Nualéla and accompanied Korohama as living proof of the transfer of rank. Despite the implication that they were slaves, Léla Onā are remembered as having been a very rich clan at that time, having brought much gold with them. When they sold any of it and received payment, the objects would return to them of themselves. According to one man (whose wife belongs to Léla Onā), this is the reason they were eventually attacked.

At any event, they settled adjacent to Korohama's people and lived in peace with them. However, a woman of Lama Kera clan was married to a Léla Onã man who treated her badly, cursed her, and so on. One day he beat her nearly to death. She ran and reported to the elders of Lama Kera and the Korohama descendants.[12] The general assumption, and especially that of the woman, was that her husband was a witch, *ata menakang,* and she raved and raged at her kinsmen for forcing her to marry a witch. I have been told by villagers that people thought the members of the clan were witches, and that they thought that they were being bothered by them at night.

The elders decided to kill every man in Léla Onã, their wives, and their children. They went from house to house in the village looking for members of the clan and killing them. When the villagers were slaughtering the members of Léla Onã, they took all of their valuables. According to some accounts, only two people got away. One was a man named Pati Mangun[13] who fled to a thicket in the upper village and disappeared into thin air. At full moon he still stalks this area with his iron walking stick, which shines and makes a sound like a bell when it strikes the ground. When people hear that sound they know that someone in the clan is going to die.

The other survivor was a boy named Boli Léla. He ran through a back way and got to a house in Klodo Onã (a section of Lama Nudek clan). The owner took a turtle shell and covered him, so when his house was searched the boy was not spotted. He was then raised by that man until he was grown. The man promised to help with bridewealth, but Boli Léla refused. He may, my source speculated, have been influenced by his ancestral spirits who told him not to enter another clan or become a slave by letting them assume responsibility. He subsequently arranged his own bridewealth and became the ancestor of the present Léla Onã.

Because of these events, to this day members of Léla Onã and Lefo Tukan clans are not supposed to enter each other's clan temple, *lango bélā,* or to intermarry.[14] Bata Onã also states that they are not permitted to address each other.

The senior figure in Lefo Tukan clan referred to above said that he was unable to get clear details of the dispute. His questions always produced the answer, "We don't know." The people are afraid to talk about it. There was something about a witch and someone was killed, but he does not know whether it was his clan or the other that was the offender. Today his people are still afraid to enter the nearby *lango bélā* (clan temple) of Léla Onã.

The missionary Bernardus Bode tried to mediate between the two clans. Léla Onã did in fact present Lefo Tukan with an elephant tusk. This took place when the father of the last Kakang (district head) was in office, thus in the 1920s or very early 1930s. This man gave the tusk to the church, and Bode buried it in the foundation of the church, which was erected in 1932.

Peace, however, was not achieved, despite the efforts of both clans to go along. The reason it did not come off, a man from Léla Onã clan told me, was that blood had appeared when elders of the two clans were attempting to reach a settlement. I asked what this last comment meant. Well, if two persons sit across a table and try to make up a split, blood may suddenly appear on the table. This will be a sign that the efforts at peacemaking will not succeed. This happened in the case being discussed. It was Pati Mangun who frustrated these efforts.

In 1950 or 1951, Bode announced that he was going to cut the dead wood in Pati Mangun's sacred grove and build a chapel there. It is prohibited to cut dead wood in this grove. Preparations were underway when Bode suffered the nervous collapse that ended his mission in Lamalera. Villagers interpret his illness as being due to his efforts to place a chapel in the grove. On my most recent visit to Lamalera (in 1987), I overheard a snatch of conversation in which it was remarked that Pati Mangun is too strong to permit a resolution of the disagreement between Léla Onã and Lefo Tukan. He was too strong even for Bode. Certainly he is ever present, as are in a way the tragic events of legend. The marriage plans between members of the two clans mentioned above fell through because of the resistance of the old people of Léla Onã.

The Return of a Lost Kinsman

Another of several occasions when legend came to life in my direct experience occurred during my visit in 1979. Once, in the distant past, a man of the Lango Fujo clan, members of which are the lords of the land of Lamalera B, committed adultery with a widow of Lefo Léin clan, which at that time was very warlike. The Lefo Léin clansmen captured the malefactor and took him to Timor where they sold him as a slave. Today Lefo Léin exists as a single household headed by Petrus Koli Lefo Léin.

Father Arnoldus Dupont's assistant in 1979 was a brother named Yohanis Napon Lango Fujo who had been in service in Lamalera for about three years. He came from Timor and had specifically asked for the posting in Lamalera because of his family's tradition that they had come from the village. He was keen to establish ties with Lango Fujo and Lefo Léin clans. He got on well with the head of the only remaining household in Lefo Léin, but because of their history he was not permitted to enter that house nor could they exchange betel quid or tobacco. This condition could be lifted only by holding a small ceremony in which each would provide palm wine, which they would mix before drinking it. Brother Yohanis kept asking when they could hold the ceremony so that he could visit, but he was put off. He was told to wait a while longer to make sure that things would be all right. A member of Lango Fujo clan told me that Brother Yohanis had asked to enter the clan temple of Lango Fujo, but that a ceremony would be required before he could be called back into the clan. I do not know that he was ever successful in entering either house, and a short time later he was transferred elsewhere.

A Migration Charter and the Discovery of a New Home

The clans of Lamalera, like so many other Lamaholot speaking villages, may be divided into those whose ancestors migrated to the village following a natural disaster at Keroko Pukan or Lapan Batan to the east of Lembata, those whose ancestors emerged on Lembata itself, and a residual category of clans whose ancestors came to Lamalera from elsewhere at different times in the past. Obviously, clans of this latter group have individual histories that stand by themselves. However, the legend that serves as what Malinowski (1926:29) called the charter myth for much of the village relates to the flight of a group that accompanied Korohama from Lapan Batan. As will be seen, Korohama acquired this name only much later, but today villagers refer to him by this name at all times in the legend.

This legend resembles Trobriand myths in which, according to Malinowski (1926:44–45), "the very foundation of such mythology is flagrantly violated."

> This violation always takes place when the local claims of an autochthonous clan, i.e., a clan which has emerged on the spot, are overridden by an immigrant clan. Then a conflict of principles is created, for obviously the principle that land and authority belong to those who are literally born out of it does not leave room for any newcomers. On the other hand, members of a subclan of high rank who choose to settle down in a new locality cannot very well be resisted by the autochthons. . . . The result is that there come [sic] into existence a special class of mythological stories which justify and account for the anomalous state of affairs.

He also comments (1926:57–58) that myths serve to cover certain inconsistencies created by historical events, rather than to record these events exactly, and that myths cannot be sober history, since they are always made *ad hoc* to glorify a certain group, or to justify an anomalous status.

The anomalous status of the core clans of Lamalera lies in precisely this claim to rank coupled with dependency on autochthonous peoples for land.[15] The central legend does glorify and justify to an extent the leading clans, although there are references to humiliations. Above all it accounts for the relative positions of various village groups. It also belongs to a widely spread cycle of migration legends, about which Vatter (1932:71) wrote that "despite their mythological garb and many purely fairy tale features, the stories . . . nevertheless almost always reveal an historical core."

History begins in Luwuk-Bélu, which is generally identified with Luwuk in central Sulawesi, although the pairing possibly combines both Luwuk and the Belu or Tetun regions of Timor. There is not enough room here to sort out all the issues involved in interpreting the toponyms in the legend. Suffice it to say that upon leaving Luwuk the ancestors sailed east to Ambon and Seran and then down through the south Moluccas back west until they reached Lapan and Batan (in local pronunciation Lepā Batā), which today survive as two tiny islands in the strait between Lembata and Pantar. There an old, childless woman found an eel while gathering shellfish at the shore, which she brought home and raised until it was grown, whereupon it disappeared into a hole in a tree in the forest. Every day when the adults went to work, the eel came out and ate a child. Eventually the adults discovered the eel and killed it by sticking a red-hot iron rod into the hole. When this happened the sea began to rise, inundating the village. Those who survived fled. Most of the refugees came west in the boat named *Kebako Pukā* (which also carried the unfinished boat named *Bui Pukā*)[16] until they reached Kédang on the eastern end of Lembata where they went ashore and stayed briefly. The old woman who had caused the disaster followed on land.

It just so happens that the sinking of this island figures in a similar way in the legends of many clans on the islands of the Lamaholot-speaking area.[17] Often it is called Keroko Pukan, but although Lamalera lists Keroko Pukā as the stop just before Lapan Batan, many people of the area identify Keroko Pukan with Lapan Batan and place it to the east of Lembata (Vatter 1932:71; Arndt 1938:24; Suban Leyn 1979:12, 20). For example, the clan Hayong in Menanga, Solor, derives from Lapan and Batan, "also called Keroko Pukan Kemahan Niron" (Suban Leyn 1979:34).

Another thing that such groups have in common is an association with the toponym Lewo Hayon, sometimes abbreviated as Lohayon. Generally they relate difficulties in their journeys as they fail to be received with hospitality at places where they wish to come ashore, until finally they are accepted in their present location. Acceptance also often means having their own claims to rank recognized. The earliest mention of this tradition that I have come across concerns the "Second" Raja of Larantuka, Flores, who lives in the Larantuka ward called Lohayon and descends from refugees from Keroko Pukang who were given the position of "Second" Raja by the ancestors of the Raja of Larantuka (Heynen 1876:73, 77–78; Couvreur 1907; Seegeler 1931:80; Dietrich 1984:321). Arndt (1940:56, 58, 59, 81–83, 85, 87, 163, 167–68) mentions such groups in eastern Flores and Adonara. In Pamakayo, Solor, the clan Lewo Hayong came from Keroko Puken (Arndt 1940:216–18). The village of Lohayong, Solor, was the site of a Portuguese/Dutch fort, which played a central role in local politics since the mid-sixteenth century (Boxer 1947, Barnes 1987b). Although its present clans appear to have lost the Keroko Pukan tradition (being later comers from the Moluccas), it lives on in the closely linked neighboring village of Menanga (Suban Leyn 1979:28–29, 34; Dietrich 1984:320, 324, n.13).

The refugees sailed further along the south coast of Lembata, while the old woman followed along the shore crying. They did not stop for her because they were angry (Keraf 1978:236–37). When they got to the peninsula marked on some modern maps as Bela Galeh, but locally referred to as Gafé Futuk, Lefo Kumé, or Noor Futu, the currents were too strong, so they threw overboard a golden bench named Kuda Belaung (Golden Horse) in order to "buy" the currents and wind (*hopi enā angi*).[18]

Having calmed the sea, they rounded the peninsula and approached the shore at Bobu. There they requested of people who were in the tops of lontar palms (i.e., tapping palm wine) permission to come ashore and settle. These people answered that there was no more room, the place was full, the land packed, "just look at us who have to live in the tops of palms."

Tricked in this way they sailed on until they got to the Ata Déi (Standing Person) Peninsula, where again their safety was endangered by winds and currents, which prevented them from passing.[19] A young woman named Somi Bola Deran (wife of the ancestor who later assumed the name Korohama) gave birth to a son in the boat. The baby cried very much, so they decided again to "buy" the wind and the current and threw a chain of gold links (Sora Kai) into the sea. The wind and current became calm again and the party sailed through and anchored at Luki Lewobala (the village Labala and the Luki Point in the Labala Bay). Sora Kai was swallowed by a porpoise, which was subsequently harpooned and the chain recovered. I have been shown this object. At Ata Déi, the old woman who had been following them on shore turned to stone, giving the peninsula its name (Keraf 1978:236). She can still be seen there.[20]

They went into the mountains above Labala and settled in a hamlet named Lefo Hajo.[21] Here they lived comfortably for a time, but once a woman from the clan Tana Krofa was pounding maize when unexpectedly the stamper slipped from her grasp and the mortar tipped over, crushing a baby chick. Tana Krofa, which until then had held the leading rank, was unable to pay the price demanded by the local owner of the chick. Korohama then gave what was demanded, reported to be a gold chain or a gold ring, whereupon they left. As a consequence of this payment, Korohama assumed leadership from the Tana Krofa clan. When they reached the shore and started to board their boat, the head of Tana Krofa told Korohama to sit on the deck: "we will sit on the thwarts and row—we will stand in the bow and all the way to the stern in order to push you forward."

From there they sailed the short distance across the bay to Doni Nusa Léla (the village Nualéla and today's marketplace Fulan Doni). At Doni Nusa Léla they were well received and permitted to settle. Here there was a momentous exchange of crafts and functions. Lamalera's ancestors received iron harpoons in exchange for their own brazilwood harpoons. They reciprocated by giving Nualéla the skills to make earthen pots (and *fato faka,* the stone and the wooden paddle used for shaping the clay), which today remains their monopoly. The Kebelen Nualéla then gave his rank of *kebelen* [great person] of the earth to Korohama, who, announcing that he was now of the same standing as the leader of Nualéla, assumed the name Korohama ["I with him am the same"] for the first time. Through this act the office of *kakang,* district leader, passed into the hands of Lamalera.[22]

They dwelled comfortably at Fulan Doni for a long time. However, when they went to sea the winds and currents frequently carried them far to the west, bringing them ashore in the land of Libu Lama Mau[23] and Gesi Raja, or Gesi Guan Bala Bata Balamai.[24] Since they felt that this was a more convenient situation for their fishery, they requested permission to settle there. At first it seems this permission was refused (at least that is what members of Lango Fujo, Gesi Raja's clan and the principal lords of the land, say). However, agreement was eventually reached, and a site in the upper village was purchased from Lama Mau clan with five brass bracelets and a chain. Because the chain always fled back to Korohama, they built a boat to replace it and gave it to Lama Mau clan. According to Dasion (n.d.) this boat was named *Baka Fai,* although it is now called *Baka Téna.*[25]

Once this transaction was completed, they moved to the upper part of the present village. This location they first called Lefo Hajo, until Korohama changed the name to Lama Lerā (Sun Plate).[26] According to Dasion (n.d.), Korohama's eldest son remained in his father's house in the clan Lefo Hajo/Beliko Lolong.[27] He became the ancestor of Beliko Lolong ('top of the stone wall') clan. The second son was told to settle in the lower section of the village to guard the beach and assist passersby. His descendants became the Bata Onā clan and its various offshoots (Bata For, Sula Onā, and Bedi Onā). Bata Onā ('within the Bata') takes its name from the stone elevation Bata Bala Mai that they found there. The youngest son was ordered to the center of the village to guard the peace. From him descends the clan Lefo Tukan ('center of the village'), and through this line has passed the office of *kakang*. These three clans are *lika,* like the three stones that make up the hearth.

This history tells us the following about the ancestors of present clans of Lamalera. All clans descended from Korohama have a comparable status as core clans, but in particular Beliko Lolong, Bata Onā, and Lefo Tukan are in a sense the aristocrats. Korohama and the ancestors of Tana Krofa clan and Lama Nudek clan traveled together from Lapan Batan and shared a common history and fate, although Tana Krofa lost its original position of leadership to Korohama. The ancestor of Lama Nudek clan was the master builder of the boat they traveled in, Kebako Pukā, and today members of this clan still claim the position of premier master builders.[28]

In the very earliest days, then, Korohama and his family, his dependents in Léla Onā, as well as Tana Krofa and Lama Nudek, lived closely together (as indicated by the locations of the various clan temples) and not far from the indigenous lord of the land in Lama Mau clan (and possibly at that time the ancestors of Tufa Onā clan, who latter supplanted members of Lama Mau as lords of the land). When the ancestor of Bata Onā clan moved down to the beach, he established himself on land associated with a different lord of the land in the clan now called Lango Fujo. This second lord of the land is today the more significant of the two, at least in the sense that he has certain powers over success in the fishery.

Relationships among these groups were further defined by the establishment of ties of marriage alliance. The social organization of Lamaholot-speaking communities of Lembata and neighboring islands presupposes asymmetric marriage alliance among patrilineal descent groups. Each such group takes women from certain wife-giving groups, and gives women to different groups. The asymmetry of the system means that each group's wife-takers must be kept separate from their wife-givers. According to Yosef Bura

Bata Onā's history, Korohama and the ancestors of Lama Nudek clan became wife-givers to the ancestors of Tana Krofa, Lama Mau, and Tufa Onā. He writes that later, as the village grew larger, the pattern changed somewhat, so that there were three sections, composed as follows, arranged by clan.

I.	II.	III.
Beliko Lolong	Lefo Tukan	Lama Kera
Lama Nudek	Tana Krofa	Lefo Léin
Léla Onā	etc.	Lama Nifa
Bata Onā		Tapo Onā
Bedi Onā		Olé Onā
Bata For		Eba Onā
Sula Onā		

Group I were wife-givers to group II, who gave wives to III, who in turn gave wives to I. Bata Onā's history does go on to acknowledge that in recent times the patterns have become more complicated. Nevertheless, his list corresponds exactly with a somewhat more detailed listing I was given independently by Kakang Petrus Bau Dasion.[29]

Village history does not stop at this point, of course, but, just as Bata Onā's history does, I am beginning to drift off from narrative into ethnographic description. Essentially we have succeeded in crossing the bridge between myth and oral history, and the recent end of this history is firmly linked to modern chronology. Indeed, Bata Onā prefaces his history with a detailed chronology going back to the fourteenth century. Allowing for a hundred-year slippage because he misunderstands what terms like *fourteenth* or *nineteenth century* mean,[30] his chronology is in fact highly plausible, although, of course, it is based on pure guesswork.

I regard the account given above as an authentic version, in the sense that it would be recognized locally with no more quibbles or demurrals than would be directed at any of my sources. It is not just a conflation of different versions. I received accounts of all of the principal events, either in answer to specific questions, in volunteered comments, or in overheard conversations. The oral narratives thus depend on this shared understanding. Admittedly my recitation of this history has included more events and presented a clearer sequence than any of my sources. Especially, the course of events in passing the two peninsulas has been presented more clearly than in any single source, some of which conflate the two peninsulas into one. There are also discrepancies about which object was dropped where. On the other hand, like each version of the narrative I have referred to, I have been constrained by the space available and the attention of my readers to compress the account. A complete "scholarly" treatment of the issues raised by the narrative could easily take up many times more pages.

Bata Onã's history is the product of the man, a local schoolteacher with religious training. It goes on for many pages, drawing together many different stories and events in history that are deemed important locally. He garnishes the basic narrative with speculations about the states of mind of the ancestors at various moments in the past, ethnographic digressions, rhetorical questions and answers, notes, and Christian moralizing. The whole contains many interesting ethnographic dimensions.

At the beginning of his account Bata Onã says that in July 1939 the head of the village called together the elders to discuss the histories of each of the clans. After a period of preparation they met and each took his turn. Some recitations were brief, others long and complete. Unfortunately, no one was delegated to write them down. Among those present was an elder from Bata For clan, who recited the above history as he had heard it from his mother's brother, Kialakatana of the clan Beliko Lolong. Bata Onã records this narrative as he purportedly remembers it (in the early 1970s). Kialakatana's story is a strictly oral performance, using parallel phrasing and binary pairs in a familiar eastern Indonesian pattern (Fox 1988). It is almost little more than a laundry list of places visited as the ancestors traveled, and can only be understood by those who know the story. Of course, such explanation is always readily available to people living in the village.

Of the other sources, the one that comes closest to that by Kialakatana is Petronela Ema Keraf's school report, demonstrating that the oral form lives on within even the context of modern educational institutions.[31] For comparison, I give a brief section of each.

Kialakatana (in Bata Onã, n.d.)	Ema Keraf (school report)
1. *Goé pipé usulk-asalk rang² lau Luwuk—Lau Belu.* I will now tell our story as they went [from] Luwuk—Belu.	*Usal asalk lamanéku Lau Luwuk, lau Belu.* Our ancestry from Luwuk, from Belu.
2. *Haka kaik téti Seran—téti Gorang.* I went to the east to Seran and Gorang.	*Haka kaik téti Séra téti Gora.* I went to the east to Seran and Gorang.
3. *Téti Abo—téti Muan.* To Ambon and to the Moa Archipelago.	*Lodo hauk téti Abo téti Mua.* I descended to Ambon and Moa.

4. *Hauk téti Vato Béla Lako.*
I returned to the Watu Bela Archipelago.

Hauk téti Fato Béla.
I returned to Wato Bela.

5. *Hauk téti Kroko Tawa—Triagéré.*
I returned to where the Kroko tree [*Calotropis gigantea*] grows—the Tria rises.

Hauk téti Kroko Tafa Teria Géré.
I returned to where the Kroko grows, the Teria rises.

6. *Hauk kiangk téti Lapan—téti Batan—(blébu lébu ékan).*
I returned and rested at Lapan at Batan—(where the inundation covered the earth).

Hauk téti Lepa, téti Bata.
I returned to Lapan and Batan.

Dasion's version is a concise account of some twelve handwritten pages, which follows a pattern very close to that of Kialakatana, while being by no means so terse. It is intended after all to explain the history to an outsider. It is also the most complete and well ordered in covering the entire sequence of events. The least complete version is that given in Keraf's appendix I, perhaps because it takes the form of a dialogue between Keraf and the narrator, and Keraf's frequent requests for clarification interrupt the narrator's train of thought and lead to digressions. For the outsider, understanding is hampered because the narrator alludes to, rather than explicitly states, information that he knows Keraf has. Also, the purpose of the appendix is to illustrate the use of the language not to provide a polished account of village history.

There are certainly many inconsistencies, contradictions, and reversals in this tale. At the beginning a sterile old woman raises a creature of the sea as her own child, which then nearly annihilates the children of the village. When the adults destroy this creature, the land itself is inundated. On their journey they are rebuffed in various ways when they seek a place to settle. They are forced to display wealth and then destroy it or give it away in order to secure their safety at sea (twice) and then on land (once). Twice they have to purchase a calm sea. At Lewo Hajo, above Labala, the injury of an insignificant animal forces another large payment. There are two changes of rank, at Lewo Hajo and later at Nualéla. At Nualéla, this transfer of rank is accompanied by an exchange of primary subsistence skills, establishing the specialties that to this day remain basic to the economic contract between the villages. All of these important transformations occur before they move to their present village where, although powerful, they remain virtually landless and dependent upon the indigenous clans not only for such land as they managed to purchase but also in important respects for the very success of the fishery upon which they depend for a living. The *tana alep,* or lord of the land, in Lango Fujo clan played, and to some extent still plays, a crucial role in the annual fisheries ceremony.

There are a variety of, if not proofs, then evidences of authenticity, which villagers resort to in relation to this history. Important among them are the heirloom items that came from Lapan Batan, which, if they are properly disposed toward a visitor, they may bring out and show him. Both in the narratives and today these heirlooms demonstrated and demonstrate wealth and status, and in doing so follow a widespread Southeast Asian pattern. Part of the test the ancestor of the "second" Raja of Larantuka was put through involved just this act of demonstrating the precious objects that he owned and had brought with him in his boat (Seegeler 1931:80–82). Other material evidence of veracity includes the petrified standing woman and the other topographical features referring to known places in the local region. Then there is the comparative dimension. When scholars such as Vatter, Dietrich, and myself emphasize that the story of the disaster at Lapan Batan is widely shared and often associated with the name Lewo Hajon, we are doing no more than what locals do themselves. These points are also made by Dasion and Bata Onã (also Kumanireng 1990:45). They lend authenticity to aspects of the story and demonstrate its power to take over even foreign narrators. Like Vatter, I am convinced that there is a historical truth in the narrative, but the "proofs" available to me are no sounder or different than theirs.[32]

It should also be acknowledged that, whatever historical validity these legends may or may not have, various of their features belong to stereotypical patterns of wide distribution. The theme of a sea creature found inland, representing a categorical confusion and leading to disaster and dispersal, is one (Barnes 1974a:35). McKinnon (1991:55) notes that in the Tanimbar Archipelago the dispersion and migration of people is related most often to a catastrophe such as a flood, plague of mice, or epidemic illness. Categorical oppositions of broad cultural importance such as that between land and sea, and land people and sea people, play a prominent part in the legends, just as they do in everyday life. Furthermore, the shores of eastern Indonesia are liberally dotted with petrified boats and people.

Genealogies and Historical Chronologies

Unlike their neighbors in Kédang and some other Lamaholot-speaking groups, the people of Lamalera are not great genealogists. Certainly there is nothing currently available like the fifteen names, from Raja Lorenso Usi Dias

Vierra Godinho to his ancestor Patigolo (who came from Timor) in the short history of the descent of the rajas of Larantuka, Flores, which Raja Usi wrote for his son, the later Raja Servus, during Raja Usi's exile in Java following his deposition by the Dutch in 1904, nor the twenty-three generations that the raja's agent in Riang Koli, Flores, was able to recite for Vatter (1932:135). Happenstance of colonial history generally left Lamalera out of the purview of European contact and interest until the 1880s, so there is virtually no way of comparing local traditions with European records. For these two reasons, it is not possible to perform the sort of test that permitted Fox to demonstrate that nine of the thirty-two ancestors in the royal genealogy of Termanu, Roti, were historical individuals, carrying the genealogy back three hundred years, and to link the genealogies to historical events, as well as showing that it was not intended as a purely historical record but as an assertion of a moral order and a claim to the right of succession (Fox 1971:67–68). Kakang Petrus Bau Dasion's genealogy goes back only to his great-grandfather, Atakebelaké Dasi, and this lack of depth is currently typical for the village, although it is possible that this situation is the result of Bode's suppression of the native priesthood, which may have been responsible for preserving the extended genealogies of at least the more prominent lineages.

Fox (1971:70–71; see also 1979b) links Rotinese royal genealogies to a form of historiography common throughout the Malayo-Polynesian world, from the Merina Kings of Madagascar to the chronicles of the rulers of Java and the simple folk histories of the chiefs of Tikopia, which serve to extol rulers. The internal chronology of the genealogical sequence, he says, superficially resembles western chronological historiography. To the best of my knowledge, the Lamalera counterpart to this tradition is now lacking. Bowen (1989:676) comments on the Gayo legends:

> The orientation of the stories is lateral (toward the entry of newcomers and the exit of the older rulers) and not toward the succession of an established line. No genealogies link the first generation of Lingë rulers to those of recent times.

For whatever reason, it appears that in Lamalera there was no attempt to rework origin accounts into genealogical form, as Bowen said was done for the first time by the Gayo rulers of Bukit and Ciq in the 1930s (Bowen 1989:678–79).

Conclusion

I would agree with Hubert that ritual and legend may serve to make certain events belonging to different periods of time contemporaneous in a metaphorical sense. I resist, however, the notion that the underlying category or medium of time (or time-space) changes. It seems to me that something must remain constant while we happily get on with describing cultural or psychological variations in representing the passage of events. Lamalera has available a wide range of modern methods of measuring, classifying, and characterizing time and its passage, while it also retains older means.

There is really no end to the discussion of a people's historical consciousness. That of the people of Lamalera is inevitably changing and is too varied to be adequately described in a brief essay. Perhaps it can be said of traditional history that it provides a normative framework that serves to govern the debate about the past, and that it permits new forms of action while allowing cultures to regulate social change (Appadurai 1981:218). As we have seen, however, villagers do not always feel as free of the constraints of the past as this optimistic generalization would suggest. Nevertheless, it is certainly true that the last thing Lamalera historiography describes is an unchanging past, or even a social structure fixed in an unchanging mold. Malinowski denied that the purpose of myth was to explain anything. The account of the journey from Lapan Batan to Lamalera is what anyone will be told if they ask how the core clans of the village got there. That is an explanation of sorts. Nevertheless, perhaps Malinowski is right, in the sense that this story, and others of its kind, are less satisfactory as explanations than they appear at first. Not only do they constantly suggest that things could easily be different but also that the present may not be quite all that it claims to be. Above all they convey a sense that the resolutions of the central structural challenges, which broke down so often in the past, may not be stable in the present either.

NOTES

1. Full names of Lamalera people today consist of a Christian name, e.g., Petrus, a village given name, e.g., Hidang, and a surname of recent coinage, usually, but not always, based on a clan name, e.g., Beliko Lolong. Some individuals bear titles such as Kakang (district head), Guru (teacher), or Professor, thus Professor (title) Gregorius (Christian name) Prafi (village given name) Keraf (surname). In the village, people generally are known by their Christian name, followed by their village given name or

by an honorific such as Bapa (father), followed by their village given name. People who play a role in public life, on the other hand, are generally known by their Christian name followed by their surname, their village given name being neglected.

2. See Davis (1989:107) for a similar list of written resources for Zuwayi oral history.

3. Professor Gregorius Keraf once objected to my reference to the "traditional village calendar," saying that they had no calendar until they were given the Gregorian calendar. His point, I think, is that the villagers lacked a formally organized calendar. Anthropology, however, commonly uses the word more loosely to refer to conventional patterns of annual time reckoning, as do I in this essay.

4. The full name of *lerā* is *lerā matā*.

5. The rainy season has these divisions. In November and December is *urā sedā tana*, 'rain sounds [drums] on the earth', early showers wetting the soil. Then comes *tenika* in late December and early January, indicating that the dry season is over. According to Keraf (1978:196) *tenika* is a device for dividing, setting aside. It is also the period of change from the dry season to the rainy season and from the rainy season to the dry season: *Kofa pé legur-legur bolo pi pé tenika lépé* 'The clouds keep sounding, perhaps now is the season of change' (pp. 197, 205). The third period is *nalé datāng onā,* or the bad sea worms, from January through February, when the wind and rain are so bad that you cannot go outside.

6. *Nalé saré* 'good sea worms'.

7. *Temakataka* is the harvesting period and seems to be composed of *temaka* 'to steal' and *takā* 'ripe', but also note *taka* 'to steal' (Keraf 1978:79, 111).

8. More precisely, of the three arrows of time, the thermodynamic and psychological arrows may not be reversed, although possibly the cosmological arrow can be. For the purposes of anthropology, it is enough to know that the first two cannot change.

9. "We do not hold that the religious and magic notion of time always differs from the common notion; it is enough that they differ frequently and that the differences in question are regular" (Hubert 1905:8).

10. Cyclical chronological systems did, of course, exist in parts of what is now Indonesia. The most accessible brief English summary of the Javanese calendars and chronology, and the transformations they underwent during colonialism, is in Ricklefs (1978:223–38). Aveni (1990:101) has recently attributed weekly cycles to Kédang, but he has confused Kédang with Bali. It was Geertz (1966:45) who wrote that the Balinese had ten kinds of cycles, ranging from one to ten days. Nothing of the sort occurs on Lembata.

11. There are some nineteen named patrilineal descent groups in the two halves of Lamalera of quite varied sociological significance, twenty if the lineage of the lord of the land of Lamalera B is counted. (Lamalera is divided into two villages for administrative purposes: Lamalera A is higher and lies to the west, while Lamalera B is lower and lies to the east. Each has a different lord of the land.) Some of these consist of no

Time and the Sense of History in an Indonesian Community / 263

more than a family or two. Others are divided into large subdivisions, which are virtually independent in their own right. In the traditional marriage culture, these decent groups are linked by what anthropologists call asymmetric marriage alliance. Catholicism and modern life have much modified this marriage system, but it is still central to family relationships. The legend discussed here concerns the founding descent groups of the village.

12. Carolus Sinu Ebang said she went to the elders of Lama Kera and Beliko Lolong, but it is uncertain whether Korohama's descendants had yet split into the present clans. The consequences of the dispute today affect only Lama Kera, Lefo Tukan, and Léla Onã but not, so far as I know, Beliko Lolong. Beliko Lolong and Lefo Tukan descend patrilineally from Korohama.

13. This personal name is still passed down within the clan.

14. Guru Yosef Bura Bata Onã says that the marriage prohibition also exists between Lama Kera and Léla Onã. However, my genealogical record shows that five men of Léla Onã have married women of Lama Kera. There are no marriages between Léla Onã and Lefo Tukan.

15. Lamalera's central legend is related to the stranger-king theme widespread in the Austronesian-language world (see the references cited in Bowen 1989, and Barnes n.d.). For the Gayo of Sumatra, Bowen (p. 676) writes that in the stories of origin, "Many of [the] subsequent founding events are ascribed to an outsider who is superior to an older ruler and in some cases displaced him." These stories are the converse of Negri Sembilan legend in which "the legitimate claimant to the throne [is] supplanted by a usurper (and a foreigner's puppet at that) [but] who, after a year of humiliation, wins through and gains his heritage" (de Josselin de Jong 1975:298).

16. Kakang Petrus Bau Dasion gives the name of this boat as *Bui Pukā/Keroko Pukā*. *Pukā* means the trunk of a tree. *Bui* is a form of cork tree, perhaps *Alstonia Scholaris* or *Sonneratia acida*, Linn. I have had no other indication that this boat bears two names.

17. Lembata lies on the band of volcanic activity circling the globe known as the Ring of Fire. The island, like its neighbors, is the product of volcanism and tectonic upheaval. A new volcano appeared at sea just south of the Lerek Peninsula in 1974. In July 1979, the combined processes of earthquake, volcanic eruption, seafloor subsidence, and tsunami wiped out the villages of Wai Teba and Sara Puka on the south Lembata coast, with a loss of more than five hundred lives, including sixteen from Lamalera and neighboring villages. There is a long-standing record of volcanic activity on these islands (see West and Barnes 1990:144–45), and it would be surprising if there were no local traditions about such events. In fact, there are several such traditions.

18. According to Petrus Hidang Beliko Lolong, this bench was a buffalo in the form of gold with a place to sit (*karbau bentuk emas dengan tempat duduk*). *Belaung* is usually an earring, but it has been translated variously as silver (Vatter 1932:51),

bronze (Keraf 1978:237), and gold (Müller 1857:310). Petrus Hidang specified gold, as opposed to Keraf's bronze.

19. The currents at Ata Déi continue to pose a danger to overloaded boats. In recent years both the *Demo Sapang* and the *Léla Sapang* have sunk there for this reason.

20. Two people told me that the Standing Person is Somi Bola Deran.

21. The rajadom of Labala also has the tradition of coming from Lapan Batan, which was connected to Lembata until it was partly destroyed. The mountain people then fled overland to the beach at Wato Téna where they went into the mountains. The coastal people fled by boat to Luki Point where their boats turned to stone. Wato Téna are some rocks just offshore to the west of Luki. The name means "boat stones," although they do not much look like boats (*wato/fato* 'stone', *téna* 'boat'). They built the village of Lewo Raja, which they had to flee because of a war with Nualéla. The walls of this village were still visible in 1910. Warfare between Nualéla and Labala used to break out from time to time whenever there was a drought. The blood of the fallen was needed to asperse on the sacred stones, *nuba nara*, in order to bring rain (Beckering 1911:187–88; 1921:136).

22. It is significant that, according to Munro (1915:5), the commander of the Dutch company that pacified the island in 1910 mistakenly divided the Lamalera district into two kakangships, Lamalera and Nualéla. This mistake was rectified in 1911. There are ten such districts in the old realm of the Raja of Larantuka who would have been able to provide confirmation of the correct situation.

23. Kakang Petrus Bau Dasion renders this clan name as Lamaau, but everyone else has it as Lama Mau. Lama Mau was the clan of the lords of the land of the upper village of Lamalera (Téti Lefo, or "up at the village"). It has died out and this function was passed on to Tufa Onā clan.

24. Gesi Raja was the lord of the land of the lower site of Lamalera (Lali Fata, or "down at the beach") where there was a structure of layered stone that served as a place for the elders to sit. This was called Bata Bala Mai.

25. *Baka Ténā*, which has long been disused, belongs to Tufa Onā clan. I have been told that its former name was actually *Baka Walang*. In this respect, it is interesting that the clan Lewo Hajon, of Pamakayo, Solor, relates that when in Keroko Pukan they lived on an island named Nusa Toné Tobo Baka Walang Pai. *Baka Walang* is said to be a morass (Arndt 1940:216).

26. I was mistaken when in 1974 I wrote that the clans had come from Lohayong, Solor, but I was right in saying that the first clan was Lohayong (Lefo Hajo), as does Vatter (1932:205), perhaps better written Lefo Hajo for Lamalera (Barnes 1974b:140). My mistake has been cited as fact in a local publication (Wignyanta 1981:21).

27. This tradition contrasts with the practice in Termanu, Roti, where the youngest son inherits his father's house, and the oldest has the right to replace his father in the ceremonies of former affines, while in the succession to political office all sons are considered equal (Fox 1971:53).

28. What this story does not explicitly state is that at some point in their travels the

ancestors acquired a new language, form of social structure, and local culture; that in Lamalera the population eventually grew as more clans came in or split off; and that the fleet expanded over the centuries. It also fails to mention that another industry that may have been introduced by the refugees from Lapan Batan is the weaving and dyeing of cloth, although when Vatter was in Lamalera he was told that groups of Lapan Batan origin had introduced it (1932:205).

29. These reports conflict with Vatter's statement (1932:205) that marriage is strictly forbidden among all the clans that came from Lapan Batan in that Lefo Tukan and Tana Krofa are wife-takers from group I.

30. He thinks this century is the nineteenth and that the fourteenth century covers the years 1400–1499. This application of dates to legends of migration is, of course, not an isolated example. The Fetor of Abi, Bonji, and Lasu, Timor, dated events as early as 1460 in his account of Timorese migrations written in December 1901 (Middelkoop 1952:182, 191).

31. Petronela Ema Keraf was thirteen years old when she wrote this report, in September 1982, for her class in the local junior high school.

32. Finnegan (1970:200) has written that "The notion that when using 'oral tradition' one can suspend many of the normal critical canons of historical research is, despite the caution of more experienced historians, surprisingly prevalent." While written sources are open to distortions due to prejudice, personal interests, aesthetic forms and so on, "Oral sources are in many ways even more open to such factors than written ones." The nature of the oral performance, its audience, the social setting, and the relationships between the audience and the performer all have their effects (see also Finnegan 1988). Nevertheless, there are constraints of a sort on oral history, which is independent to a degree from individual performances of it. Furthermore, as these traditions become written, the written forms show many of the characteristics of the oral versions.

REFERENCES

Appadurai, Arjun. 1981. The Past as a Scarce Resource. *Man* 18:2, 201–19.
Arndt, Paul. 1938. Demon und Padzi, die feindichen Brüder des Solor-Archipels. *Anthropos* 38:1–58.
———. 1940. *Soziale Verhältnisse auf Ost-Flores, Adonara und Solor*. Anthropos, Internationale Sammlung Ethnologischer Monographien, vol. 4. Münster i. W., Germany: Aschendorffsche Verlagsbuchhandlung.
Aveni, Anthony. 1990. *Empires of Time: Calendars, Clocks, and Cultures*. London: Tauris.
Barnes, R. H. 1974a. *Kédang: A Study of the Collective Thought of an Eastern Indonesian People*. Oxford: Clarendon.
———. 1974b. Lamalerap: A Whaling Village in Eastern Indonesia. *Indonesia* 17 (April): 137–59.

———. 1986. Educated Fishermen: Social Consequences of Development in an Indonesian Whaling Community. *Bulletin de l'Ecole Française d'Extrême-Orient* 75:295–314.

———. 1987a. Anthropological Comparison. In Ladislav Holy, ed., *Comparative Anthropology*. Oxford: Basil Blackwell.

———. 1987b. Avarice and Iniquity at the Solor Fort. *Bijdragen tot de Taal-, Land- en Volkenkunde* 143:2/3, 208–36.

———. 1989. Méi Nafa: A Rite of Expiation in Lamalera, Indonesia. In C. Barraud and J. D. M. Platenkamp, eds., *Rituals and Socio-Cosmic Order in Eastern Indonesian Societies*. Part 1: *Nusa Tenggara Timur. Bijdragen tot de Taal-, Land- en Volkenkunde* 145:4, 539–47.

———. N.d. The Power of Strangers in Flores and Timor. Paper given at the Colloquium on Timor held in Lisbon in December 1989.

Bata Onã, Yosef Bura. N.d. *Sejarah Lamalera*. Typescript.

Beckering, J. D. H. 1911. Beschrijving der Eilanden Adonara en Lomblem, Behoorende tot de Solor-Groep. *Tijdschrift van het Koninklijk Nederlandsch Aardrijkskundig Genootschap*, 2e Serie 28:167–202.

———. 1921. De Bevolking van de Timor-groep. In J. C. van Eerde, ed., *De Volken van Nederlandsch Indië*, Deel 2. Amsterdam: Elsevier.

Beding, Alex. 1986. *Seratus Tahun Gereja Katolik Lembata*. Ende-Flores: Percetakan Arnoldus.

Bloch, Maurice. 1977. The Pasts and the Present in the Present. *Man* 12:2, 278–92.

Boxer, C. R. 1947. The Topasses of Timor. *Koninklijke Vereeniging Indisch Instituut*, Mededeling no. 73, no. 24.

Bowen, John R. 1989. Narrative Form and Political Incorporation: Changing Uses of History in Aceh, Indonesia. *Comparative Studies in Society and History* 31:4, 671–93.

Cohn, Bernard S. 1961. The Pasts of an Indian Village. *Comparative Studies in Society and History* 3:3, 241–49.

Couvreur, A. J. L. 1907. Verslag van eene dienstreis benoorden Larantoeka, 18–28 April 1907; Beschrijving van het rijk Larantoeka en verslag van eene reis rond Larantoeka, 13–23 May 1907. Algemeen Rijksarchief, the Hague, the Netherlands, LeRoux Collection, no. 7. Typescript.

Dasion, Petrus Bau. 1978. *Singkatan Usul Asal Dari Mojang Kami Bernama Korohama Lewohajo*. Typescript.

Davis, John. 1989. The Social Relations of the Production of History. In Elizabeth Tonkin, Maryon McDonald, and Malcolm Chapman, eds., *History and Ethnicity*. ASA Monographs, no. 27. London: Routledge.

———. 1991. *Times and Identities*. Inaugural Lecture delivered before the University of Oxford, 1 May 1991. Oxford: Clarendon.

Dias Vieira Godinho, Raja Usi. N.d. Tjeritara pendek dari Ketoroenan Radja di Tanah Larantoeka. In Couvreur 1907. Verslag van eene dienstreis benoorden Larantoeka, 18–28 April 1907; Beschrijving van het rijk Larantoeka en verslag van eene reis rond Larantoeka, 13–23 May 1907. Algemeen Rijksarchief, the Hague, the Netherlands, LeRoux Collection, no. 7. Typescript.

Dietrich, Stefan. 1984. A Note on Galiyao and the Early History of the Solor-Alor Islands. *Bijdragen tot de Taal-, Land- en Volkenkunde* 140:2/3, 317–34.

Farriss, Nancy M. 1983. Remembering the Future, Anticipating the Past: History, Time, and Cosmology among the Maya of Yucatan. *Comparative Studies in Society and History* 25:566–93.

Finnegan, Ruth. 1970. A Note on Oral Tradition and Historical Evidence. *History and Theory* 9:2, 195–201.

———. 1988. *Literacy & Orality.* Oxford: Basil Blackwell.

Fox, James J. 1971. A Rotinese Dynastic Genealogy: Structure and Event. In T. O. Beidelman, ed., *The Translation of Culture: Essays to E. E. Evans-Pritchard.* London: Tavistock.

———. 1979a. The Ceremonial System of Savu. In A. L. Becker and Aram A. Yengoyan, eds., *The Imagination of Reality: Essays in Southeast Asian Coherence Systems.* Norwood, N.J.: Ablex.

———. 1979b. 'Standing' in Time and Place: The Structure of Rotinese Historical Narratives. In Anthony Reid and David Marr, eds., *Perceptions of the Past in Southeast Asia.* Asian Studies Association of Australia, Southeast Asian Publications Series, no. 4. Singapore: Heinemann.

———, ed. 1988. *To Speak in Pairs: Essays on the Ritual Languages of Eastern Indonesia.* Cambridge: Cambridge University Press.

Geertz, C. 1966. *Person, Time and Conduct in Bali: An Essay in Cultural Analysis.* Cultural Report Series, no. 14. New Haven: Yale University.

Hawkings, Stephen. 1988. *A Brief History of Time: From the Big Bang to Black Holes.* London: Bantam.

Heslinga, T. 1891. Larantoeka op het eiland Flores: Ter Nagedachtenis van den Eerw. Pastoor Conr. ten Brink. *Berichten uit Nederlandsch Oost-Indië voor de Leden van den Sint-Claverband* 3:46–84.

Heynen, F. C. 1876. *Het rijk Larantoeka op het Eiland Flores in Nederlandsch Indië.* Studiën op Godsdienstig, Wetenschappelijk en Letterkundig Gebied, 8:6. The Hague: Van Gulick.

Howe, Leopold E. A. 1981. The Social Determination of Knowledge: Maurice Bloch and Balinese Time. *Man* 16:2, 220–34.

Hubert, H. 1905. Étude Sommaire de la Représentation du Temps dans la Religion et la Magie. *Annuaire de l'École Pratique des Hautes Études, Section des Sciences Religiuse,* 1–39.

Josselin de Jong, P. E. de. 1975. The Dynastic Myth of Negri Sembilan (Malaya). *Bijdragen tot de Taal-, Land- en Volkenkunde* 131:2/3, 277–308.

Keraf, Gregorius. 1978. *Morfologi Dialek Lamalera.* Ende-Flores: Percetakan Arnoldus.

Kumanireng, Piet. 1990. *Himpunan Cerita Rakyat Suku Demong Pagong di Flores Timur, Nusa Tenggara Timur.* Surabaya, Indonesia: Penerbit Mingguan ASAS.
Leach, E. R. 1961. *Rethinking Anthropology.* London School of Economics Monographs on Social Anthropology, no. 22. London: Athlone.
McKinnon, Susan. 1991. *From a Shattered Sun: Hierarchy, Gender, and Alliance in the Tanimbar Islands.* Madison: University of Wisconsin Press.
Malinowski, Bronislaw. 1926. *Myth in Primitive Psychology.* New York: W. W. Norton.
Middelkoop, P. 1952. Trektochten van Timorese Groepen. *Tijdschrift voor Indische Taal-, Land- en Volkenkunde* 85:2, 173–272.
Müller, Salomon. 1857. *Reizen en Onderzoekingen in den Indischen Archipel, Gedaan op Last der Nederlandsche Indische Regering, Tusschen de Jaren 1828 en 1836, nieuwe uitgave,* vol. 2. Werken van het Koninklijk Instituut voor Taal-, Land- en Volkenkunde van Nederlandsch-Indië, Tweede Afdeeling, Afzonderlijke Werken. Amsterdam: Frederik Muller.
Munro, W. G. 1915. Nota over Lomblen (Afdeling Flores). Typescript.
Ricklefs, M. C. 1978. *Modern Javanese Historical Traditions: A Study of an Original Kartasura Chronicle and Related Materials.* London: School of Oriental and African Studies.
Seegeler, C. J. 1931. Nota van Toelichting betreffende het zelfbesturende landschap Larantoeka. Museum van den Tropen, Amsterdam. Typescript.
Suban Leyn, Gabriel. 1979. Sejarah Daerah Pulau Solor. Laporan Kegiatan Penilik Kebudayaan Kandep. Menanga: P & K Kec. Solor Timur. Typescript.
Vatter, Ernst. 1932. *Ata Kiwan, unbekannte Bergvölker im Tropischen Holland.* Leipzig: Bibliographisches Institut.
West, Janet, and R. H. Barnes. 1990. Scrimshaw by William Lewis Roderick: A Whale Bone Plaque Dated 1858 showing the Barque *Adventure* of London Whaling off Flores and Pulau Komba in the Indian Ocean. *The Mariner's Mirror* 76:2, 135–48.
Wignyanta, Thom. 1981. Ceritera dari Lamalera (1): Lalilola. *Dian* 8:15, 17, 21.

Ruins of Time: Estranging History and Ethnology in the Enlightenment and After

Peter Hughes

I

> And into that, from which existing things come to be, they also pass away according to necessity; for they suffer punishment and pay retribution to one another for their wrongdoing according to the ordinance of time.
> —A Fragment of Anaximander

Distinctions and differences that now separate historical from anthropological discourse are deep and widespread. Johannes Fabian, in his challenging study *Time and the Other,* has shown that anthropology owes its origin to Enlightenment fascination with human nature, whose temporality was the principle of the "new science" advanced by Vico, Montesquieu, and Herder. But he has also shown how that shared origin has since been denied, most vehemently in Lévi-Strauss's erasure of time and revulsion from history.

Lévi-Strauss' attitude toward Time is firmly rooted in nineteenth-century notions of natural history, a fact which casts considerable doubt on his claim to be the legitimate heir of the eighteenth century. Admittedly, Enlightenment thinkers were interested in history for "philosophical" reasons. Above all they saw history as the theater of moral principles ultimately traceable to "constant laws of nature." But nature was decidedly human nature and the challenge of the historian was to show the temporal unfolding of its principles. The radical distinction between contingent human history and necessary natural history was drawn in the nineteenth century. To maintain, as Lévi-Strauss does, that anthropology tout court belongs to natural history is to deny the Enlightenment origin of our discipline.[1]

We will return to and account for the larger anguish about time that led Lévi-Strauss to rouse his readers (and his own discourse) from what James Joyce a generation earlier had called "the nightmare of history."[2] But the fear of time that first led Lévi-Strauss to geology and away from historical chronology later led him to Brazil and the taxonomy of myth and away from France and the taxidermy of defeat following its agony in 1940. The rise of structuralism owes much, as we shall see, to the fall of France.

Distinguishing between the worlds and languages of history and ethnology may mask or conceal what will be the subject of this essay—their shared origin in an experience of time expressed through ruin and revolution. What I want to show is that the historical revolution, in all the meanings of that term, shares origins and originators with the rise of anthropology. I also want to show how that shared origin lay in a sense of time as duration, which by a paradox revealed itself through things and artifacts that had not endured—objects and codes that survived through the half-life of ruins and allegories. We could say that metaphors are the ruins of language, repeating Walter Benjamin's ratio: allegories are to thought what ruins are to architecture. What histories and ethnologies had in common during their shared beginnings in the eighteenth century was an attempt to fasten temporal discourse to the weight and force of actions, archives, and artifacts: folk customs, texts, and charters that had to be read as palimpsests, abstract language telescoped back to its force of metaphor through etymology, and everywhere ruins and meditations on ruins—of kingdoms and cities, temples and castles, statues and amphorae. Even that fascination with Roman antiquity in the French Revolution or with Homer's epic poetry in Vico's *New Science* views these objects of interest and desire as defaced models, material paradigms of class struggle and overthrow.

From the Enlightenment to the present, the renewal of history and the birth of anthropology have depended on a sense of a sudden disruption in the temporal order, not just or only in the sense of a new cycle, or "revolution of the times," but in the more drastic sense of a sudden break with the past—time seen as a broken pillar. Time seems caught in a revolutionary shift of pace or temporal perspective that violently estranges the world we live in, making the present seem suddenly primitive, ruined by events. Most of my discussion will circle or cycle around the expression of this temporal trauma and *Traum* by revolutionary and romantic voices, focused on Vico, Coleridge, and Michelet. But I want to begin and end with modern voices that most of us would identify with the magisterial voice in historical and cultural theory.

Let me then begin with Marc Bloch's attempt to grasp the meaning and causes of the fall of France in his strangely neglected book *Strange Defeat*. The book's title reflects, as a *mise-en-abyme* on several levels, the estrangement that concerns us here. The author's reflections on France's ruin were written while he was still in a state of shock that records his thoughts more clearly and bleakly than hindsight might have made possible.

> The privations resulting from war and defeat have had upon Europe the repercussions of a Time Machine in reverse. We have been plunged suddenly into a way of life which, only quite recently, we thought had disappeared for ever. I am writing these lines in my house in the country. A year ago, when I and the local tradesmen had all the petrol we needed, the country town, on which the economic life of the district centers, seemed close at hand. Now, when bicycles are the quickest means of transport available, and heavy loads have to be carried in donkey-wagons, every trip to market becomes a major expedition. We have gone back thirty or forty years! The ruling idea of the Germans in the conduct of this war was speed. We, on the other hand, did our thinking in terms of yesterday or the day before. Worse still: faced by the undisputed evidence of Germany's new tactics, we ignored, or wholly failed to understand, the quickened rhythm of the times. So true is this, that it was as though the two opposed forces belonged, each of them, to an entirely different period of human history. We interpreted war in terms of assagai *versus* rifle made familiar to us by long years of colonial expansion. But this time it was we who were cast for the role of the savage!

Coming from an historian revealed by this memoir as ardently French and yet committed to the intellectual life, the thoughts that lead into these conclusions are even more ominous: "Our leaders, or those who acted for them, were incapable of thinking in terms of a new war. In other words, the German triumph was, essentially, a triumph of intellect—and it is that which makes it so peculiarly serious."[3]

During these months and years of war and occupation, the period covered by Bloch's memoir, the philosopher Simone Weil was engaged in a study of the ways in which catastrophic violence destroys any stable sense of time and closes off that sense of a future without which history cannot be imagined, let alone written or lived. From 1939 until 1941 she was at work on a long essay on Homer's *Iliad* as "The Poem of Force." It is both a reading of the poem as the epic of suffering and an allegory of her experiences in Civil War Spain and Occupied France. In an unpublished passage, which may have been deleted by the censors, she meditates on Hector's taunting of the defeated Greeks.

> Once the enemy is present it is too late; to capitulate is to be massacred. It is necessary to accept (since one must suffer it) this life where thought cannot pass from one day to the next without death's evocation. The mind is then stretched as it can only be for a little while, a single day; but after that, another day comes, requiring an equal tension; and another after that, until weeks, months, years roll on. This situation is impossible for the soul, but it *is*—the soul must do itself violence to bow to it. It must void itself of all hope, of all thought of the future, of every notion of an end near or far.[4]

It is easy to understand why these Æsopian words may have been censored. They give the lie to the Vichy delusion of a contented present and a shining future. Just as death turned Homer's warriors into carrion, "dearer to the vultures than they are to their wives," the forces of war turn human beings into slaves, or even into things, existing without time and outside time.

The ease with which the *Blitzkrieg* deconstructed France struck Marc Bloch as much more than a military defeat, and it evoked in the veteran historian many of the same unsettling responses that we have recently found in the collaborationist journalism of the young Paul De Man. Nazi policy and propaganda deliberately tried to degrade its enemies as primitives and reactionaries, especially in creating a contrast between a new era of youthful revolution and a dead world of the past. This message and contrast had more impact on Bloch and De Man and a multitude of others than we are usually led to believe. Later, in his examination of conscience (to use his own phrase), Bloch shows this link between renewal and ruin in a comparison between the French Revolution and this new force.

> We judge revolutions to be admirable or hateful according as their principles are or are not our own. All of them, however, have one supreme virtue which is inseparable from the vigor out of which they grow: they do thrust the young into positions of prominence. I detest Nazism, but, like the French Revolution, with which one should blush to compare it, it did put at

the head, both of its armed forces and of its government, men who, because their brains were fresh and had not been formed in the routine of the schools, were capable of understanding "the surprising and the new." All we had to set against them was a set of bald-pates and youngish dotards.[5]

Meditating upon ruins in the past, if we think of Vico, or Gibbon, or Volney, was a way, even a method, of confirming the revolutionary function of time: that force of time, we shall see, to advance political arguments all the more deeply felt because usually masked by the Enlightenment and its rhetoric. What a contrast it is to follow Bloch's troubled conclusion that the strangest aspect of this *Strange Defeat* is its inversion and perversion of time and modernity. François Mitterand, who was jolted by Occupation and Resistance from royalism to socialism, has presidentially defined France in terms of its "modernity," and I have suggested how much that cherished assertion owes to the shattering denial it suffered in 1940. And how much, by a paradox that our twofold approach to time should expose, that "modernity" owes to the "pensées sauvages" and "tristes tropiques" of those others among the vanquished who strike out of a decadent and decrepit Europe toward strange worlds of myth and metaphor. This flight from history and ruin should also be seen as an impulse toward ethnology and the primitive, but always as a counterhistory that seeks to abolish time, to pass from temporality to taxonomy, from line to grid through the synapse of cycle.

II

> To estimate a man like Vico, or any great man who has made discoveries and committed errors, you ought to say to yourself—"He did so and so in the year 1720, a Papist, at Naples. Now, what would he not have done if he had lived now, and could have availed himself of all our vast acquisitions in physical science?"
>
> —Coleridge, *Table Talk*

Marc Bloch's shocked feeling that the fall of France had created a parody of the agon of colonialism, turning him and his culture's sense of time and place into displaced persons staring at the camera of history like the refugees in contemporary news photos, is a modern form of that encounter with the ruins of time that provoked in the Enlightenment the need for a revised view of our relations to both the temporal past and its cultural analogue of the "primitive." This double sense of renewal and ruin repeats the vision of Vico that inaugurated both historical and ethnological notions of time we still live and think with. And Vico, even more revealingly than Bloch, writes himself and the decadence of Naples into his attempt to envision the world of temporality as the faded and disordered design whose original and material power can only be perceived through temporal allegories and etymologies. We know from Augustine that there can be no time without change.[6] From Vico we learn that all change is for the worse, or at least for the weaker.

His sense of time might be described as a declension down into the present of a physical world that can only be grasped in that present through an imaginative return, a *ricorso* that is also an ascension. The paradox, which explains why this theory of human society has been at the root of both histories and ethnographies, is that the primal and physical origin can only be grasped by the modern and belated imagination, through an individual awareness whose temporal capacity is memory. When he offers the principle of his new science, even darkness is made material as well as visible.

> In the night of thick darkness which envelops the earliest antiquity, so remote from ourselves, there shines the eternal and never failing light of a truth beyond all question: that the world of human society has certainly been made by men, and its principles are therefore to be found within the modifications of our own time and mind.[7]

This declaration of principles then goes on to pull us away from the Newtonian and Cartesian world of nature ("which, since God made it, he alone truly knows") to that of human society. But the strange inversion of his argument lies in the way it weights the human spirit in the same passage that lightens the physical universe. He speaks of "that infirmity of the human mind, by which, immersed and buried in the body, it naturally inclines to take notice of bodily things, and finds the effort to attend to itself too laborious." This remark about the condition of modern man takes on greater force when we know that Vico elsewhere says that language itself emerges from the body of our primal ancestors.

The difficult practice of "attending to itself" is made possible by philology, which starts from the premise that the language we know is the ruin of a lost language in which every word is physical and no metaphor is figurative but plain truth. The world of rocks and tides is presented as timeless, and even abstract, while the world of language and artifacts is presented as weighted by the past, by the burden every word and art object bears into the present. Taken to its extremes, taken as far back as etymology can go, language returns to the physical and material order of sounds. This includes those that imitate or recreate things through onomatopoeia, gestures—which Augustine describes in the opening pages of his *Confessions* as "the first language of mankind"[8]—and inscriptions or incisions such as the laws and charters whose effaced meaning and defaced wording Vico struggles to interpret. The diachronic meanings of a word are established through the estranged history recounted by etymology. Even the most abstract of legal terms is shown to have a metaphorical weight and value.

Arguments or proofs derived from etymology bear a curious relation to time and memory. They are clearly temporal and historical to the extent that they depend upon derivation and sequence, but they lack any precise chronology or dates. Even stranger, etymology discovers or invents origins that have no causal relation to the words that are their effects. The word *charm,* for example, takes us back etymologically to the Latin *carmen,* but the way back is intuitive rather than logical, a *ricorso* rather than a causal relation. And it tells a story its teller can't remember: it is a history without any sense of the past. If Vico's *New Science* presents etymology as theory, Joyce's *Finnegans Wake* presents its practice. Its endless wordplay was meant, according to Joyce, to offer a history of mankind, a retelling of the narrative implied by Vico's book, which is what he urged his baffled admirers to read when they said they found *Finnegans Wake* unreadable. Samuel Beckett, who read Vico more carefully than anyone else trying to account for *Finnegans Wake,* later offered the best description of the etymological method common to both books. "In the beginning," said Beckett, "was the pun."[9]

This is a joke that is itself a pun, an allusion to and parody of St. John, but it is also a hint at the enigma at the center of ethnographic notions of time. These are reducible to the notion that certain stories are first or primal stories and the ethnographer's task is to find, record, and interpret them. But any story, even if it accounts for the Creation, always *re*counts or retells another story in terms of narrative pattern or plot. Historians, at least since the appearance of Hayden White's *Metahistory,* have been more alert than anthropologists to the twist that plotting imparts to verbal representations of temporality, the spin that *res gestae* owes to *historia* or *Geschichte* to *Historie.* There is always an earlier story. The original is always to some extent a copy or even a fake.

This does not mean that nothing is true but rather that the chief value of a historical account, as Paul Ricoeur argues in his great book *Temps et récit,* its power to tell us something worth knowing about our temporal world, is its ability to constitute a *narrative.* When that attempt succeeds, it creates what

Ricoeur calls temporality, "the structure of existence that reaches language as narrativity." This power to *re*count is reciprocally defined as "the language structure that has temporality as its ultimate referent."[10] And this implies that the truth of historians and ethnographers depends upon a sense of time and duration quite different from that of a chronicler or archivist but similar to that of the novelist or dramatist. I don't mean that historians tamely follow fictional models (some actually tried to *outdo* novelistic plots, as Michelet tried to outdo Scott and Chateaubriand, but that's another story), rather that they achieve, or at least attempt, an account that conveys what it means to live in a world that combines sequence and simultaneity. Such an account conveys the value and intricacy of the ways in which some things happen *after* and others *before,* while other events or scenes interrupt sequence to form tableaux or ensembles of groups and societies. This account of narrative deliberately says nothing about causality—not because it does not exist but rather because any narrative dominated by causal concerns tends toward the perfection of the whodunit or the propaganda leaflet, in both of which the unanswered questions of the one, like the unquestioning answers of the other, tend to eliminate and reduce the mingling of sequence and simultaneity in favor of simpler solutions.

Narrative time might best be described as the ruin of experience. Insofar as narrative puts things in sequence, one after the other, it no sooner erects a structure of events than the property is condemned to a falling down or falling away (in the further active sense of the Latin *ruina*) out of the reader's or listener's awareness and into the solitary confinement of memory. And yet the extent to which narrative repeats and varies events, or portrays and centers itself around a character or characters, is the degree to which it projects presence and aesthetic order. The Romantics grasped, in ways we will turn to later, that what is a ruin or fragment to one observer may be a torso or vision to another. A person from another culture unfamiliar with our aesthetic category of torsos or busts might take whole galleries of the British Museum or Louvre to be junkrooms accidentally left open to the public, and Romantic poets were obsessively ready to describe their works as "fragments."

That is what Coleridge called his *Kubla Khan,* even smuggling the word into the poem's text (so that the ruin can be identified by the epitaph RUIN), not because the poem is incomplete but rather because it is different from another poem he wanted to write or even from yet another poem written by Wordsworth. His elliptical poem could also be read as an example of the tension in narrative I have just described because it begins with a story about (or around) a figure and ground identified as Kubla Khan and Xanadu; but this narrative sequence then shifts into a visionary trance in which simultaneity displaces sequence. Its sense of time also shifts from historical (or Æsopian) allusions to the French Revolution in the first part to ethnographic rites and charms in the second.

The poem, like some of Ezra Pound's *Cantos,* begins in a narrative mode and then shifts into a lyric mode: it goes in a few lines from *corso* to *ricorso.* The famous opening lines, "In Xanadu did Kubla Khan/A stately pleasure-dome decree," historicize and date what they narrate through the slightly archaic past of "did . . . decree." This historicity is further increased by the strangeness of "decree"-ing the pleasure-dome, bringing it into existence with a word or a document. Used in this sense, *decree* sounds like a translation from the French, turning into English the verb that opens so many Revolutionary decrees read and referred to by Coleridge in other more overtly historical contexts. The narrative mysteriously announces the coming of war and then fades away, like the dome itself. Recalling and reviving the pleasure-dome can only be accomplished through a lyrical conclusion that slides, like a Vichian *ricorso,* back into the rhythm and imagery of a magic spell or folk-rhyme: "And all should cry Beware! Beware! / His flashing eyes, his floating hair! / Weave a circle round him thrice, / And close your eyes with holy dread. . . ." The poem accomplishes in its compressed scale and force the double movement aimed at in much longer meditations on ruins and vanished monuments. This movement is not only forward out of the past through historical time but also backward through memory and through the imagination that Coleridge described as the repetition of divine creativity in the human world.

Coleridge, like another figure we will soon come to, was fascinated by Vico and planned to write his life as the biography of an intellectual hero. This project is no less interesting because Coleridge never carried it out, or even because Vico himself already had. It matters because Coleridge recognized more clearly than any other Romantic the narrative crossing where the ethnographic subject, which he called "Superstition," merged with the psychological life of the individual. He recognized as well that Vico's works opened a door into a world of language that offered an alternative to the abstract world of *Naturphilosophie* or pure physics in which he despaired of creating his "Opus Maximum." And that order of language was also an order of time in which he could transform his own sense of self-betrayal and ruin into a *Biographia Literaria.* Just as Vico argued that the decadent present could only understand the "allochronic" prehistorical world by and through an exercise of memory and imagination (*ricorso*), precisely because the individual recapitulates the *corso* of human history, Coleridge thought of himself as someone whose life could serve as the pattern for a metaphysics that could include ignorance and superstition as well as the more usual modes of thought. The shame that he said dominated his life began to modulate into a new confidence that he would be the first, unless that honor was Vico's, to turn the problem of one's life into its solution as the pattern for the development of theory—or "the growth of the poet's mind," to recall Wordsworth's description of his poetic autobiography, *The Prelude*. At the center of both Wordsworth's poem and Coleridge's memoir there is not only the autobiographical attempt to constitute a narrative and, in Ricoeur's terms, to create a truth, but also the attempt to come to terms with the historical truth that both Coleridge and Wordsworth had in their earlier life been ardent supporters of the French Revolution and of English Radicalism, and then, as Byron put it, "turned out Tory at last."

Coleridge's uneasy relations with history and with his own past are matched, as we have seen, by his attempt to get back before history into a world of myth, folktale, and symbol. The most brilliant of these attempts, *The Rime of the Ancient Mariner,* offered to the reader as a traditional ballad with learned marginal glosses, is entirely the invention of Coleridge and Wordsworth. It is both original and much closer to the style of a folk-origin than the broadside and period. But the original is a fake, and Coleridge was as inspired in his faking of ethnographic or ancient origins as Vico was in his searching of Roman law and Homeric poetry for the material origins and etymologies of an older world defaced by abstraction and ideology. Coleridge's motives or impulses in this estrangement of history are entangled in his anguished and tantalized attraction to superstition and evil. Coleridge's *Notebooks,* an astonishing revelation of his "effort to attend to himself," are also an attempt to rescue himself through visions and reveries from waking and historical time, from the ruin of his own life and times. He wants to live instead in the dream time of an opium trance, a state in which he would imagine the creation of poems that exceeded his powers or willingness to concentrate. He once noted that his greatest pleasure would be to *dream* the whole of Milton's *Paradise Lost,* and both *Kubla Khan* and *The Rime of the Ancient Mariner* are ascribed to "reveries," by which Coleridge means daydreams in his public statements but nightmares in his private *Notebooks*.

The connection with Vico's theory of the *ricorso* is striking: both are ways of creating a mode of strange time, a sense of duration purged of history and chronology, through which the individual can imagine an age of fable by reading or even writing the runes of the psyche's or memory's deepest fears and, to use Coleridge's own phrase, of the individual's sense of "holy dread." What Coleridge expresses with much greater clarity than Vico, though enveloped in the obscurity of his *Notebooks,* is a deep revulsion from time itself, from history as Vico's "the course the nations run." Coleridge went so far as to confess that he found creation itself to be an anticlimax: when he reflected on the creative power of God, the Creation itself struck him as a great disappointment! Here the romantic poet asserts a sense of time that is profoundly antihistorical, treating the rise of history as a fall into time. The Fall of Adam and Eve, according to his mythic or ethnographic romanticism, is merely the aftershock of a primal ruin. The prehuman origin of this Fall, which predates what John Henry Newman suggestively called an "aboriginal calamity," strengthens the bond between ruins and the ethnological other.

Before turning to Vico's and Volney's treatment of this *topos,* we might first consider how close to home, and to Vico's own time in Italy and Naples, strikes the theory of time exposed and expressed in the *New Science.* It is actually a threefold theory of which two parts were risky business for anyone who explored them. Vico passes as quickly as possible over providential time, the theory of time that still governed both the sacred and profane powers in Naples, and to which they expected consent or at least lip-service from their subjects. The Inquisition was still active in Naples during Vico's lifetime, and little could be written about time and history that trespassed on notions of providence without attracting its dangerous interest. Hence the careful limits set by the title of Pietro Giannone's *Civil History of the Kingdom of Naples.* When it was published in 1723, despite this cautionary step, "It was promptly put on the Index and translated into English."[11] Gibbon, himself a great meditator on ruins, learned "the use of irony and criticism on subjects of Ecclesiastical gravity" from his reading of Giannone, and the light touch of irony in the choice of *gravity* shows how Gibbon's generation could dismiss the claims of providence.[12]

The second theory of time, which takes up much of Vico's study, historicizes time by making available to the philological method even the most obscure myth or allegory. Even the strangest time becomes intelligible, temporal, and political when examined to discover its place in the course of human history. One of Vico's most basic axioms is that human beings can only know what they have made; and time as memory and duration is a human and cultural construct. Time as chronometric measure is another matter, and Vico's argument that measuring is not knowing had even greater force in the eighteenth century when even finding the longitude with precision was not possible until after his death. Fixing the longitude was one of the decisive victories of the European colonizers over time. Another victory, the irony of which a Gibbon would have grasped but hardly enjoyed, was that these same navigators, especially those from Catholic countries, gave passage to missionaries who renewed a providential sense of time, or at least brought it into conflict with other theodicies, at the very moment when enlightened Europe was turning away from providence for what seemed like the last time. This irony shows that the third and ethnological notion of time was being exalted by Vico in Europe as the conversion of the native peoples, and the repression of their cultures, was trampling it under in the East and the Americas. His enlightened *ricorso* had been turned into the *corso* and triumphant course of European history.

The civil historian in Vico saw both the thesis and the writing of his *New Science* as a revolutionary gesture, an awakening of corrupted and fallen Naples, divided and decadent Italy, against the judgment of the world. Few themes in his *Autobiography* are more painfully clear than the parallel between his own struggles, not just against the ruling Cartesianism of the age but also against the favoritism and cowardice of his culture, and the struggles of the early Romans to overthrow tyrants. Part of the analogy suffers from the academic pettiness of his defeated attempts to be appointed and promoted at the Royal University: everything about them is ludicrous except the anguished intelligence of the victim. And yet the northern slant of most studies of Enlightenment and Revolutionary Europe should not lead us to forget that Naples was, with all its defects, the largest and one of the most splendid of European cities. Vico was forced to work against both the hostile forces of reaction within Naples and the double disdain for the south within Italy and for Italian scholarship within Europe. Vico agreed to write his *Autobiography* in part because the request came from Venice and the north. A letter of praise from Jean Le Clerc in the Netherlands stirred Vico so deeply that he quotes it in full.

> I saw many excellent things, both philosophical and philological, which will give me occasion to show our northern scholars that acumen and erudition are to be found among the Italians no less than among themselves; indeed, that more learned and acute things are being said by Italians than can be hoped for from dwellers in colder climes.[13]

Vico, who throughout speaks of himself in the third person, accepts the compliment of this and lets the condescension pass. But not so his many envious enemies in Naples.

They said it was impossible that as a result of this work of Vico's Le Clerc should be willing to retract what he had been saying for nearly fifty years, namely that Italy produced no works that could stand comparison for wit or learning with those published in the rest of Europe.[14]

Vico insists on his incorruptible determination to read Homer in the light of his philological theories, unswayed by the conventional wisdom on both sides of the Alps. This is essential to his *New Science,* which finds in Homer a double time-sense. Homer is revealed on the one hand to be historically true, not only in the sense demonstrated in the nineteenth century through the finding of Troy but in the wider sense that the poem reveals a stage in heroic culture. Homer read in this way refutes, on the other hand, the academic attempt to familiarize the poems and the culture that produced them. Vico, like Fustel de Coulanges in a later generation, estranges the classical world by denying its classical status and insisting on its closeness to the moody tribesmen that Gibbon later saw in the Tartar and Kalmuck horsemen who posed the only threat, in his eyes, to polite Europe and its stable expectations. Coleridge was fascinated by this interplay in Vico between ancient and modern, historical and ethnological. In the course of an eloquent discourse among his friends, including the Italian scholar and exile Prati, who had introduced him to Vico:

> Coleridge referred to an Italian Vico who is said to have anticipated Wolf's theory concerning Homer [which Coleridge says was his at College]. Vico wrote *Sur une nouvelle Science,* viz. Comparative History. Goethe notices him in his Life as an original thinker and great man. Vico wrote on the origin of Rome. Coleridge drew a parallel between West India planters and the negroes, the subjection between them and the condition of the plebs of Rome toward the patricians; but when I inquired concerning the origin of the inequality Coleridge evaded giving me an answer.[15]

The converse of this reading, in both Vico and Coleridge, emerges from the way it leads to a new way of writing history, of constituting a narrative that does not separate itself from the past but rather tries to write in its own language, which is the language of poetry. In his continuation to the *Autobiography,* written six years after the first version of the *New Science* appeared in 1725, he goes further, suggesting a style for history that will partly abandon modernity itself, whose emblem is prose, and aim instead at a style between poetry and prose. He finds this historical style in his own work, especially in an oration meant to praise the heroic virtue of a noble judge, a contemporary whose judgments restored Naples to a greater past.

> He made a digression in style half way between that of poetry and that of prose. (Such indeed should be the historical style, in the opinion of Cicero, as indicated in the brief but pregnant remark he makes concerning the writing of history. For history should use, he says "words for the most part taken from the poets;" perhaps because the historians were still clinging to their ancient possession, for, as is clearly demonstrated in the *New Science,* the first historians of the nations were the poets.)[16]

The final stage of the process by which style becomes the historian's *ricorso* occurs (or recurs) once again in his own life as it draws toward its end. He describes himself as sinking down toward death, a ruin at last restored by the heroic revenge of his writings.

> He however blessed all these adversities as so many occasions for withdrawing to his high impregnable citadel, to meditate and to write further works which he was wont to call "so many noble acts of vengeance against his detractors."[17]

What is striking in this conclusion to the life is its merging into the work. Michelet, who resurrected Vico three generations later, used to say of himself, "I am not a man, but a book," and it is remarkable that some of the most subtle early autobiographies are by historians—Vico, Gibbon, Michelet—who finally offer the work as an alternative to the life.

III

> Michelet is known in the Quartier Latin as "Monsieur Symbole." His way of writing history reminds me of a remark once made by my great teacher Hegel. He said that if some one were to write down all the dreams that people dreamed during a certain period, the record of those dreams would constitute the best history of the age.
>
> —Heine, *Augsburger Zeitung*

The significance of this alternative (and alternation) is that the temporal structure of narrative and the source of its truth, the alternating awareness of sequence and simultaneity, appears in both the history and the autobiography. The *corso,* or "course the nations run," oscillates into and reverses itself in the *ricorso* by which the imagination returns through memory to a "strange time" that can be peopled by the child, the primitive, or the outlandish. Just as Coleridge writes metaphysics as his life, the historians renew and estrange history by rewriting it as *their* lives. Introspection, what Vico called "the effort to attend to itself," leads the historian's quest to the discovery that the self offers a sense of time and narrative much more capable of shifting between tenses than a history chained to chronicle or causality. The self, embodied in stories with many versions, is constantly revising itself, doing and undoing itself. Vico memorializes several stages in his life, Gibbon's palimpsest is rightly entitled *Autobiographies* in the plural, and Michelet's *Journal* presents a polymorphous figure whose thoughts and dreams become a bilingual discourse. What happens is that the traditional way of judging the validity of a historical text—its relation to other texts such as archival documents—is subverted through this autobiographical subtext, which is closer to fiction in that it establishes its validity through its relation to human experience, to what has had a long genealogy since Aristotle called it probability. Until that moment in historical discourse when anthropology separated itself from history, or was driven out as too superstitious, a moment that occurred around the time of Michelet's death, the historian could recognize, even if he rarely reported, the other, or primitive, within himself. Michelet imagines his sorceress and his Joan of Arc, even his insects and mountains, as we can tell from parallel imaginings in his *Journal,* in aspects and possibilities perceived in himself. Marc Bloch lived well after those polymorphous days for professors of history, and his shock at being transformed into the primitive other, the savage, of the German conqueror, is all the greater because it is such a relic out of the ethnological past of historical writing.

The last stage of this coexistence within the historian of both chronological time and psychic time was reached in the life of Michelet and the upheavals of the French Revolution. The frequency with which romantic visions of the past, ruins in estranged settings, and (counter)revolutionary issues run together in his life and work suggests that they may be linked by some deeper reasons than those of imitation. Michelet was by no means alone in deriving historical method from ruins and tombs, though he was, as we shall see, ready to put them at the very center of his sense of the past.

A generation before he published his intensely personal translation, or transformation, of Vico's *Autobiography* and *New Science,* Volney's *The Ruins* pulled together the impact of the French Revolution and the shock of exotic desolation to argue that ruins, like the fallen statue of Ozymandias, confirm radical ideals. The book's invocation begins,

Hail solitary ruins, holy sepulchres and silent walls! . . . When the whole earth, in chains and silence, bowed the neck before its tyrants, you had already proclaimed the truths which they abhor; and, confounding the dust of the king with that of the meanest slave, had announced to man the sacred dogma of Equality.[18]

Volney's traveler surveys the ruins of lands in the Middle East, now part of the Ottoman Empire, as part of his concern for "the happiness of man in a social state." Here again ruins are both historical and political—all the more so when we recall the standing analogy, the Enlightenment code in which the Turkish or Persian empires stand for France and its absolutist corruption. The material weight of all this wreckage offers a way of overcoming the absence of the past when it is limited to paper and ink or condemned to fading memory.

Often I met with ancient monuments, wrecks of temples, palaces and fortresses, columns, aqueducts and tombs; and this spectacle led me to meditate on times past, and filled my mind with serious and profound contemplations.[19]

Passages such as these, which reappear in many other works or meditations on ruins, give a new third dimension to one of the Enlightenment's basic codes. Just as the Ottomans can be identified as the Bourbons, so that the exotic other can be seen to be here and in power, a reader of *The Ruins* would be expected to recognize another axiom in the Enlightenment's strategy toward the other. This was the principle, put clearly and perhaps for the first time by Racine in the preface to his tragedy *Bajazet,* that remoteness in space could be taken as equivalent to remoteness in time. For historical and allegorical purposes, the Ottomans may be substituted for the Romans. Volney sets his ruins in both of these contexts that make time imaginable in terms of space. The solidity of stones and pillars resists the erosion of ages, and the third dimension of ruins is lacking in charters and dated documents. The distance to Xanadu equals the centuries that separate us from Imperial Rome, and these two separate time-space equations could be combined into a single equation with two unknowns. Coleridge's *Notebooks,* for example, confirm that Kubla Khan is in some respects linked to Napoleon, and his newspaper *The Watchman* creates an allegory that narrates wars involving khans in Central Asia; but these are to be read against Pitt's censorship as a second narrative of Bonaparte's campaigns with the Army of Italy. Coleridge takes

Vico's etymological sense of time into inner worlds as well. His fascination with evil validates and makes serious the wordplay of Tartary in Asia and the infernal Tartarus of mythology. This may seem farfetched, but it could be shown, "had we but world enough and time," that it in fact understates an obsessive analogy between the ruins of time and the being of the traveler or historian. As Volney's other and observing self meditates on the glory that was of these Levantine ruins, he reflects, "And now a mournful skeleton is all that subsists of this powerful city!"[20]

Ruins call Volney back to the flourishing state of France a few years before the Revolution. This moment of suspended time, in which France seemed to possess historical dominance and poise, came "in 1782, at the close of the American war." But the estranged perspective of Syria and Egypt reminds Volney's traveler that French dominance depends upon the unending movement of power from east to west, and this in turn provokes the thought that Paris and London may soon be one with Ninevah and Tyre. The cycle or wheel means that European ascendancy is poised over a temporal abyss.

> Who knows, said I, but such may one day be the abandonment of our countries? Who knows if on the banks of the Seine, the Thames, or the Zuyder Zee . . . some traveller, like myself, shall not one day sit on their silent ruins and weep in solitude over the ashes of their inhabitants, and the memory of their greatness?[21]

This question had turned into the response of the Revolution before Volney actually asked it. *The Ruins* was written and published in the early stages of revolutionary overthrow, and Volney is commenting rather than prophesying. What is unusual about his bleak reading of the ruins is that he was himself an advocate and supporter of the Revolution that its opponents and survivors made the necessary cause of our modern historical sense. Michelet was one of the first to see that the Revolution had broken the temporal thread that joined the past to the future, and that, from the point of view of historical thinking, modern times began in the attempt to overcome that sense of abandonment— the sense that the past itself was a ruin. R. W. Southern has argued that Romantic fascination with the historical past was the fascination of the dispossessed, of those who hungered after a past that the French Revolution seemed to have distanced and even destroyed.[22]

The Romantic ratio between psychological time and primitive awareness, put most simply in Shelley's *Defence of Poetry* ("the savage is to ages what the child is to years") underlies Michelet's early fascination with Vico. Although Michelet became in later years an ardent republican, he began to read and translate Vico when still a royalist and a Catholic. He was even a tutor to the royal princes of the Restoration court, and his later attempt to constitute a coherent narrative for himself is every bit as problematic as Coleridge's or Vico's. Michelet should have emancipated himself from early superstition to become such a heroic defender of revolutionary and republican values. But he was actually born into a radical printer's family, was never baptized, spent his childhood under Napoleon, and turned to both Catholicism and Reaction in young manhood. He came to Vico by way of a very different historian. His constant reading as he first began to teach history was Gibbon's *Decline and Fall*, and his first historical publication is a *Tableau* of modern history that runs precisely from, as its subtitle puts it, *Depuis la prise de Constantinople par les Turcs, jusqu'à la Revolution Française, 1453–1789*.[23] He turned out radical at last, but his grand vision of history as a resurrection (a view inscribed on his gravestone at Père-Lachaise) could be understood as a rewriting and revision of his own redemption from royalist superstition.

We recall how Gibbon in his *Autobiographies* traces back to his meditations among the ruins of Rome, among its classical temples whose only echo was that of chanting friars, the origins of his *Decline and Fall*. This may lie behind Michelet's reconstruction of his own childhood as one of silence and emptiness in the desolation of the Quartier St Antoine. Now this memory is on one level astonishing, even incredible: he is recalling one of the most crowded sections of Paris during the tumultuous years around 1800. And yet, like Gibbon's ruins or Vico's Naples, Michelet's Quartier St Antoine is a place that is being displaced into a strange time of the historical imagination. It was the birthplace of Revolutionary action, and Michelet declares that he entered the world at the moment of the Revolution's death, brought about by Napoleon and kept in the tomb through all the years of the Empire. Napoleonic France, celebrated or abhorred for its monumental splendor and power, always figures in Michelet as a house of death built on boredom and void. The text that Michelet returns to, obsessively and eloquently, to express the desolation of that time is itself a fragment or shard, the prose sketch of an epic poem by the otherwise obscure writer Grainville, *Le Dernier homme*. Like several of Coleridge's imaginary or "intentional" epics, this prose work calls attention to its ruined state by incorporating ruin into both its subject and its style. Its opening words echo an early passage in Volney's *The Ruins*, published a few years before the writing of Grainville's. But this epic attempt is much more "timely," in the double sense that it refers directly to

Napoleon's campaign in Egypt and because it is directed toward the future rather than backward into the past or around him in the present: "I was in Syria, not far from the ruins of Palmyra. There gaped a great cavern, where no one ever entered and lived to see the light. Our brave troups from the Army of Egypt risked the adventure and never came back."[24] The tense of Grainville's vision shifts forward toward the future, a future that closes with the end of history. And the figure of the sole survivor of the human race, which recurs in later romanticism in Mary Shelley's novel *The Last Man,* might even be said to replace the Enlightenment fascination with ruins. Both Grainville and Shelley close with a vision of death that means the end of the human world: *The Last Man* by plague, *Le Dernier homme* by the resurrection of the dead that Michelet so often proclaimed.

And yet, ruins—however much they may seem to offer an aesthetic consolation or escape from history—pull the historical imagination toward perceiving its own task, which is to make present through narrative what is, like ruins, *there* as decline and fall but *not there* as rise or triumph. The revolt against what Nietzsche describes as "monumental history," and in great part the growth of modern historical studies, can be traced back to the enlightened fetish of ruins and catastrophe. Since the revolt or *ricorso* begins in the individual imagination, it is also true that the ruinous constitutes a "savage" time in which the ethnographic impulse of the writer coincides with the writer's own misgivings or estrangement, both about the self and about the story the self needs to tell.

We may assume that history is in love with time and its narratives, but I would suggest in conclusion that the shock of ruin reveals to historians their anxiety or revulsion when confronted with temporality. Marc Bloch's traumatic response when he found that defeat had pushed him into a primitive past cannot be overcome, and his *Strange Defeat* is also the defeat of his own historical method. Michelet recognized a century before that narrative is the most agonizing task of the historian, while the tableau or scene is the most ecstatic. Presence and simultaneity are actually far more the time scale of history than absence and sequence. Ruins, taken with the artifacts and events that together shape them are at the center of "the history of things" so probingly discussed by George Kubler in his *The Shape of Time.* "Such things," he observes, "mark the passage of time with far greater accuracy than we know, and they fill time with shapes of a limited variety."[25] The ties that bind the ruins of the archaeologist with those of the autobiographer are the bonds of attraction and revulsion. Nietzsche once said that the chief function of art was to enable us to look on the world without desire, but Nietzsche was wrong. Art, and our response to it, enable us to see how much we desire in the world and disable us by revealing how much of art we lack and how much of it is lacking in our world. Kubler expresses this power of presence in his opening axiom.

> In effect, the only token of history continually available to our senses are the desirable things made by men. Of course, to say that man-made things are desirable is redundant, because man's native inertia is overcome only by desire, and nothing gets made unless it is desirable.[26]

We began by seeing how the fall and ruin of France, the end of a certain view of enlightenment and revolution, deconstructed Marc Bloch's historical awareness and led him to question both his own perceptions and those of his culture. Those same shattering events could be said to have renewed the time sense in French fiction, as in Claude Simon's *The Flanders Road,* and even more to have driven so many thinkers into the strange time of folktale and ethnological system. The radical changes of narrative and drama, obvious in writers such as Claude Simon and Samuel Beckett, whose *Waiting for Godot* reflects his own encounter with the boredom and servitude of Occupied France, suggest that the temporality exalted by Ricoeur may have other manifestations than those of novelistic realism. The narrative of history may need to reflect and register the skepticism about sequence, the revulsion from time, the fascination with the synchronic and syncopated in modern culture. The example of Claude Lévi-Strauss has already been mentioned. An even more startling example was that of Simone Weil, all the more so because her flight into that world is marked by a fragmentary paper trail of anthropological jottings and possibilities that are in total contrast to her previous historical and philosophical writing. Her *Notebooks,* as remarkable in their way as Michelet's, turn in her New York and London exile of 1941–43 away from history and against time itself. Sometimes the best witness is the most hostile, and her witness is telling: "it is time that tortures us. Man's whole effort is to escape from it, that is to say, to escape from past and future by embedding himself, in the present, or else by inventing a past and a future to suit himself."[27] The ruins of time are the encounter between the presence that underlies all understanding and the absence that underlies desire. Scientists and positivists have argued against historians and anthropologists that we can demonstrate only what is present, only argue about what is absent. Ruins of time, in the senses proposed here, are what make present the absent world of history and ethnology: present in bringing them to light and present in making them part of our time.

NOTES

1. Johannes Fabian, *Time and the Other: How Anthropology Makes its Object* (New York: Columbia University Press, 1983), 57.
2. James Joyce, *Portrait of the Artist as a Young Man* (London: Jonathan Cape, 1956/1914), 233.
3. Marc Bloch, *Strange Defeat,* trans. Gerard Hopkins (London: Oxford University Press, 1949), 36–37.
4. Simone Weil, *La Source grecque* (Paris: Gallimard, 1969/1941), 35.
5. *Strange Defeat,* 161.
6. St. Augustine, *The City of God,* xi.6.
7. G. Vico, *New Science,* para. 331.
8. St. Augustine, *The Confessions,* i.2. Wittgenstein, who takes this passage as the epigraph for his *Philosophical Investigations,* perceived the significance of this Augustinian and Vichian principle that the first human language is of gestures rather than words: that we live in a world of appearances that are real, in no way inferior to an abstract world of causes.
9. Samuel Beckett, *Disjecta* (London: John Calder, 1983), 27.
10. Paul Ricoeur, "Narrative Time," *Critical Inquiry* 7 (1980): 169.
11. Max Harold Fisch and Thomas Goddard Bergin, eds. and trans., *The Autobiography of Giambattista Vico* (Ithaca: Cornell University Press, 1944), 31.
12. E. Gibbon, *Autobiographies,* ed. John Murray (London: John Murray, 1896), 235.
13. G. Vico, *The Autobiography,* 159.
14. Idem.
15. *Table Talk,* ed. Carl Woodring, 2 vols. (Princeton: Princeton University Press, 1990), 1:565–66.
16. Vico, *The Autobiography,* 180.
17. Ibid., 200.
18. Constantine Francis Chassebeuf de Volney, *The Ruins,* trans. Joel Barlow (Philadelphia, 1800), 11.
19. Ibid., 21.
20. Ibid., 23.
21. Ibid., 25.
22. R. W. Southern, "Aspects of the European Tradition of Historical Writing: The Sense of the Past," *Transactions of the Royal Historical Society,* 5th ser., 23 (1973): 127.

23. Jules Michelet, *Tableau chronologique de l'histoire moderne*, in *Œuvres complètes*, ed. Paul Viallaneix, 21 vols. (Paris: Flammarion, 1971–82), 1:64, 71.

24. Given by Michelet as an appendix to his *Histoire du XIXe siecle*, in *Œuvres complètes*, 21:655.

25. George Kubler, *The Shape of Time* (New Haven: Yale University Press, 1962), 1–2.

26. Ibid., 1.

27. Simone Weil, *First and Last Notebooks*, ed. and trans. Richard Rees (London: Oxford University Press, 1970), 328.

Contributors

R. H. Barnes, a University Lecturer in Social Anthropology at the University of Oxford, is the author of a monograph on a people of eastern Indonesia entitled *Kédang* and is completing another on Lamalera, the subject of his essay in this volume.

Bernard S. Cohn, an India specialist and a friend of *CSSH* since its inception, has been exploring the borderland between what he calls Historyland and Anthropologyland as long as *CSSH* has—most recently in his collection *An Anthropologist among the Historians and Other Essays*. He is in the Departments of Anthropology and History at the University of Chicago.

Nancy M. Farriss, Walter Annenberg Professor of History at the University of Pennsylvania, is author of *Maya Society under Colonial Rule: the Collective Enterprise of Survival* and many other writings on Mayan history.

Anthony T. Grafton, Professor of History at Princeton University, has published extensively on the history of scholarship and education in Renaissance Europe. His most recent book is *Forgers and Critics: Creativity and Duplicity in Western Scholarship*.

Diane Owen Hughes, a specialist in the history of medieval and renaissance Italy, has written on a variety of cultural topics from kinship patterns to clothing codes and is finishing a book entitled *Inexpressible Grief: Restraint and Display in Premodern Italy*. She is a member of the History Department at the University of Michigan.

Peter Hughes is Professor of English at the University of Zurich. His area of research is intertextual relations in literature and culture. Recent publications include a book on V. S. Naipaul, studies of allusion, and a work-in-progress on intellectual biography in Vico, Coleridge, and Michelet.

Maria Minicuci is director of the Istituto di Storia dell'Arte and Professor of Anthropology in the University of Messina. One of the Calabrian villages she compares in her article for this volume is the subject of a major anthropological study of its occupants and its emigrants to Argentina, entitled *Qui e altrove: Famiglie di Calabria e di Argentina.*

Peter Rigby is a citizen of Uganda and a member of the Anthropology Department of Temple University. His research on the Ilparakuyo Maasai eventuated in his 1983 *CSSH* article reprinted here and in two books: *Persistent Pastoralists: Nomadic Societies in Transition* and *Cattle, Capitalism, and Class: Ilparakuyo Maasai Transformations.*

Thomas R. Trautmann of the University of Michigan holds appointments in the Departments of History and Anthropology, specializing in ancient India. He is writing a book on the British Sanskritists and their ethnologies, called *Aryans and British India.*

Jonathan Wylie, a graduate of Reed College and Harvard University, has done ethnographic field work in fishing villages in the Faroe Islands and Dominica. Twice a Fulbright scholar, he has taught anthropology both in this country and in Denmark. He is currently affiliated with the Anthropology/ Archaeology Program at the Massachusetts Institute of Technology.

Index

Acosta, Joseph de, 143, 144, 155
Africa, 10, 13, 15, 61, 202, 210, 212, 238
  concepts of time in, 206, 209–10, 234
  philosophy in, 203, 234, 236, 238, 239
Alberti, Leon Battista, 140–41
Álvabøur, 31–32, 41–51, 52, 53, 56, 61, 63
  founding of, 42–44
  genealogy in, 32, 41
  history of, 42–49
  people of, 32, 54
  religion in, 49–50
Alphabetic script, 108, 110, 130
Althusser, Louis, 213, 214, 238
Americans. *See* United States
Anaximander, 269
Annius of Viterbo, 154–55
Anthropology, 1, 2, 4, 7, 10, 15, 16, 29, 201, 202, 206, 210, 211, 234, 235, 236, 238, 240, 245, 247, 262, 263, 270, 275, 283, 288
  birth of modern, 1, 269, 270
  consciousness of, 7
  discourse of, 3, 15
  French, 209
  imagination in, 203
  Marxist, 234, 239
  physical, 10
  positivist, 235, 239
  social, 10, 203
  structural, 1
Arab, 180, 189
Arabization, 208

Archaeology, 45, 155, 185, 287
Argentina, 8, 72, 74, 80, 86, 87, 89, 99
Aristophanes, 145
Aristotle, 7, 148, 149, 283
Arya Samaj, 26, 27, 28, 29
Ashanti, 208
Asia, 177, 180, 182, 284
Astrology, 113, 114, 131, 189
Astronomy, 9, 12, 132, 140, 145–46, 158, 170, 184, 189
Augustine, St., 1, 2, 109, 148, 155, 177, 191, 274, 275
  view of time, 177
Authority, 117, 153, 180, 186, 187, 193
  Indian concepts of, 188
  municipal, 118
  political, 45
  religious, 179, 187
  royal, 128
Autobiography. *See* Biography
Awareness, 4, 72, 86, 99, 208, 274, 276, 285
  historical, 14, 15, 86, 288
  of historical change, 208
  of sequence, 282
  temporal, 2, 5, 7, 15
Aztec, 143, 144, 149
  calendar, 149
  religion, 144–45

Bali, 2, 15, 112, 115, 131, 248, 262
Barbaig, 212
Barnes, R. H., 53

293

Bateson, Gregory, 228
Beckett, Samuel, 275, 288
Beidelman, Thomas, 206, 207, 208
Benjamin, Walter, 270
Berosus, 154, 159, 164
Bible, 50, 144, 148, 152, 155, 156, 157, 164, 176, 177, 178, 180, 181, 183
Biography, 81, 119, 126, 189, 190, 277, 282, 283
al-Biruni, 179
 view of Indian time, 178
Bloch, Maurice, 2
Bloch, Marc, 5, 246, 271, 272, 273, 274, 283, 287, 288
Bodin, Jean, 144, 152–55
 ideas of progress, 154
Bohannan, Paul, 206, 208, 210
Boucher de Perthes, 14
Bourdieu, Pierre, 101, 102, 212, 226–27
Brahmans, 24, 25–26, 27, 29, 180, 181
Brazil, 270
Brixham Cave, 14, 178, 185, 192
Buddha, 183, 187
Buddhism, 171, 186, 187, 188, 189
Burckhardt, Jacob, 139
Busbecq, Ogier, 143

Calabria, 7, 9, 71, 72, 73, 79, 86, 100, 101
Calendar, 3, 111–14, 117–19, 122–23, 130, 132, 139–40, 262
 agricultural, 150
 ancient, 143, 149
 Aztec, 143, 148–51
  European criticism of, 148–51
 ecclesiastical, 140, 141, 149
 European, 122
 Greek, 149
 Gregorian, 217, 262
 Indian, 150
 Javanese, 262
 Julian, 123, 143, 149
 lunar, 201
 Mayan, 110–12, 125–26, 130

 Mesoamerican, 151
 reform, 142
 religious, 247
 ritual, 248
 solar, 145
 study of, 145
 western, 150, 245
Calvisius, Seth, 147, 148
Canek, 126–29, 133
Capitalism, 208, 227, 228
Caribbean, 31, 32
Cartesianism, 274, 280
Casaubon, 158
Casse, 31–32, 35–40, 51, 58–66
 genealogy in, 32
 landed elite in, 59–60
 reinvention of the past in, 38
 view of race, 35–36
Caste, 11, 23, 24, 25, 27, 29
 mythological origin of, 24
Cato, 154
Certeau, Michel de, 236
Chamars, 24–25, 29
 origin myth of, 24–25
Change(s), 4, 5, 9, 21, 72, 123, 173, 207, 262, 274, 285
 economic, 25, 55, 61
 historical, 71, 208
 laws of, 212
 pattern of, 4
 recognition of, 5
 social, 51, 55, 61, 211, 261
 technical, 11
 temporal, 2, 226
 and time, 177
Chaos, 12, 13, 115, 123, 145
Chilam Balam, books of, 110, 111, 115, 117, 120, 121, 122, 124, 126, 129, 131, 132
China, 108, 146, 174, 178, 181, 186
Christ, 13, 113, 141, 144, 176
Christianity, 61, 113, 176, 177, 178, 182, 247, 248, 257. *See also*, Church
 conversion to, 124, 125, 243, 244
 missionaries of, 49–50, 51, 111, 121, 185, 249–50

moralizing of, 257
movable feast in, 151
sense of time in, 142
understanding of history in, 145
Chronicle, 111, 116, 121, 126, 132, 260
   Andean, 120
   Byzantine, 158
   clan, 11
   of Eusebius, 144, 146, 148, 158, 159, 184, 193
   Mayan, 111, 116–17, 121 (*see also* Chilam Balam)
   Medieval, 146
   Mesoamerican, 14
   of Moses, 10
   Nuremberg, 146
   world, 146
Chronology, 1, 3, 14, 111, 112, 117, 139, 144, 146, 151, 155, 157, 158, 161, 193, 207, 217, 218, 219, 220–21, 222, 224, 228, 243, 256, 262, 275, 278. *See also* Time
   biblical, 10, 144, 176, 177, 188, 192
   Christian, 145, 146, 248
   chronometric order of, 3
   controversies over, 148
   European, 157
   European transformation of, 145
   external, 14
   Greek, 156
   Hindu, 188, 191, 192
   historical, 259, 270
   Indian, 184
   linear, 117, 118
   links secular and sacred, 145–46
   long term, 144
   mark of civility in, 143, 147
   merits of, 148
   modern, 3, 153, 160, 256
   Mosaic, 10, 184, 185
   Neapolitan, 280
   power of, 141
   providential, 10
   reconciliation of biblical and Indian, 181–86
   Renaissance, 12, 160
   sequential, 112
   technological, 142, 145
   topological, 223–24
   used to attack Christianity, 178
   world, 184
Chronologers, 147, 148, 151, 152
Church, 74, 75, 84, 85, 123, 148, 245
   Assembly of God, 49–50
   Catholic, 141, 142, 243, 244, 263, 280, 286
   Fathers of, 109
   Mass, 113
   Lutheran, 49
Cicero, 282
Civilization, 143, 183, 186
   European, 172
   non-European as embodiment of pre-modern, 167
   "superiority" of, 172
   Indian, 172, 173, 185
Clock, 3, 12, 139, 140, 151
   effect on European culture, 142
   index of time, 12
   Renaissance, 12
   of Strassbourg, 12, 139, 141, 147
Coleridge, Samuel Taylor, 271, 273, 276–79, 281, 283–84, 286
Colonialism, 235, 273. *See also* England
   British, 10
   European, 208
   Spanish in Mesoamerica, 110, 112, 116, 118, 120–30
Columbus, 10, 13, 146
Comestor, Petrus, 157, 159–60
Commoditization, 227, 228
Communication, 60, 108, 236
   mass, 29
   non-literate modes of, 108
   oral, 108
Copernicus, 10, 140
Cosmogony
   Hindu, 11
   Judeo-Christian, 11

Cosmology, 13, 107, 114, 121, 125, 262
  Christian, 122
  historicist, 113
  modern scientific, 176
  of time, 210–11
Cosmos, 10, 109, 114, 116, 130, 176
Creation, 122, 141, 144, 158, 168, 176, 275, 279
Culture(s), 2, 5, 8, 22, 32, 39, 52, 53, 55, 56, 61, 62, 64, 65, 73, 101, 121, 130, 145, 194, 202, 206, 207, 228, 239, 246, 247, 276, 288
  biblical, 178
  black, 36, 40
  definition of archaic and non-progressive, 167
  folk, 84
  heroic, 281
  literate, 110, 243
  local, 35, 36, 57, 59, 60, 73
  modern, 236, 288
  nonwestern, 143
  primitive, 236
  shock, 35
  western, 14, 108, 227, 237
  white, 36
Cuneiform, 177
Cycle
  astronomical, 149
  conception of, 112
  lunar year, 216
  structure of, 111
  of nature, 173
  pattern, 116, 117, 118, 119, 121
  of rebirth, 172
  of renewal, 247
  ritual, 123
  seasonal, 82, 206
  temporal, 225
Cyclical thought, 109
Cyclicity, 113–14, 176, 246

Darwin, Charles, 10, 14
Ibn Daud, 164
Davis, John, 102, 247

Dasypodius, Conrad, 139–40, 141, 142, 147, 161
De Man, Paul, 272
Denmark, 31, 32, 42, 52–55, 61, 62
Devil, 59, 144, 149, 150, 151
Diachrony, 117, 120, 208, 211, 212, 213
Discourse, 71, 208, 283
  anthropological, 3, 15, 204, 236, 240, 269
  ethnographic, 4
  historical, 3, 9, 240, 269
  Marxist, 240
  temporal, 4, 9, 270
Dominica, 31, 35, 37, 39, 40, 51, 56–64
  history of, 33–34, 36–37
Dohrn-Van Rossum, Gerhard, 141
Douglas, Mary, 203
Dow, Alexander, 179–81
Duran, Diego, 143, 150
Durkheim, Emile, 205

Egypt, 151, 158, 176, 177, 178, 185, 285, 287
Eliade, Mircea, 109, 167, 190
Engels, Fredrick, 209, 229, 239
England, 31, 33–38, 56–57, 60–64, 173, 182
  conquest of Bengal, 167, 171, 179
  rule in India, 173
Enlightenment, 5, 11, 14, 15, 108, 178, 179, 269, 270, 273, 280, 284, 287
Epic, 11, 170, 187, 219, 286. *See also* Homer, Mahabarata
  Homeric, 270
  Indian, 22–23, 187
Erasmus, 154
Eschatology, 113
  Christian, 210
Eternity, 12, 114, 115, 140, 148, 171
  Greek ideas of, 175
  historical definition of, 148
Ethnology,
  Mosaic, 180, 182, 183, 184, 193

revolution in ethnological time, 185
structural, 9
Etymology, 270, 278, 285
as history without past, 275
Euhemerism, 159–60
Euhemerus, 157
Europe, 36, 40, 63, 73, 110, 121, 139, 142, 144, 172, 177, 178, 181, 184, 185, 193, 273, 280, 281. *See also* Enlightenment, Reformation, Renaissance
civilization, 61
consciousness, 171
Early Modern, 13
EEC, 52
Medieval, 130, 141, 142, 146
Eusebius. *See* chronicle
Evans-Pritchard, Edward, 203, 206–8, 227
Events
ceremonial, 201
chain of, 108, 129
cyclical, 208
demographic, 92
economic, 201
historical, 2, 3, 4, 73, 85, 110, 189, 218, 260
in human history, 159
legendary, 3
meteorological, 101
mythical, 3
narration of, 88
natural, 51
patterns of, 4, 115
secular, 117
sequence of, 6, 16
temporal, 1
unrepeatable, 109, 112
Evil, 13, 14, 66, 115, 116, 285
randomness of, 13, 115 (*see also* Chaos)

Fabian, Johannes, 4, 10, 15, 167, 229, 235, 236, 269
Fabricius, Laurentius, 146
Farriss, Nancy, 246

Faroe Islands, 15, 16, 31, 32, 41, 46–53, 59–63, 65
Fentriss, James, 9
Feuerbach, 206, 212
Finnegan, Ruth, 265
Firth, R., 71
Fitili, 7, 8, 72–102;
genealogical memory in, 88–89
marriage practices in, 75–77, 92, 102
naming system, 91
view of past, 81–82
Flood, 144, 158, 184
Folk culture, 22, 52, 53, 55, 56, 59, 60, 62, 65, 84, 87, 270, 278
Fortes, Meyer, 208
Fosbrooke, Henry, 201–2, 218
Foustel de Coulanges, Numa, 281
Fox, James, 260
France, 5, 33, 36, 37, 60, 61, 64, 273, 284, 285, 288
Fall of, 5, 9, 270, 271, 273
Vichy, 272
French Revolution, 5, 139, 184, 270, 272, 277, 278, 283, 285
Freud, Sigmund, 238
Functionalism, 206, 235
Fundamentalists, 184
Future, 37, 58, 72, 99, 107, 110, 129, 141, 177, 208, 210, 215, 225, 226, 227, 272, 285, 287, 288
event, 204
prediction of, 150
sense of, 8

Galen, 146, 148
Gandhi, Mahatma, 11, 27, 28, 29
Garibaldi, Giuseppe, 86
Gayo, 260
Geertz, Clifford, 2, 15, 227, 246, 262
Genealogy, 7, 8, 11, 25, 32, 37, 41, 45, 51, 55, 73, 75, 77, 78, 83, 88, 89, 90, 91, 93, 95, 96, 99, 101, 102, 111, 112, 116, 117, 126, 128, 157, 183, 208, 217–22, 224, 256, 259, 260, 270, 283
Germany, 15, 271, 272

Giannone, Pietro, 279
Gibbon, Edward, 6, 155, 273, 279, 280, 281, 282, 283, 286
God, 32, 37, 39, 51, 57, 61, 84, 98, 101, 154, 164, 177, 178, 274
God(s), 117, 118, 156, 170, 187
Goethe, Johann Wolfgang von, 281
Gogo, 222
Gomara, Francisco Lopez de, 143
Grainville, 286–87
Greece, 108, 141, 145, 146, 151, 154–60, 172, 176, 185, 186, 189, 193, 272. *See also* Language(s), Religion, Time
 cyclical sense of time, 175
 knowledge of India, 174–75
 notions of eternal world, 176
 paganism of, 145, 185
Grenada, 35
Guadeloupe, 33, 35, 39, 40

Habermas, Jürgen, 237–38
Haguelon, Pierre, 148
Halhead, Nathaniel, 179, 181–82
Hallowell, A. I., 227
Hartley, L. P., 16
Harvey, David, 238–39
Harvey, Gabriel, 153
Hawkings, S., 247
Hegel, G. W. F., 167, 172, 173, 178, 185, 190, 206, 212, 213, 228, 238, 282
Herder, J. G., 269
Heresy, 180
Hermeneutics, 147
Herodotus, 154, 156, 157
Hieroglyph, 5, 110, 112, 177
Hinduism, 10, 27, 28, 168, 171, 180, 185–87
 antiquities of, 179
Hippocrates, 148
History
 ancient, 142, 156
 biblical, 145, 160, 176
 clan, 244, 253–57
 colonial, 260

 conflation of, 143
 consciousness of, 4, 111, 130, 201, 219, 223, 224–25, 239, 261
 construction of, 121
 creation of, 73, 77
 criticism of, 154, 157, 160, 161
 cyclical view of, 129–30
 discourse, 3, 9
 end of, 146
 family, 83, 97
 folk, 260
 genealogical, 206
 human, 10, 13, 14, 159, 175, 178, 185
 idea of, 71
 ideological conception of, 214
 knowledge of, 79
 lack of, 172, 173
 law of progress view of, 171
 local, 78, 79, 87, 112, 224, 244, 248
 long term, 144
 maya, 117, 125, 126
 meaning of, 148
 methods of, 287
 medieval, 142
 modern, 140, 286
 Mosaic, 19
 national, 183
 natural, 178
 Near eastern, 147
 oral, 45, 117, 212, 265
 periodization of, 163
 personal, 38
 philosophy of, 177
 political, 119
 providential, 145, 164
 recent, 142
 sense of, 243
 universal, 183
 views of, 130, 285
 western concepts of, 116
 world, 142, 145, 147, 288
 writing of, 282
Historiography, 38, 55, 61, 65, 161, 217, 202, 218
 bourgeois, 207

oral, 55
western chronological, 260
Hollis, A. C., 216–17
Homer, 160, 270, 272, 278, 281
Hoskins, Janet, 16
Howe, Leopold, 2, 112, 246–47
Hubert, H., 243, 247, 248, 261
Huet, P. D., 160
Hume, David, 205
Hyppolite, Jean, 228

Iceland, 52
Identity, 2, 80, 90, 96, 97, 99, 102, 225, 226
 collective, 52
 ethnic, 80, 98
 regional, 79, 80
 national, 79, 80
Ideology, 1, 10, 21, 113, 122, 205, 213–14, 217, 227, 228, 278
 capitalist, 210
 millenarian, 131
Ilarusa, 223
Ilparakuyo, 201, 202, 210, 215–17, 218, 219, 220–21, 222, 223, 224, 225, 226, 227, 228, 229, 234, 235, 239, 240
 ceremonies, 225
 notions of temporal units, 215–17
Imagination, 149, 274, 277, 286, 287
India, 7, 10, 11, 13, 21, 24, 28, 29, 108, 170–75, 179–86
 civilization, 167
 culture, 11, 167, 171
 divisions of time, 167–71
 knowledge of Europe, 182
 learning in, 189
 Persian history of, 179
 view of history, 181
 views of time, 167
Indonesia(n), 2, 16, 243, 244, 257, 259. *See also* Bali, Lamalera
Iraqw, 206, 212, 217, 228
Isampur, 223
Isidore, St., of Seville, 157

Islam, 24, 27–28, 109, 177–78, 180, 193, 247
 Thakurs versus, 23
 understanding of history in, 145
Israel, 109
Italy, 71, 72, 73, 74, 79, 86, 140, 279, 280, 281, 284

Jainism, 171, 186, 187, 188, 189, 191
Jamaica, 35
Jerome, 146, 148, 159
John, St., 275
Jones, Sir William, 179, 182–84
Joyce, James, 275
 nightmare of history, 270
Judaism, 142, 145, 146, 157, 178, 180

Kaguru, 205–8, 222
Kalpa, 169, 170, 175
Kant, Immanuel, 2, 212
Kambyle, 226
Katun, 10, 13, 110, 111, 113, 117, 118, 119, 120, 121, 123, 124, 125, 127, 130, 131, 132
Kedang, 246, 247, 252, 259, 262
Kenya, 201, 225, 227
Keraf, Gregorius, 244, 262
Kimbu, 222
Kinship, 28, 45, 47, 65, 73–76, 88–89, 91–99, 208
 area in Zaccanopoli and Fitili, 92–95
 degrees of, 46
 marriage, 49, 73, 75, 89, 91, 94, 95, 96, 98, 99, 100, 101, 102, 250, 255, 256, 263
 structures, 99, 100
Kingship, 119, 124
Kisongo, 222, 223
Koran, 28
Kubler, George, 287, 288

Labatwi, 32–34, 36, 38, 39, 40, 60, 63
Lamalera, 2, 243, 244, 247, 248, 250, 251, 254, 255, 259–65
Language(s), 14, 15, 35, 36, 39, 52, 54, 55, 56, 64, 80, 81, 110, 116, 117,

Language(s) (*continued*)
130, 177, 204, 243, 244, 270, 274, 277
Arabic, 171, 178
Bantu, 210, 217, 222
Bengali, 181
Celtic, 184
Danish, 31
English, 23, 31, 36, 64, 279
Faroese, 31, 52–54, 54
and political independence, 53
French, 31, 33
Gothic, 184
Greek, 154, 155, 172, 184
Hebrew, 171
Indo-European, 184
Kalenjin, 223
Lamaholot, 243, 247, 251, 252, 255, 259
Latin, 146, 156, 159, 184, 276
Maa, 201, 202, 215, 216, 217, 222, 223
Patois, 31, 33, 36, 58, 59, 62, 63, 64
Persian, 171, 179, 181, 184
Sanskrit, 171, 173, 174, 179, 180–85, 189, 194
Last Judgment, 113
Law(s), 1, 14, 40, 63–64, 143, 189, 190, 269, 275
Koranic, 28
process of, 40
literature of, 189
Roman, 278
Le Clerc, Jean, 280, 281
Leach, Edmund, 10, 71, 209, 228
Lefebvre, Henri, 212–13
Lembata, 2, 244, 246, 252, 253, 263, 264
Lenin, V. I., 209
Lévi-Strauss, Claude, 5, 8, 14, 269–70, 288
Lévy-Bruhl, Lucien, 203, 204–6, 208, 209, 210, 214, 227
Lineage, 47, 111, 118–20, 222, 260
Thakur, 23, 36

Linear,
progression, 112, 131
thought, 109
Linearity, 246
Linguistics, 189, 193, 225, 270
Literacy, 4, 27, 29, 85, 86, 107, 108, 110, 129, 130, 243
rate of, 110
London, 34, 56, 61, 285, 288
Luther, Martin, 154

Maasai, 10, 15, 201–2, 215–40
conceptual time of, 202
historical time linked to spatial, 226
notions of temporal units, 201–2, 215–17
social formation, 218, 226, 235
MacCannell, Dean, 236–37
Machiavelli, 153
Madagascar, 260
Magic, 87, 247, 262
Magus, Olaus, 151
Mahabarata, 11, 22, 171, 179, 180, 187, 189
Mahayuga, 169, 170
Malayo-Polynesian world, 260
Malinowski, Bronislaw, 261
Manetho, 158, 164
Manu, 168–71, 188, 189
Manvantara, 170
Martinique, 33, 35, 37
Marx, Karl, 2, 10, 15, 167, 173, 185, 205, 206, 209, 212, 213, 239
Marxism, 10, 209, 234, 235, 239
Mathematics, 189, 190
Matthew, St., 190
Mauss, Marcel, 205
Maya, 9, 12, 13, 14, 107, 110–32, 248
hieroglyphic writing, 110, 121
language, 110
perception of time, 110–16
religion, 114, 116, 118–19, 123–24
time, 128
Mayer, A. C., 102
Mbiti, John, 209, 210

Melanchthon, Philip, 147–48
  on importance of chronology, 147
Memory, 2, 4, 7, 8, 14, 16, 50, 61,
  71–73, 77–79, 81, 86–90, 93, 95,
  97–100, 107, 116, 117, 141, 144,
  246, 247, 274, 275, 277, 278, 283,
  284
  genealogical, 9, 78, 80, 81, 83, 84,
    88, 89, 92, 95
  historical, 83–86
  human, 119
  power of, 5, 237
  social, 9, 83
Mentality, 203
  civilized, 203
  primitive, 203, 204, 205, 208
Merton-Williams, Peter, 211–12
Merker, M., 216
Mesoamerica, 13, 108, 110, 112–16,
  121
Mexico, 7, 107, 111, 121, 127, 128,
  133, 143
  Spanish Conquest of, 124
Michelet, Jules, 271, 276, 282, 283,
  285–88
Middle Ages, 140
Migration, 60, 72, 74, 75, 78, 111,
  117, 224, 226, 251–56, 259, 265
Milan, 8, 142
Mill, James, 167, 172, 173–74, 177,
  185
  attack on non-European time, 178
Millenarian movements, 126, 131
Miller, Arthur, 119
Milton, John, 278
Mitterand, François, 273
Moctezuma, 124, 128–29
Modernity, 5, 14, 140, 160–61, 167,
  185, 186, 191, 237, 243, 273, 281
  Europe's experience of, 139
  master narrative of, 167, 185, 186,
    191
Modernization, 10, 21, 22, 25, 26, 73,
  157, 235, 243
  of ancient world, 157
  process of, 14, 29

Mol, Frans, 216, 218
Monotheism, 185
Montaigne, Michel de, 143, 153
Montesquieu, Baron de, 269
Montezuma. *See* Moctezuma
Morton-Williams, Peter, 213
Motolinia, Toribio de, 148, 149–51
  analogy of Indian and classical calendar, 148–49
Mtango, 222
Mudimbe, Valentin, 234–39
Mumford, Lewis, 228
Munz, Peter, 3
Myth(s), 8, 9, 24–25, 33, 36, 38, 51,
  84, 129, 133, 146, 156, 157, 160,
  207, 208, 218, 248, 251, 261, 278,
  279
  age of, 158
  concept of, 8
  cyclic, 9
  discourses of, 9
  of genealogical accounts, 7
  as history, 159, 160
  influence on daily life, 243
  lack of, 217
  mythical expectation, 124
  of origin, 7, 24, 84, 111
  served as history, 46
  as sources of knowledge, 9
  strange time of, 1
  taxonomy of, 270
  topography of relationships, 8–9
  understanding of, 9
Mythology, 10, 22, 24, 38, 187, 206,
  235, 243, 251, 285
  Greek, 159
  Indian, 24, 174

Naples, 274, 277, 280, 282, 286
Napoleon, 284, 286–87
Narrative, 1, 3, 4, 5, 6, 88, 100, 167,
  257, 259, 275–78, 283, 287, 288
  analysis, 1
  anthropological, 2
  ethnographical, 5, 6, 256
  historical, 2, 4, 5, 6, 14, 288

Narrative (*continued*)
  Indian Creation, 168–71
  of modernity, 167, 185, 186
  Mosaic, 182, 183
  of progress, 167
  temporal structure of, 282
Nationalism, 11, 31, 48, 50–52
  movements of, 27–28
Nazism, 272
Needham, Rodney, 203, 205, 209
New Spain, 150
New Year Ceremonies, 118, 123
New World, 178
Nietzsche, Friedrich, 287–88
Nigeria, 206
Norway, 42, 52, 53, 54, 65
Nuer, 206–7, 217, 239
  time according to, 206–7
Nusa Tenggara Timur, 246

Old Testament, 109, 121
  linear theodicy of, 113
Orality, 4, 51, 79, 91–92, 108
Order, 5, 7, 8, 120, 130
  chronometric, 3
  cosmic, 13, 109, 112, 114, 115, 118, 121, 123
  divine, 145
  economic, 227
  of events, 246
  historical, 10, 148
  of language, 275, 277
  moral, 66
  of nature, 32, 50, 52, 149
  political, 115, 118
  sequential, 1, 111
  social, 11, 32, 51, 61, 127, 227
  of sounds, 275
  spatial, 10, 15
  temporal, 15, 127, 270
  of things, 39
Orientalism, 167, 171, 172, 173, 179, 182, 184, 191
  Indian-centered, 172
Ottoman Empire, 143, 153, 154, 284
Ovid, 141

Palmyra, 287
Parcus, David, 155–56
Paris, 285, 286
Past(s)
  as another country, 16
  classical, 157
  construction of, 72
  continual reinvention of, 38
  cyclical view of, 123, 127
  distinct from present, 153
  ethnological, 283
  European notions of, 120, 122
  Hindu fathomless, 180, 187
  historic, 11, 22–29, 245
  human, 141
  Judeo-Christian system of, 120
  legendary, 23, 24, 26
  local historic, 22
  "modern," 11
  moral judgments based upon, 23, 55
  narrative of, 6–7, 11
  plurality of, 29, 244
  and primitive, 273, 287
  representation of, 186
  sense of, 8
  traditional, 11, 22–29
  uses of, 152
Persia, 180, 284
Petavius, 158
Petrarch, 146, 159
Phenomenology, 239
Philology, 14, 51, 274, 281
  Renaissance, 160
Philosophy, 3, 8, 203, 208, 209, 210, 213, 229, 238, 269, 288
  Christian, 190
  classical, 205, 206
  Greek, 229
  of history, 177, 237
  Indian, 26, 27
  moral, 152
  western, 208
Pitt, William, 284
Plato, 149, 155
Pliny, 148
Postmodernism, 188, 238

Pound, Ezra, 277
Power, 117, 118, 120, 153, 235, 236
  divine, 127
  of past over present, 238
  political, 25
  shift to West, 285
  spiritual, 24
Progressivism, 123, 167, 185, 190
Prophecy, 107, 110, 124–26, 133, 217
  apocalyptical, 141
  of Daniel, 153, 156
  historical, 121, 124, 129
  katun, 124, 125, 127

Racine, Jean, 284
Radcliffe-Brown, Alfred, 211
Radicalism, 278
Ramayana, 11, 22
Ram Lila, 22–23, 26
  description of presentation, 22–23
Ranke, Otto, 1
Ray, Benjamin, 210
Reformation, 144
  effect on ecclesiastical time, 142
Religion, 10, 26, 28, 122–23, 152, 167, 170, 187, 217, 224, 243, 247, 259, 262
  natural, 185
Renaissance, 2, 139, 140, 144–46, 157, 161, 172
  chronologers, 149
  English, 152
  French, 152
  Oriental, 172, 185
  science, 142
Repetition, 6, 72, 113, 114, 115, 246, 248, 277
  of the past, 225
  repetitiveness, 100, 123, 124
  sequential, 5
Revolution, 272, 280, 286
  historical, 153, 270
  Industrial, 139
Ricoeur, Paul, 7, 115, 237, 275–76, 278, 288

Ritual(s), 2, 26, 27, 28, 96, 100, 122, 123, 130, 223, 261
Rolewink, Werner, 146
Romanticism, 276, 279, 283, 285
Rome, 141, 146, 154, 157, 158, 185, 270, 280–81, 284, 286
  fall of, 5, 6
Roseau, 31, 33, 36, 59, 60
Ruin(s), 5, 10, 269, 270, 272, 273, 274, 276, 285, 286, 288
Rulership, 111, 120, 128, 132
  cyclical model of, 119, 120

Saga, 54
  Icelandic, 42, 53
Sahagún, Bernardino de, 149–50
St. Thomas (island), 35
Sampur, 225
Samsara, 187
Sankan, S. S. ole, 216
Scaliger, Joseph, 143, 147–48, 154–61
  defense of comparative chronology, 156
  reconciliation of biblical and classical history, 156
Scandinavia, 52, 54, 61, 66, 140
Scholar(s), 25, 139, 155
  Arabic, 179
  medieval, 157, 159
  modern, 158
  Persian, 179
  Renaissance, 151, 158, 161, 281
  western, 11
Science, 181, 193, 212, 274, 288
  Greek, 176
  Indian, 189
  modern, 141, 176
  occult, 39
  scientific revolution, 108
Secularization, 108
Semite, 108, 109
Senapur, 11, 12, 21–28
Sequence, 1, 4, 5, 6, 8, 114, 246, 256, 276, 288
  awareness of, 282
  chronological, 89

Sequence (*continued*)
  ethnological, 5
  historical, 4, 5–6
  linear, 111, 112, 116, 119, 121, 123, 247
  long term, 117, 132
  narrative, 277
Shastras, 26
Shelley, Percy Bysshe, 285, 287
Shils, Edward, 21, 29
Sidney, Philip, 152
Simone, Claude, 288
Smith, Jonathan Z., 8
Socialism, 273
Society, 4, 14, 16, 21, 25, 27–29, 39, 52, 62, 64–66, 71, 72, 75, 94, 98, 100, 108–20, 129, 139, 171, 212, 219, 224, 226, 228, 237, 255, 261, 265
  Afro-Caribbean, 32
  agrarian, 113
  archaic, 190
  Asiatic, 173
  bourgeois, 211
  capitalist, 217
  categorization of, 21, 29, 88
  colonial, 61
  construction of, 51
  effects of history in, 83
  egalitarian, 46
  formation in, 202, 207–19, 224–29
  human, 274
  inferior, 203
  local, 47, 51, 61
  mercantile, 140
    importance of time in, 140
  modern, 14, 21, 29, 167
    definition of, 29
  non-western, 205
  organized, 143
  precapitalist, 228
  primitive, 15, 155, 190, 210, 246
  Scandinavian, 32, 55
  static, 71
  traditional, 21, 28, 29, 71, 73, 167, 190
  tribal, 71
  village, 11
  western, 173
Sociology, 3, 15, 205, 206, 262
  comparative, 152
Southern, R. W., 5, 285
Space, 2, 8, 10, 15, 71, 79, 86, 87, 98, 100, 101, 112, 127, 151, 205, 224, 228, 235, 236, 238, 239, 284
  Copernican, 10
  social, 15
  spatial relations, 226, 284
  spatial relations in language, 204
  spatial-temporal relations, 2
Spain, 110, 112, 116, 120, 121, 122, 123, 124, 143, 272
  conquest of Mesoamerica, 113, 121
  ideas of history, 128
  rule in Mesoamerica, 125, 126
  revolt against, 126–29, 133
Stephens, Walter, 154
Structural totalization, 203, 208
Structuralism, 9, 107, 173, 209, 211, 212, 214, 228, 236, 270
Symmetry, 164
Synchrony, 209, 211, 212, 213, 214
Syria, 285, 287

Talensi, 208
Tanzania, 201, 206, 219, 222, 226, 227
Tatian, 148
Taxonomy, 1, 4, 273
Technology
  modern, 141
Teleology, 108–10
Telis, 24, 27–28
Temporality, 1, 5, 7, 12, 240, 269, 273, 274, 275, 276, 287, 288
  alternate, 16
  anthropological, 8
  definition of, 4
  historical, 8
  human, 9
Thakur, 11, 21, 22–25, 26, 27, 28, 29
  genealogical connections of, 23–24
Theodicy, 109, 113

Theology, 9–10, 185, 187, 193, 203
Thompson, E. P., 227
Thornton, Robert, 206, 212
Thucydides, 154, 155
Time
 anthropological, 14, 15
 awareness of, 5
 beginning of, 189
 biblical frame, 10, 145, 177, 178, 179, 180, 185
  defense of, 177
 "big," 11
 biographical, 16
 calendrical, 160
 Christian, 123, 141, 167, 176 (*see also* Chronology)
 chronometric measure of, 6, 279
 civilized, 204
 computation, 146
 concept of, 117, 204, 238
 consciousness of, 3, 16, 139, 142, 174, 176
 constructing of, 9
 cosmic, 12, 114, 115, 116, 117, 118, 119, 121, 124, 206
 cyclical, 3, 80, 100, 107, 108, 109, 110, 112, 113, 115, 117, 119, 120, 121, 123, 124, 125, 126, 129, 130, 171, 172, 175, 176, 178, 246, 247
 dual conception of, 120
 durational, 2
 ecological, 206
 effect on modern culture, 141
 end of, 82, 146
 ethnographic notions of, 275
 ethnological, 5–6, 13, 14, 185, 274
 European fascination with, 141, 144
 experiencing, 13, 245
 genealogical, 206, 207, 208
 historical, 3, 5–6, 13, 16, 84, 100, 113, 115, 116, 117, 118, 119, 120, 145, 157, 182, 206, 210, 211, 213, 214, 226, 274, 276, 278
 human, 10, 12, 115, 171, 178
 Indian, 11, 167–71, 172, 174–86, 189, 193
 European views of, 171–74, 179–86
 irreversible, 10
 legendary, 3
 linear, 2, 3, 4, 80, 100, 107, 108, 110, 111, 112, 113, 114, 118, 119, 120, 121, 131, 176, 178, 246, 247
 local, 11, 14
 Mayan, 128
 measureless, 13
 measuring of, 12, 13
 misuses of, 235
 modern, 3
 mythical, 84, 160, 209, 215
 nothing geographical about, 209
 pagan views of, 177
 passage of, 3–4, 110
 perception of, 16
 power of, 5, 139
 problem of, 1–2, 148
 process of, 2–3
 profane, 12
 prophetic, 10
 providential, 10, 279, 280
 Renaissance, 12
 repetitive, 10
 representation of, 16, 245
 ruins of, 273
 sacred, 12, 146
 secular, 141, 146
 sense of, 31, 32, 36
 sequential, 4
 shamanic, 13
 structural, 3, 10, 206, 208
 universal, 144
 western sense of, 2–3, 10, 171–76
Timelessness, 114
Trithemius, Johannes, 157
Tiv, 206, 208, 210
Tovar, Juan de, 143, 144
Tradition, 12, 28, 29, 58, 71, 72, 112, 139, 170, 190, 207, 237, 260, 263, 264
 anthropological, 204
 biblical, 180
 Brahman, 25–26, 27, 28

Tradition (*continued*)
definition of, 72
historical, 178, 222
Indian, 25, 193
   flood narrative, 183
   sacred, 22, 25
intellectual, 189
Judaic, 109
Judeo-Christian, 10, 109
national, 177
oral, 64, 110, 116, 212, 243, 257, 258, 265
philosophical, 206, 212
western, 113, 145, 227
Transcendentalists, 185
Transformation
   historical, 226
Transmigration, 187–88, 190
Tribe, 202
Troy, 155, 159, 281
   fall of, 144
Truth, 1, 6, 14, 39, 187, 188, 210, 235, 282
   biblical, 181
   eternal, 189
   historical, 39, 259, 278

Uduk, 247
United States, 29, 33, 61
Universe, 109, 216, 247, 274

Varro, 157
Vatter, Ernst, 265
Vedas, 26, 180, 181, 185, 186, 190, 194
Vico, Giambattista, 14, 15, 160, 269, 270, 271, 273, 274, 275, 277–82, 283, 285, 286
   mockery of chronology, 152
Viking, 42, 52, 61
Volney, Comte de, 273, 279, 283–85, 286
Voltaire
   attack on Christianity, 182
   mockery of chronology, 152
   rejection of Indian time, 182
Vossen, Rainer, 225
Vossius, G. J., 160

Weil, Simone, 5, 271–72, 288
West, 108, 144, 179, 181, 189
   historicism of, 109
   thought of, 10, 109, 114, 130, 205
Whickam, Chris, 9
White, Hayden, 6, 275
White, John, 153
Witchcraft, 39, 60
Wittgenstein, Ludwig, 205
Wordsworth, William, 276, 278

Young, Robert, 167
Yucatan, 107, 110, 116, 118, 124, 126, 133
Yugas, 169, 172, 175, 179, 180, 181, 184, 185

Zaccanopoli, 7–8, 72–102
   genealogical memory in, 88–89
   marriage practices in, 75–77, 92, 102
   naming system, 90–91
   view of past, 82–83